AF353001

Remote Sensing of Savannas and Woodlands

Remote Sensing of Savannas and Woodlands

Editor

Michael J. Hill

MDPI • Basel • Beijing • Wuhan • Barcelona • Belgrade • Manchester • Tokyo • Cluj • Tianjin

Editor
Michael J. Hill
University of North Dakota
USA

Editorial Office
MDPI
St. Alban-Anlage 66
4052 Basel, Switzerland

This is a reprint of articles from the Special Issue published online in the open access journal *Remote Sensing* (ISSN 2072-4292) (available at: https://www.mdpi.com/journal/remotesensing/special_issues/savannas_woodlands).

For citation purposes, cite each article independently as indicated on the article page online and as indicated below:

LastName, A.A.; LastName, B.B.; LastName, C.C. Article Title. *Journal Name* **Year**, *Volume Number*, Page Range.

ISBN 978-3-0365-1957-9 (Hbk)
ISBN 978-3-0365-1956-2 (PDF)

Cover image courtesy of Michael J. Hill

Contents

About the Editor

Michael J. Hill is Emeritus Professor of Earth System Science in the Department of Earth System Science & Policy at the University of North Dakota (UND), Grand Forks, USA. He has a background in grassland agronomy and ecology but has worked with spatial information and remote sensing of grassland and savanna systems for the past 30 years. He has published widely on agronomy, ecology, biogeography and production of grasslands, radar, multispectral, and hyperspectral remote sensing of grasslands, and vegetation dynamics of Australian and global rangeland and savanna systems. He retired from UND in 2016. He now resides in Australia and is a Visiting Scientist at CSIRO Land and Water in Canberra, Australia. His current interests involve assisting CSIRO scientists with global remote sensing of vegetation fractional cover, and the development of global vegetation cover products and the GEOGLAM RAPP project to enhance productivity of global rangelands and pastures.

Preface to "Remote Sensing of Savannas and Woodlands"

Savannas and woodlands are one of the most challenging targets for remote sensing. However, they require increased attention, especially in Africa and South America since they are prime candidates for agricultural conversion, important resources for livestock production and subsistence of indigenous communities, and could play a significant role in signalling vegetation shifts driven by the interaction of climate change and rising atmospheric $CO2$ concentrations. The sub-Saharan region of Africa played a key role in exemplifying the value of early low resolution polar orbiting satellites and introducing the public to large scale regional dynamics of climate driven vegetation growth via time series of the Normalised Difference Vegetation Index. The 21st century has produced a revolution in remote sensing with both public and private initiatives rapidly addressing the historical trade-offs associated with spatial resolution, temporal frequency and spectral resolution that includes the active LiDAR and Radar domains. These trade-offs are particularly important in savanna and woodland systems where over-story, mid-story and under-story vegetation strata have equal importance in structure, function and dynamics. The suite of papers provides a current snapshot of the geographical focus and application of the latest sensors and sensor combinations in savannas and woodlands.

Michael J. Hill
Editor

Editorial

Remote Sensing of Savannas and Woodlands: Editorial

Michael J. Hill

Department of Earth System Science and Policy, University of North Dakota, Grand Forks, ND 58202, USA; michael.hill4@und.edu

Citation: Hill, M.J. Remote Sensing of Savannas and Woodlands: Editorial. *Remote Sens.* **2021**, *13*, 1490. https://doi.org/10.3390/rs13081490

Received: 6 April 2021
Accepted: 12 April 2021
Published: 13 April 2021

1. Background

Savannas and woodlands represent one of the most challenging targets for remote sensing. They provide a complex gradient of woody and herbaceous plant species that varies widely in spatial arrangement, plant density and height, and canopy persistence (evergreen to deciduous) of the woody plant component. The understory component also varies from annual to perennial grassland, and pure grassland to a complex mix of grasses, forbs, palms, cycads and small shrubs. Savannas and woodlands are geographically associated with grasslands, shrublands and dry forests at their dry margins, and temperate and tropical forests at the wet margins. Savannas are generally regarded as an intermediate state between grassland and forest maintained by herbivory and wildfire and influenced by climatic and edaphic characteristics.

The grasslands, savannas and woodlands of sub-Saharan Africa played a key role in development of the remote sensing of vegetation cover and dynamics [1–7]. This vegetation provided a fertile target for the earliest applications of the Normalized Difference Vegetation Index (NDVI) and introduced the public to large-scale regional dynamics of climate-driven vegetation growth. Although there is wide geographical variation in the density and seasonality of woody canopies, savannas and woodlands are associated with highly seasonal climates, and the pattern of vegetation greening and browning detected by space-borne sensors is largely driven by the behavior of the understory. Despite this, much of the attention on remote sensing of savannas and woodlands has historically focused on the woody component.

Since the 1980s, savannas and woodlands have continued to be at the forefront of developments in remote sensing. Savannas and woodlands were the target for international projects aimed at understanding ecosystem behavior such as the Hydrological Atmospheric Pilot Experiment (HAPEX) Sahel [8], the Southern Africa Regional Science Initiative (SAFARI 2000) [9] and continued studies along the Northern Australian Tropical Transect [10], all of which have included major remote sensing components especially utilizing the Advanced Very High Resolution Radiometer (AVHRR) and the MODerate resolution Imaging Spectroradiometer (MODIS). African savannas have played an important role in assessing the potential of radar backscatter for retrieval of, for example, herbaceous biomass with the European Remote sensing Satellite (ERS) wind scatterometer [11] and woody biomass with the Advanced Land Observing Satellite Phased Array L-band Synthetic Aperture Radar (ALOS PALSAR) [12]. Since wildfire is such an important feature of savannas, these ecosystems have been, and continue to be, at the forefront of the development of fire products such as hotspots and burned areas [13]. In the absence of a satellite-based imaging spectrometer with a high signal-to-noise ratio, especially in the short-wave infrared range, airborne spectrometers alone and in association with full-waveform lidar have been used to characterize the biochemistry [14], floristics [15], structure [16,17] and species composition [18] of savannas at the landscape scale.

Savannas and woodlands, especially in Africa and South America, deserve increased attention for remote sensing studies since they are prime candidates for agricultural conversion, important resources for livestock production and subsistence of indigenous communities, and could play a significant role in signaling vegetation shifts driven by the interaction of climate change and rising atmospheric CO_2 concentrations. This Special Issue sought to provide an overview of the application of the latest sensors and sensor combinations to retrieval of quantitative properties of savannas and woodlands. It was hoped that contributions would illustrate improvements in retrievals of attributes of cover components and component dynamics and relate these attributes to the wide diversity of issues faced by savanna and woodland systems globally.

2. The Papers

The papers provide a very current perspective on remote sensing of savannas and woodlands and reflect both methodological trends, and geographical imperatives driven by threats. An analysis of word frequency in the abstracts from the 12 papers, using Wordclouds.com, revealed several specific points of focus (Figure 1). The word cloud is presented with six types of words. The most common "ecosystem" words were *tree, soil* and *water*. *Grassland, forest* and *species* were present but less common in keeping with a continued emphasis on the woody component of savannas and woodlands. Among "climate-related" words, *season* and *dry* were most common matching the broad climatic characteristics of savannas and woodlands involving distinct periods of growth, senescence and dormancy. Among the "process" words, *spatial, trends, phenology, distribution, change* and *time* were most common reflecting interest in the arrangement of vegetation, the interaction of climate and growth pattern and the driving forces affecting savannas and woodlands. The "method" words were dominated by terrestrial laser scanning (*TLS*) and related acronyms (*ULS*—unmanned aerial vehicle laser scanning; *MLS*—mobile laser scanning), and by *NDVI* and *indices*, indicating the legacy of those initial remote sensing studies discussed above and the enduring value of the NDVI. More generic words such as *sensor, images, mapping, analysis, area, difference, scale* and *monitoring* were also common. The papers also used an array of space-borne sensor systems including Landsat, MODIS (various products), Sentinel 1, Sentinel 2, ALOS PALSAR and Compact High Resolution Imaging Spectrometer (CHRIS). Several studies used combinations of sensor types including optical, SAR and LiDAR. The "geography" words were dominated by *cerrado* and *Brazil* with *biome* and *global* next most common. However, the 11 research papers provide a relatively balanced coverage of savanna and woodland geography, with four articles on South America, two articles on Africa, two articles on Australia, one each on North America and Europe, and one article with global coverage. Finally, the "attribute' words included *types, structure, AGB, biomass, woody,* and *density* reflecting the fundamental characteristics of savannas and woodlands that need to be measured and that make them a complex target for remote sensing.

The Special Issue is nicely balanced in terms of geography and technology. It has two feature topics: the Cerrado and bordering ecosystems; and laser scanning as the technology of the moment for characterizing savanna vegetation structure. The Special Issue is conceptually organized in the first instance by geography. The lead article [19] covers the global extent of grasslands and savannas using the MODIS sensor and provides global geographical context. This is followed by the feature review on TLS in savannas [20]. The articles then follow global geography starting in the southern hemisphere dealing with TLS in the Australian tropical savanna [21,22]. Next come four articles on various aspects of savanna and woodland vegetation in Brazil [23–26]. These are followed by two articles on African environments that include a mix of savannas and woodlands and other vegetation [27,28]. The final two articles address functional ecological aspects of two different types of northern hemisphere oak savannas in North America [29] and Europe [30].

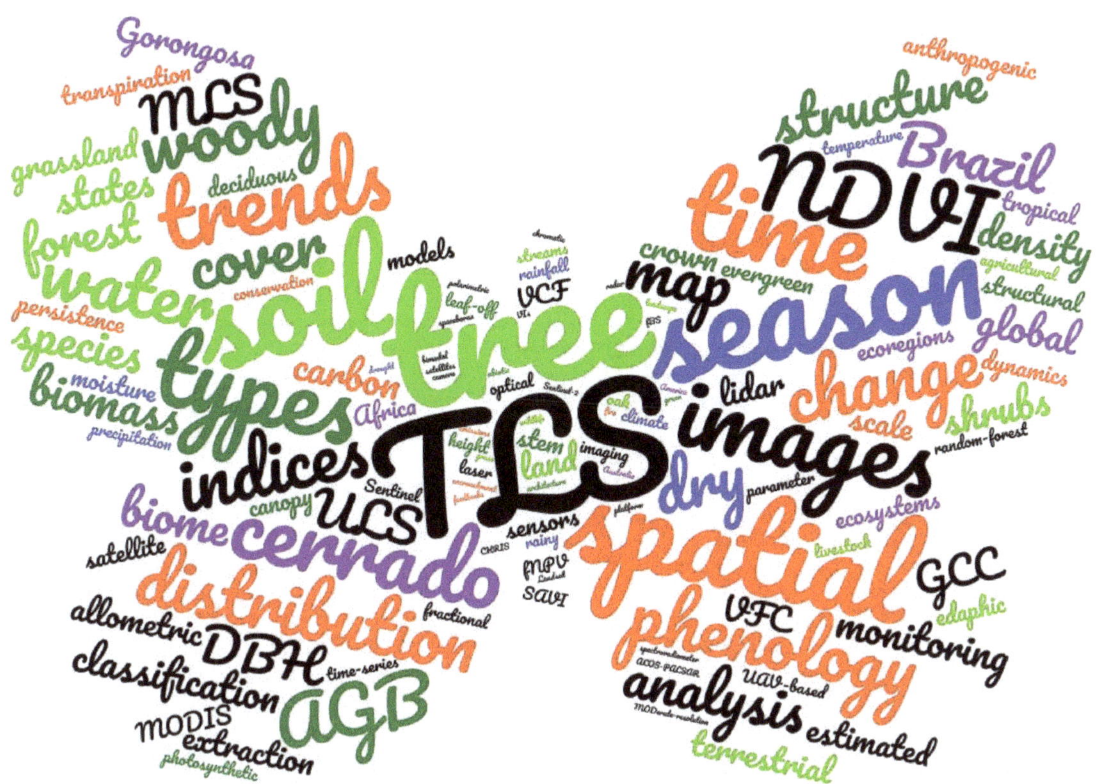

Figure 1. Word cloud based on the article abstracts (two occurrences or more). The terms *savanna*, *woodland*, *remote* and *sensing* are excluded since they are the topic of the Special Issue. Words represent all versions of the term or concept, e.g., season includes seasonal and seasonality. Only place names and acronyms are capitalized. Color scheme: light green—ecosystem words; blue—climate -related words; red—process words; black—method words; purple—geography words; and dark green—vegetation attribute words. Word cloud produced on http://wordclouds.com (accessed on 6 April 2021).

The group of four papers on South America span the ecotone from the heavily altered Mato Grosso forests in Amazonia through the savanna of the Cerrado to the dry forests of the Caatinga. These papers explore the mapping of vegetation types [23], the measurement of woody above-ground biomass (AGB) [24], the long-term change in savanna vegetation [25] and water use [26]. The emphasis on this continuum from tropical rainforest through savanna to dry forest is timely and important as this environment is subject to massive anthropogenic change and complex political forces within Brazil that influence the balance between conservation and exploitation of these ecosystems. The paper by de Souza Mendes [23] uses multiple satellite sources of SAR imagery with Sentinel 2 data to distinguish among forest and woody savanna types in the Cerrado. The paper by Bispo et al. [24] combines airborne LiDAR with Landsat 8 and ALOS PALSAR to map AGB for a catchment-scale study region of the Cerrado. The paper by Paloschi et al. [26] reports on the strong dependence of water use on soil moisture, and the variation in phenology, and physiology of individual species in this dry, fragile Caatinga environment. The paper by Alencar et al. [25] is a new and significant documenting of long-term change in natural native vegetation in the Cerrado using 33 years of Landsat imagery in the Google Earth Engine. This is already the most viewed and cited article in this Special Issue.

The review of the application of TLS with a focus on savannas [20] and the two papers focused on the application of TLS in savannas and woodlands of Northern Australia [21,22] form the second feature topic. The review [20] and the article by Levick et al. [22] are complementary in that the former identifies both the value of TLS in savannas, and the limited application and potential issues, whilst the latter explicitly answers issues surrounding the spatial extent and structural complexity by demonstrating the value of combining TLS, ULS and MLS. Both articles are important contributions to the literature and to the application of laser scanning in savannas and woodlands. The paper by Luck et al. [21] provides an example of retrieval of savanna structure using TLS and an explicit description of a northern Australian savanna in the wet tropical zone.

The article by Hill and Guerschman [19] describes the MODIS Global Vegetation Cover Product and characterizes trends in vegetation cover fractions between 2001 and 2018 across global savannas and woodlands as well as associated grasslands. The study documents trends in non-photosynthetic (dry or brown) vegetation for the first time and identifies grasslands and savannas with concerning positive trends in the bare soil fraction. The two articles based in Africa explore aspects of savanna and woodland vegetation structure and temporal persistence using MODIS data. Kumar et al. [27] examine apparent differences in spatial structure of woody vegetation across sub-Saharan Africa retrieved from different MODIS products. They suggest that multimodal spatial structure is regionally disaggregated, and that apparent spatial structure retrieved may be dependent upon the remote sensing product and spatial scale of retrievals. Herrero et al. [28] provide a temporal study of the Gorongosa National Park in Mozambique which explores the persistence of the MODIS NDVI signal in relation to changes in precipitation in the period since the year 2000. The study reveals a decline in vegetation as measured by NDVI persistence especially in grassland and rainforests potentially associated with declining precipitation.

Oak savannas were once much more common in temperate ecosystems forming an ecotone between grassland and forests. The two articles published here represent two major types: the perennial grass/oak savannas that stretched from southern Canada to northern Mexico between the grasslands and forests of North America; and the annual grass/oak savannas found, for example, in California and Spain. The paper by Hill et al. [29] describes the use of a rare time series from the CHRIS hyperspectral sensor to explore functional phenology of the woody canopy in a conserved but degraded post oak savanna in Texas invaded with evergreen shrubs. The results of this study hint at the value of hyperspectral data in characterization of differences in canopy properties and species mixtures that are important in the evolution of disturbed remnant landscapes. However, they also emphasis the limitations of 20-year-old low signal-to-noise imaging spectrometer data, and the gaping chasm in the capability of satellite remote sensing for quantitative retrievals of vegetation canopy biochemistry waiting to be occupied by a state-of-the-art imaging spectrometer. By contrast, Gómez-Giráldez et al. [30], examine the phenology of the tree and understory layers in relation to soil water dynamics in an annual grass/oak savanna in Spain using a Sentinel 2 time series. The study showed a high degree of synchrony between NDVI and soil moisture. However, some vegetation indices were potentially sensitive to subtle variations in chlorophyll content of the understory grassland.

The featured emphasis on TLS [20–22] and the Cerrado [23–26] points to the technology currently at the cutting edge for remote measurement, and to the focus of concern for the fate of carbon stocks and biodiversity in one of the largest savannas in the world. The adoption of Sentinel 2 [23,30], the application of big-data processing [25] and the combination of SAR, LiDAR and optical data [23,24] signal the future direction of remote sensing of savannas from space. Multi-sensor imaging and multi-scale laser scanning are necessary to truly capture the three-dimensional properties of savannas and woodlands. It notable that only two papers were focused on the understory grassland [19,30]. Nevertheless, there is still a role for polar orbiting medium resolution sensors [19,27,28] and a pressing need for a for a modern imaging spectrometer in orbit.

Funding: This research received no external funding.

Acknowledgments: I wish to all thank authors for submitting to the Special Issue and for contributing to what is a distinctive perspective on the current state of remote sensing of savannas and woodlands.

Conflicts of Interest: The author declares no conflict of interest.

References

1. Tucker, C.; Vanpraet, C.; Sharman, M.; Van Ittersum, G. Satellite remote sensing of total herbaceous biomass production in the senegalese sahel: 1980–1984. *Remote Sens. Environ.* **1985**, *17*, 233–249. [CrossRef]
2. Tucker, C.J.; Townshend, J.R.; Goff, T.E. African Land-Cover Classification Using Satellite Data. *Science* **1985**, *227*, 369–375. [CrossRef]
3. Tucker, C.J.; Justice, C.O.; Prince, S.D. Monitoring the grasslands of the Sahel 1984–1985. *Int. J. Remote Sens.* **1986**, *7*, 1571–1581. [CrossRef]
4. Justice, C.O.; Hiernaux, P.H.Y. Monitoring the grasslands of the Sahel using NOAA AVHRR data: Niger 1983. *Int. J. Remote Sens.* **1986**, *7*, 1475–1497. [CrossRef]
5. Justice, C.O.; Holben, B.N.; Gwynne, M.D. Monitoring East African vegetation using AVHRR data. *Int. J. Remote Sens.* **1986**, *7*, 1453–1474. [CrossRef]
6. Prince, S.D.; Tucker, C.J. Satellite remote sensing of rangelands in Botswana II. NOAA AVHRR and herbaceous vegetation. *Int. J. Remote Sens.* **1986**, *7*, 1555–1570. [CrossRef]
7. Townshend, J.R.G.; Justice, C.O. Analysis of the dynamics of African vegetation using the normalized difference vegetation index. *Int. J. Remote Sens.* **1986**, *7*, 1435–1445. [CrossRef]
8. Prince, S.; Kerr, Y.; Goutorbe, J.P.; Lebel, T.; Tinga, A.; Bessemoulin, P.; Brouwer, J.; Dolman, A.; Engman, E.; Gash, J.; et al. Geographical, biological and remote sensing aspects of the hydrologic atmospheric pilot experiment in the sahel (HAPEX-Sahel). *Remote Sens. Environ.* **1995**, *51*, 215–234. [CrossRef]
9. Privette, J.L.; Roy, D.P. Southern Africa as a remote sensing test bed: The SAFARI 2000 Special Issue overview. *Int. J. Remote Sens.* **2005**, *26*, 4141–4158. [CrossRef]
10. Ma, X.; Huete, A.; Moore, C.E.; Cleverly, J.; Hutley, L.B.; Beringer, J.; Leng, S.; Xie, Z.; Yu, Q.; Eamus, D. Spatiotemporal partitioning of savanna plant functional type productivity along NATT. *Remote Sens. Environ.* **2020**, *246*, 111855. [CrossRef]
11. Jarlan, L.; Mougin, E.; Frison, P.; Mazzega, P.; Hiernaux, P. Analysis of ERS wind scatterometer time series over Sahel (Mali). *Remote Sens. Environ.* **2002**, *81*, 404–415. [CrossRef]
12. Mitchard, E.T.A.; Saatchi, S.S.; Woodhouse, I.H.; Nangendo, G.; Ribeiro, N.S.; Williams, M.; Ryan, C.M.; Lewis, S.L.; Feldpausch, T.R.; Meir, P. Using satellite radar backscatter to predict above-ground woody biomass: A consistent relationship across four different African landscapes. *Geophys. Res. Lett.* **2009**, *36*. [CrossRef]
13. Campagnolo, M.; Libonati, R.; Rodrigues, J.; Pereira, J. A comprehensive characterization of MODIS daily burned area mapping accuracy across fire sizes in tropical savannas. *Remote Sens. Environ.* **2021**, *252*, 112115. [CrossRef]
14. Knox, N.M.; Skidmore, A.K.; Prins, H.H.; Asner, G.P.; van der Werff, H.M.; de Boer, W.F.; van der Waal, C.; de Knegt, H.J.; Kohi, E.M.; Slotow, R.; et al. Dry season mapping of savanna forage quality, using the hyperspectral Carnegie Airborne Observatory sensor. *Remote Sens. Environ.* **2011**, *115*, 1478–1488. [CrossRef]
15. Landmann, T.; Piiroinen, R.; Makori, D.M.; Abdel-Rahman, E.M.; Makau, S.; Pellikka, P.; Raina, S.K. Application of hyperspectral remote sensing for flower mapping in African savannas. *Remote Sens. Environ.* **2015**, *166*, 50–60. [CrossRef]
16. Chen, Q.; Baldocchi, D.; Gong, P.; Kelly, M. Isolating Individual Trees in a Savanna Woodland Using Small Footprint Lidar Data. *Photogramm. Eng. Remote Sens.* **2006**, *72*, 923–932. [CrossRef]
17. Asner, G.P.; Levick, S.R.; Kennedy-Bowdoin, T.; Knapp, D.E.; Emerson, R.; Jacobson, J.; Colgan, M.S.; Martin, R.E. Large-scale impacts of herbivores on the structural diversity of African savannas. *Proc. Natl. Acad. Sci. USA* **2009**, *106*, 4947–4952. [CrossRef] [PubMed]
18. Colgan, M.S.; Baldeck, C.A.; Féret, J.-B.; Asner, G.P. Mapping Savanna Tree Species at Ecosystem Scales Using Support Vector Machine Classification and BRDF Correction on Airborne Hyperspectral and LiDAR Data. *Remote Sens.* **2012**, *4*, 3462–3480. [CrossRef]
19. Hill, M.J.; Guerschman, J.P. The MODIS Global Vegetation Fractional Cover Product 2001–2018: Characteristics of Vegetation Fractional Cover in Grasslands and Savanna Woodlands. *Remote Sens.* **2020**, *12*, 406. [CrossRef]
20. Muumbe, T.; Baade, J.; Singh, J.; Schmullius, C.; Thau, C. Terrestrial Laser Scanning for Vegetation Analyses with a Special Focus on Savannas. *Remote Sens.* **2021**, *13*, 507. [CrossRef]
21. Luck, L.; Hutley, L.; Calders, K.; Levick, S. Exploring the Variability of Tropical Savanna Tree Structural Allometry with Terrestrial Laser Scanning. *Remote Sens.* **2020**, *12*, 3893. [CrossRef]
22. Levick, S.R.; Whiteside, T.; Loewensteiner, D.A.; Rudge, M.; Bartolo, R. Leveraging TLS as a Calibration and Validation Tool for MLS and ULS Mapping of Savanna Structure and Biomass at Landscape-Scales. *Remote Sens.* **2021**, *13*, 257. [CrossRef]
23. Mendes, F.D.S.; Baron, D.; Gerold, G.; Liesenberg, V.; Erasmi, S. Optical and SAR Remote Sensing Synergism for Mapping Vegetation Types in the Endangered Cerrado/Amazon Ecotone of Nova Mutum—Mato Grosso. *Remote Sens.* **2019**, *11*, 1161. [CrossRef]

24. Bispo, P.C.; Rodríguez-Veiga, P.; Zimbres, B.; De Miranda, S.D.C.; Cezare, C.H.G.; Fleming, S.; Baldacchino, F.; Louis, V.; Rains, D.; Garcia, M.; et al. Woody Aboveground Biomass Mapping of the Brazilian Savanna with a Multi-Sensor and Machine Learning Approach. *Remote Sens.* **2020**, *12*, 2685. [CrossRef]
25. Alencar, A.; Shimbo, J.Z.; Lenti, F.; Marques, C.B.; Zimbres, B.; Rosa, M.; Arruda, V.; Castro, I.; Ribeiro, J.F.M.; Varela, V.; et al. Mapping Three Decades of Changes in the Brazilian Savanna Native Vegetation Using Landsat Data Processed in the Google Earth Engine Platform. *Remote Sens.* **2020**, *12*, 924. [CrossRef]
26. Paloschi, R.; Ramos, D.; Ventura, D.; Souza, R.; Souza, E.; Morellato, L.; Nóbrega, R.; Coutinho, Í.; Verhoef, A.; Körting, T.; et al. Environmental Drivers of Water Use for Caatinga Woody Plant Species: Combining Remote Sensing Phenology and Sap Flow Measurements. *Remote Sens.* **2020**, *13*, 75. [CrossRef]
27. Kumar, S.S.; Hanan, N.P.; Prihodko, L.; Anchang, J.; Ross, C.W.; Ji, W.; Lind, B.M. Alternative Vegetation States in Tropical Forests and Savannas: The Search for Consistent Signals in Diverse Remote Sensing Data. *Remote Sens.* **2019**, *11*, 815. [CrossRef]
28. Herrero, H.; Waylen, P.; Southworth, J.; Khatami, R.; Yang, D.; Child, B. A Healthy Park Needs Healthy Vegetation: The Story of Gorongosa National Park in the 21st Century. *Remote Sens.* **2020**, *12*, 476. [CrossRef]
29. Hill, M.J.; Millington, A.; Lemons, R.; New, C. Functional Phenology of a Texas Post Oak Savanna from a CHRIS PROBA Time Series. *Remote Sens.* **2019**, *11*, 2388. [CrossRef]
30. Gómez-Giráldez, P.J.; Pérez-Palazón, M.J.; Polo, M.J.; González-Dugo, M.P. Monitoring Grass Phenology and Hydrological Dynamics of an Oak–Grass Savanna Ecosystem Using Sentinel-2 and Terrestrial Photography. *Remote Sens.* **2020**, *12*, 600. [CrossRef]

 *remote sensing*

Article

The MODIS Global Vegetation Fractional Cover Product 2001–2018: Characteristics of Vegetation Fractional Cover in Grasslands and Savanna Woodlands

Michael J. Hill [1,2,*] and Juan P. Guerschman [1]

[1] CSIRO Land and Water, Black Mountain, ACT 2601, Australia; juan.guerschman@csiro.au
[2] Department of Earth System Science and Policy, University of North Dakota, Grand Forks, ND 58202, USA
* Correspondence: michael.hill4@und.edu; Tel.: +61-262465880

Received: 24 December 2019; Accepted: 24 January 2020; Published: 28 January 2020

Abstract: Vegetation Fractional Cover (VFC) is an important global indicator of land cover change, land use practice and landscape, and ecosystem function. In this study, we present the Global Vegetation Fractional Cover Product (GVFCP) and explore the levels and trends in VFC across World Grassland Type (WGT) Ecoregions considering variation associated with Global Livestock Production Systems (GLPS). Long-term average levels and trends in fractional cover of photosynthetic vegetation (F_{PV}), non-photosynthetic vegetation (F_{NPV}), and bare soil (F_{BS}) are mapped, and variation among GLPS types within WGT Divisions and Ecoregions is explored. Analysis also focused on the savanna-woodland WGT Formations. Many WGT Divisions showed wide variation in long-term average VFC and trends in VFC across GLPS types. Results showed large areas of many ecoregions experiencing significant positive and negative trends in VFC. East Africa, Patagonia, and the Mitchell Grasslands of Australia exhibited large areas of negative trends in F_{NPV} and positive trends F_{BS}. These trends may reflect interactions between extended drought, heavy livestock utilization, expanded agriculture, and other land use changes. Compared to previous studies, explicit measurement of F_{NPV} revealed interesting additional information about vegetation cover and trends in many ecoregions. The Australian and Global products are available via the GEOGLAM RAPP (Group on Earth Observations Global Agricultural Monitoring Rangeland and Pasture Productivity) website, and the scientific community is encouraged to utilize the data and contribute to improved validation.

Keywords: vegetation; grassland; savanna; fractional cover; trend; ecoregion; bare soil; livestock; production systems

1. Introduction

Vegetation Fractional Cover (VFC) is an important global indicator of land cover change, land use practice, and landscape and ecosystem function [1,2]. Seasonal dynamics and long-term trends in the fractional cover of photosynthetic vegetation (F_{PV}), non-photosynthetic vegetation (F_{NPV}), and bare soil (F_{BS}) may identify changes in cropping cycles, impacts of livestock grazing, clearing or planting of woody vegetation, and ecosystem responses to climate shifts. The Australian Vegetation Fractional Cover Product (AVFCP) is derived from the 500 m MODIS (MODerate Resolution Imaging Spectroradiometer) NBAR (Nadir BRDF-adjusted Reflectance) product (MCD43A4) and has been comprehensively documented, validated, and improved over several versions [3–6]. In this study, we present the Global Vegetation Fractional Cover Product (GVFCP) based on the same methodology. Both the Australian and Global products are available through the GEOGLAM RAPP (Group on Earth Observations Global Agricultural Monitoring Rangeland and Pasture Productivity) website (https://map.geo-rapp.org). A list of abbreviations appears in Table A1.

A key feature of the GVFCP is the explicit sensitivity to F_{NPV} enabling better discrimination of dry cellulosic vegetation and bare soil fractions than is possible with the family of greenness indices. For this reason, the GVFCP is particularly important for monitoring of the non-forested and non-desert global biomes, woodlands, savannas, grassland and shrublands, where the overstory canopy is open, and understory dynamics involve the relationships between cellulosic herbaceous vegetation and/or senescent arboreal leaf litter and bare soil. The first version of the AVFCP [3] used the relationship between the Normalised Difference Vegetation Index (NDVI) and the Short Wave Infrared Ratio (SWIR32; a ratio of the MODIS 2130 nm and MODIS 1640 nm bands) in a linear unmixing method to approximate the relationship between NDVI and the Cellulose Absorption Index (CAI) developed and demonstrated by [7–11]. An improved version of the AVFCP utilizes all seven MODIS NBAR bands plus band transformations and interaction terms in a sophisticated unmixing algorithm [4]. Recent recalibration of the AVFCP with MODIS Collection 6 inputs and an updated validation data set established improvements in the accuracy of the retrieved fractions of photosynthetic vegetation (F_{PV} RMSE 0.112), non-photosynthetic vegetation (F_{NPV} RMSE 0.162), and bare soil (F_{BS} RMSE 0.130) [5].

Based on Australian studies [12–14], the AVFCP provides a consistent and verifiable estimate of the cover fractions for green canopy, senescent vegetation and surface litter, and the visible soil surface. Given the diversity of natural ecosystems across Australia that include tropical rainforest, tropical and temperate savanna, temperate grasslands, semi-arid and arid shrublands and grasslands, temperate eucalypt forests, and temperate rainforest, and the extent of agricultural and pastoral lands across Australia, it is reasonable to propose that the GVFCP should perform with similar levels of uncertainty across the diversity of global land covers and land uses, especially in the non-forested and non-desert biomes.

Many global analyses have focused on tree cover change [15–17] and land cover change [18]. The recent comprehensive study of [19] partitions cover into tall vegetation, short vegetation, and bare soil, and the greening analysis of [20] looks at trends in the annual values of the MODIS leaf area index product [21]. In addition, other studies have focused on the total cover and soil or bare fractions [22,23]. There are several global tree cover [24,25] and tree cover change [15] products at multiple resolutions and many FAPAR (Fraction of Absorbed Photosynthetically Active Radiation) [26,27] and green canopy cover or leaf area index products [28–30]. The global cover change study by [19] identifies changes in short vegetation (which includes shrubs under 5 m in height) and bare ground. The cover fractions are mapped from peak growing season canopy [19] meaning that deciduous savanna trees and shrubs are probably fully foliated, and understory grasses are actively growing. By contrast, the GVFCP at full temporal and spatial resolution follows the phenology of overstory and understory throughout the seasons and explicitly retrieves F_{NPV} representing senescent understory.

Studies focusing on global scale cover, dynamics, and change in savanna landscapes are less common (e.g., [31,32]), but a study based on persistence of GIMMS (Global Inventory Modeling and Mapping Studies) NDVI signals indicated larger areas of increased than decreased vegetation persistence [33]. However, there has been considerable discussion on the nature of stable states, methods of detection, and drivers of change in savannas (e.g., [34–37]). Due to the specific methodology used to derive it, the GVFCP provides a unique measure of global changes in vegetation fractional cover especially important for woodlands, savannas, grasslands and shrublands which are subject to heavy utilization and conversion for human food production and where cellulosic herbaceous cover is important.

It is instructive to provide context for the behaviour of remote sensing products using global datasets that are wholly or partially independent of remote sensing, and describe global vegetation in terms relevant to conservation, ecosystem function, and productivity. The Terrestrial Ecoregions of the World (TEoW; [38]) and the subset of these used to define World Grassland Types (WGT [39]) provides an effective framework within which to view the patterns of VFC for the grassy ecoregions of the world. Since these grassy ecoregions are subject to major utilization by humans for extensive and intensive agriculture, the Global Livestock Production Systems (GLPS) classification [40] provides a means to sample variation within WGT Ecoregions due to differences in utilisation between production systems.

In this study we have three objectives:

(1) To present the GVFCP (using a resampled 5 km^2 resolution version) and illustrate the long-term global geographical patterns of F_{PV}, F_{NPV}, and F_{BS} using the TEoW [38].
(2) To document and benchmark the levels and trends in F_{PV}, F_{NPV}, and F_B within the WGT Ecoregions at Formation and Division levels [39] using GLPS [40] to provide a measure of internal land use differences.
(3) To examine the levels and trends in F_{PV}, F_{NPV}, and F_B in savanna, woodland and scrub grassland (SWSG) Ecoregions of the WGT and explore VFC trajectories in selected example ecoregions where major changes are occurring.

2. Materials and Methods

The analysis undertaken here provides an overview of global VFC in map form; records benchmark average levels of VFC across production systems for Formation, Division, and Ecoregion areas of the grassy biomes of the world; and documents long-term trends in VFC from 2001–2018 for these areas. The analysis is designed to provide an overview framework and baseline for more detailed future studies.

2.1. Data

2.1.1. Global Vegetation Fractional Cover Product

The GVFCP estimates the fractions of photosynthetic (green) vegetation (PV), non-photosynthetic (non-green) vegetation (NPV), and bare soil (BS) within the pixel using a spectral linear unmixing method [4,5]. The base version of the GVFCP, available on the GEOGLAM RAPP-MAPP website (https://map.geo-rapp.org) is calculated using reflectance values from the MODIS MCD43A4 product which is a rolling daily 16-day composite weighted to retrieve the value on the ninth day of the 16-day period [41]. The GVFCP utilizes every eighth day from the MCD43A4 product starting with day one of each year with an adjustment for leap years proving 46 dates in each calendar year. The data are stored in a geographic latitude-longitude raster with a WGS84 spheroid. A second monthly composite product (also available on the website) is created by aggregating all values falling within each calendar month using a medoid function (a multidimensional median) following [42]. Pixels flagged as water, snow, or low quality in the MCD43A4 product are ignored in the calculation of the GVFCP. The current product is the result of a decade of development and the most recent version derived from MODIS Collection 6 NBAR data was calculated and validated using an updated and expanded database of 3022 ground measurements of fractional cover across Australia [5] and had a Root Mean Square Error (RMSE) of 11.3%, 16.1%, and 14.7% in the PV, NPV, and BS fractions, respectively. The GVFCP is derived from spectral unmixing of all seven MODIS optical bands from the 500 m MODIS Nadir BRDF-adjusted Reflectance Product (NBAR, MCD43A4 Collection 6; see Appendix A Table A1 for abbreviations) using previously described methods [4]. The product consists of four derived layers corresponding to the decimal values of cover fractions F_{PV}, F_{NPV}, F_{BS}, and the residual from the spectral unmixing in reflectance values (RNORM). The GVFCP is calculated from the daily NBAR data product producing a phased 16-day composite every eight days. For this overview study, the monthly version of the product was used and resampled to 0.05° resolution (approximately 5 km^2) by averaging.

2.1.2. Terrestrial Ecoregions of the World

The Terrestrial Ecoregions of the World (TEoW) provide a biogeographic realization of terrestrial biodiversity [38] (Figure 1a). They represent land areas that share species, ecosystems, seasonal dynamics, and environmental conditions, and therefore, in part, indicate boundaries for intrinsic capability or risk for human uses. This dataset is used to meet the first objective of this study by providing general context for the description of the GVFCP at TEoW Realm level. Since an important attribute of the GVFCP is the derivation of F_{NPV} and F_{BS}, additional datasets that focus on the characteristics and human utilization of the subset of ecoregions either partially or totally dominated by herbaceous vegetation were employed to examine behaviour in the grassy regions of the world.

2.1.3. World Grassland Types

The World Grassland Types (WGT; [39]) represent a subset of terrestrial ecoregions that were combined with the International Vegetation Classification (IVC; [43]) to spatially define the major grassy vegetation systems of the world (Figure 1b; see Table A1 for abbreviations). They describe 75% of the IVC grasslands using ecoregions. The remaining IVC grasslands were absent from the mapping due to their fine scale distribution [39]. The WGT Ecoregions cover an area of 35.5 M km^2. The Tropical Lowland Shrubland, Grassland and Savannas (TLSGS), Tropical Montane Shrubland, Grassland and Savanna (TMSGS), and the Warm Semi-Desert Scrub and Grassland (WSDSG) WGT Formations were selected to represent the savanna, woodland and scrub grasslands (SWSG) which are the focus of this Special Issue. These SWSGs occupy 46.9% of the WGT area (16.7 M km^2) of which 9.5 M km^2 are in the northern hemisphere in Africa and neo-tropical South America, and 7.16 M km^2 are in the southern hemisphere in Africa, Australia, and South America. The northern hemisphere contains 7.0 M km^2 of TLSGS, 0.3 M km^2 of TMSGS, and 2.2 M km^2 of WSDSG. The southern hemisphere contains 6.1 M km^2 of TLSGS, 0.25 M km^2 of TMSGS, and 0.82 M km^2 of WSDSG.

2.1.4. Global Livestock Production Systems

The Global Livestock Production Systems (GLPS) data describes the terrestrial land surface using livestock density maps, crop type maps, land cover, and information on management practices [40] (Figure 1c). The classification describes 12 production systems combining rangelands (L—grazing only), mixed rainfed and mixed irrigated production (M—grazing and cropping) with hyper-arid (Y), arid (A), humid (H), and temperate (T) environments at a spatial resolution of 0.00833° (approximately 1 km^2; Figure 1c). These are: Rangelands Hyper-arid (LGY); Rangelands Arid (LGA); Rangelands Humid (LGH); Rangelands Temperate (LGT); Mixed Rainfed Hyper-arid (MRY); Mixed Rainfed Arid (MRA); Mixed Rainfed Humid (MRH); Mixed Rainfed Temperate (MRT); Mixed Irrigated Hyper-arid (MIY); Mixed Irrigated Arid (MIA); Mixed Irrigated Humid (MIH); and Mixed Irrigated Temperate (MIT). The remaining land areas are classified as Urban or Other (such as forest, ice, rock, etc.). These classes describe land use with each WGT Ecoregion. The WGT Ecoregions are dominated by seven GLPS types with the six major production systems being: LGA (36.8%); LGT (17.6%); MRA (8.4%); MRT (14.5%); LGH (4.7%); and MRH (5.6%) and the remaining area being Other.

2.2. Analysis Method

2.2.1. Trend Analysis for World Grassland Types

The long-term trend from 2001–2018 in the monthly F_{PV}, F_{NPV}, and F_{BS} was evaluated for the WGT Ecoregions using the Mann–Kendall test [44,45] with a custom-built function in IDL. This non-parametric test detects monotonic trends in time series data where the least squares regression approach is invalid due the autocorrelation in time series data derived from remote sensing. We used the Mann–Kendall approach for trend analysis on the full monthly data set since these data provided more observations than annual means by a factor of 12. The time series was smoothed by the medoid procedure, spatially averaged from 500 m to 5 km pixels by the pixel aggregation, and completely continuous in the WGT Ecoregions. As a result, we did not consider it necessary to apply the seasonal Kendall method [46]. The significance (p-value) of the slope in each pixel was recorded. Global maps of pixels within the WGT Ecoregions where trends were significant at p < 0.1 were produced. The same spatial patterns and trends were evident at p < 0.05 but patterns were clearer and more contiguous at p < 0.1 similar to [20].

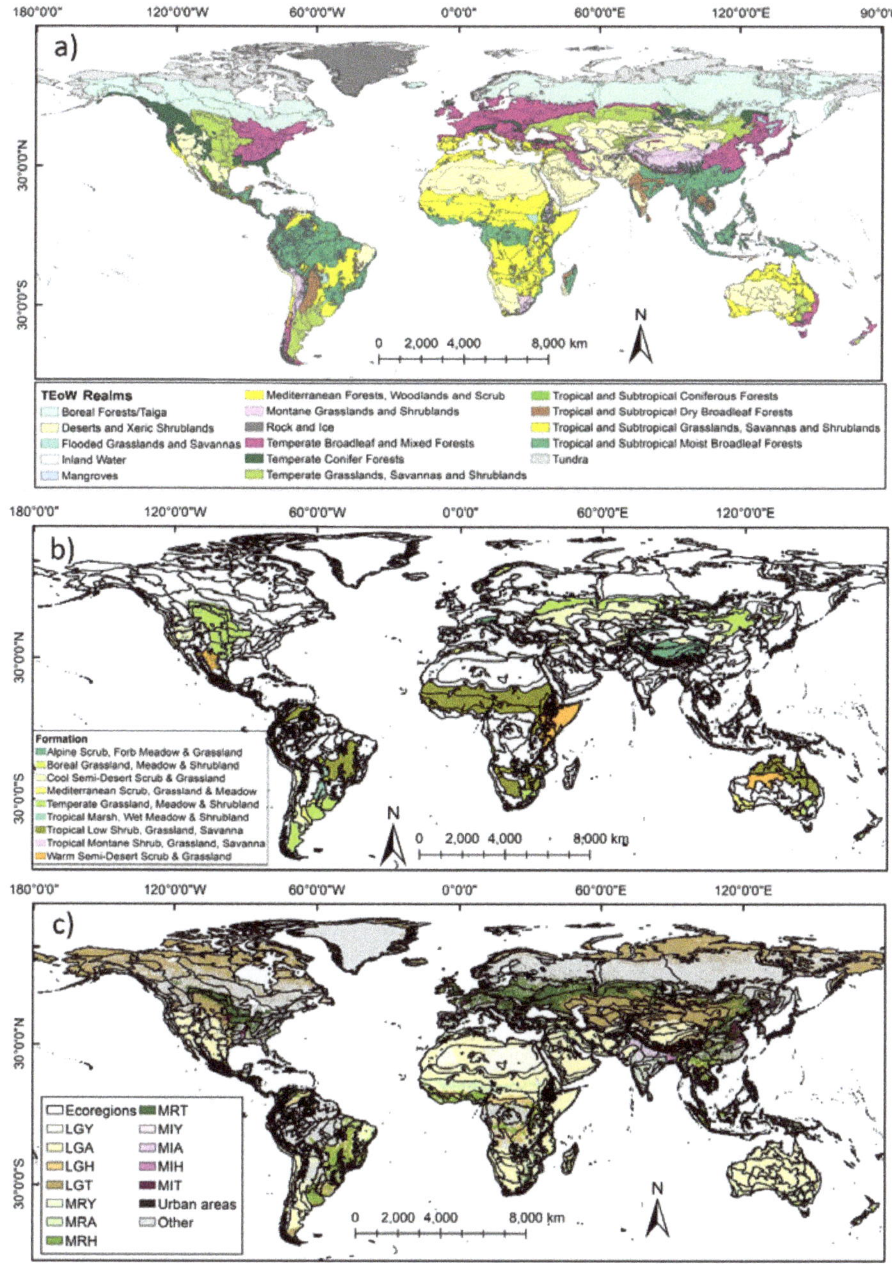

Figure 1. (**a**) Terrestrial Ecoregions of the World (TEoW) at Realm level. (**b**) World Grassland Type Ecoregions (WGT; legend shows formations). (**c**) Global Livestock Production Systems (GLPS). System codes as follows: Rangelands Hyper-arid (LGY); Rangelands Arid (LGA); Rangelands Humid (LGH); Rangelands Temperate (LGT); Mixed Rainfed Hyper-arid (MRY); Mixed Rainfed Arid (MRA); Mixed Rainfed Humid (MRH); Mixed Rainfed Temperate (MRT); Mixed Irrigated Hyper-arid (MIY); Mixed Irrigated Arid (MIA); Mixed Irrigated Humid (MIH); Mixed Irrigated Temperate (MIT).

2.2.2. Summarizing Levels and Trends across World Grassland Types and Savanna-Woodlands

The levels and trends in VFC were examined at increasing levels of detail across the WGT. The WGT and GLPS layers were combined to create a composite layer with individual classes for each GLPS-ecoregion combination. Class mean and standard deviation values for long-term average and trend in F_{PV}, F_{NPV}, and F_{BS} were extracted using ArcGIS®zonalstats. For trends, zonal statistics were collected only for slope values > 0.1 or < −0.1 using a mask that restricted extractions to only those pixels with significance at p < 0.1. The areas of significant trend by Division and Ecoregion were calculated on geographic grids using a weighted area algorithm to correct for pixel distortion with latitude away from the equator. The variation across GLPS types within WGT Divisions was displayed using the fence box plot. The fence box analysis plots the median and divides the observed values into quartiles with first and third quartiles defining the hinges, and the upper and lower fences being defined by multiplying the interquartile range by 1.5. The whiskers represent the minimum and maximum observations falling inside the fences and points falling outside this range represent outliers. The fence-box plots at Division level therefore show the range in mean values among GLPS types within that Division (whether it contains one or many Ecoregions) and the range in the standard deviation values from each GLPS type within that Division. Examination of the SWSG was carried out by the same methods at individual ecoregion level, and GLPS class behaviour was explored within selected ecoregions exhibiting large areas of significant trends.

3. Results

3.1. Global Patterns of Average VFC

Global geographical distribution of long-term average F_{PV} and F_{BS} exhibit the typical associations with forests and deserts and the gradients in between (Figure 2). The distribution of long-term average F_{NPV} tends to highlight the large grasslands of North America, South America, central Eurasia, Tibet, Mongolia, and the savannas of southern Africa and Australia (Figure 2). Based on TEoW definitions, there are 19.5 M km^2 of tropical and subtropical grasslands, shrubland, and savannas representing 13.26% of the terrestrial land surface (Table 1). In addition, the global coverage of Temperate Grasslands and Shrublands, Montane Grasslands and Shrublands, and Flooded Grassland and Savannas brings the total area of grassy biomes to 35.439 M km^2 representing 24.05% of the terrestrial land surface. Across these grassy biomes, the Montane Grasslands and Shrublands have lower average F_{PV} and higher average F_{BS} than the other grassy biomes (Table 1). Temperate and Montane Grasslands and Shrublands have higher average F_{NPV} than the Tropical and Subtropical Grasslands, Savannas and Shrublands and the Flooded Grassland and Savannas. The global latitudinal variation in average F_{PV}, F_{NPV}, and F_{BS} exhibits typical peaks in F_{PV} for tropical forests between 20° S and 20° N and boreal forests above 50° N and in F_{BS} in northern and southern hemisphere arid lands between latitudes 20 and 30° (Figure 3). However, there is a notable rise in F_{NPV} between 20 and 40° N which may reflect the large expanses of steppe environments either limited by precipitation or converted to cereal cropping. With much less land at mid- to low-latitudes, the dynamics between the cover fractions in the southern hemisphere exhibit distinct narrow latitude zones of oscillation between higher and lower levels of F_{PV} and F_{NPV} between 30 and 45°.

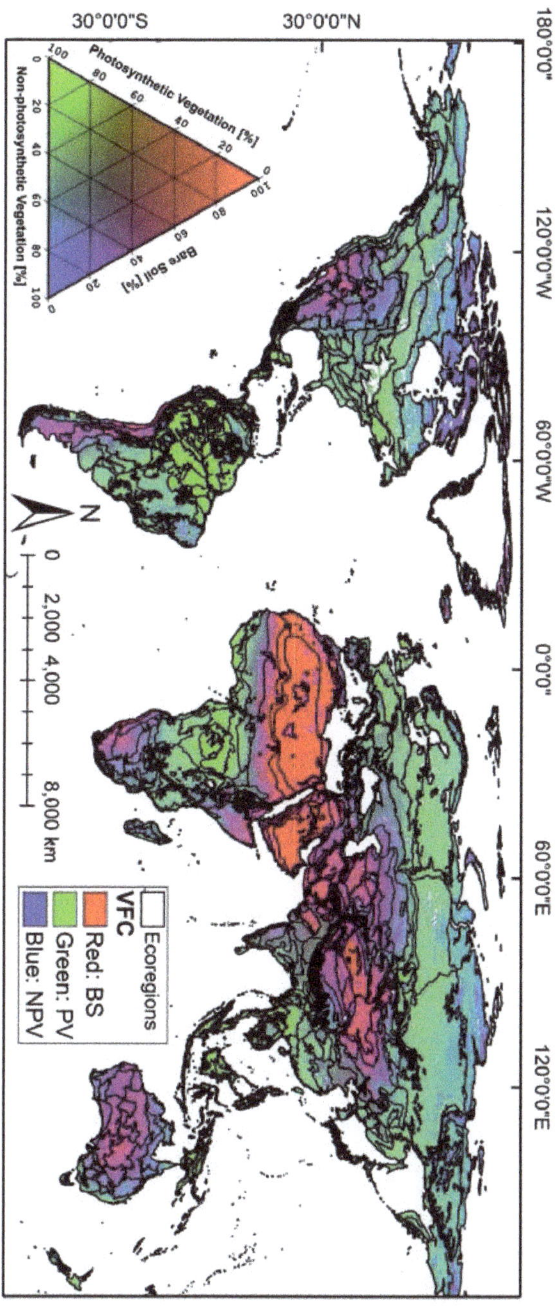

Figure 2. Global average fractional cover from 2001–2018. The ternary plot shows the correspondence between colour and the values fractions of photosynthetic vegetation (F_{PV}), non-photosynthetic vegetation (F_{NPV}) and bare soil (F_{BS}). Each ternary axis represents colours corresponding to two-factor mixtures, the greater colour intensity indicates more dominance of a single cover type.

Table 1. Mean and standard deviation of fractional cover (%) from 2001–2018 across the Realms of the Terrestrial Ecoregions of the World (TEoW). Note that overall means do not sum to exactly 100% due to temporal and spatial averaging effects.

TEoW Realms	Area (M km²)	Area (%)	F_{PV}	F_{PV} StD	F_{NPV}	F_{NPV} StD	F_{BS}	F_{BS} StD
Tropical and Subtropical Moist Broadleaf Forests	19.845	13.47	77.1	16.3	13.9	11.6	10.0	6.3
Tropical and Subtropical Dry Broadleaf Forests	3.805	2.58	53.8	18.5	29.0	11.1	16.9	10.1
Temperate Broadleaf and Mixed Forests	12.859	8.73	55.3	16.3	28.7	8.8	15.5	11.0
Tropical and Subtropical Grasslands, Savannas and Shrublands	19.531	13.26	42.3	23.4	32.1	11.8	24.9	22.0
Temperate Grasslands, Savannas and Shrublands	9.624	6.53	31.2	16.3	43.7	9.5	24.8	13.3
Montane Grasslands and Shrublands	5.189	3.52	18.3	21.1	43.0	11.7	39.0	21.5
Tundra	11.234	7.63	38.0	20.2	49.7	13.7	12.5	14.8
Mangroves	0.348	0.24	58.2	19.8	29.5	15.5	12.2	9.2
Flooded Grasslands and Savannas	1.095	0.74	43.1	22.7	31.9	11.9	24.3	21.1
Mediterranean Forests, Woodlands and Scrub	3.267	2.22	29.7	18.5	38.8	10.1	30.9	18.8
Deserts and Xeric Shrublands	27.948	18.97	6.8	12.0	27.2	18.3	66.2	25.7
Tropical and Subtropical Coniferous Forests	0.644	0.44	57.3	17.4	30.6	12.8	11.9	6.6
Temperate Conifer Forests	4.365	2.96	51.6	20.3	32.5	12.2	15.4	13.1
Inland Water	0.685	0.46	21.4	23.0	46.9	17.6	27.5	16.2
Rock and Ice	10.840	7.36	2.7	8.4	44.4	23.1	45.9	26.4
Boreal Forests/Taiga	16.042	10.89	59.0	9.2	31.6	9.0	8.2	4.7

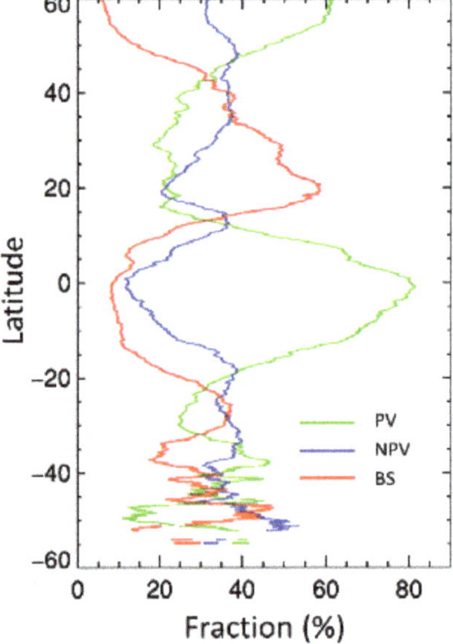

Figure 3. Global latitudinal variation in average fractional cover of F_{PV}, F_{NPV}, and F_{BS}.

3.2. Variation in Vegetation Fractional Cover within Global Livestock Production Systems

The GLPS types exhibit wide variation in mean F_{PV} across ecoregions with large ranges between upper and lower hinges and fences. There was less variation in mean F_{NPV} and F_{BS}, although the latter exhibited wide variation in the upper and lower fence values and more extreme (outlier) ecoregions than F_{PV} and F_{NPV} (Figure 4). Median and hinge values for F_{PV} increase from hyper-arid (Y) to arid (A) to humid (H) for both grazing (L) and mixed (M) land uses, while median and hinge values for temperate (T) systems were similar to arid systems for grazing and mixed land uses. The average F_{BS} also exhibits a wide range of values across ecoregions within a GLPS type with values > 70% and ≤ 5% for LGA and LGT suggesting differences in interactions between the vegetation and livestock type and management. Average F_{BS} is notably low across ecoregions for both LGH and MRH. Average F_{NPV} exhibits a lower range across ecoregions within a GLPS type, and between GLPS types than F_{PV}, with median values of ecoregion means across all GLPS types lying between 25%–45%. A gross indication of the impact of the GLPS on vegetation cover can be gained from comparing the medians and ranges of the Other class with those of the production systems. For F_{PV}, the median value is higher and the range is wider than all GLPS types except for the LGH, MRH, and MIH types where tropical pasture and cropland conversion results in higher productivity. Similarly, F_{BS} for the Other class is lower than all GLPS types except for the LGH, MRH, and MIH types.

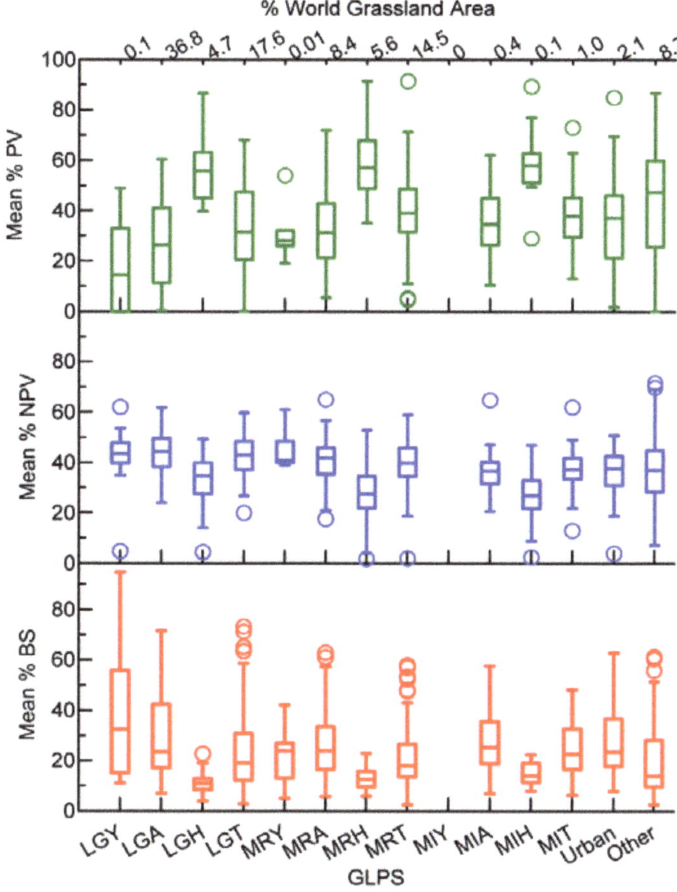

Figure 4. Fence box plots showing the medians, upper and lower hinges and upper and lower fences of mean long-term fractional cover of ecoregions over within Global Livestock Production System (GLPS) classes.

3.3. Variation in Average Vegetation Fractional Cover in World Grassland Type Divisions

The WGT Ecoregions are organized by Formations and Divisions (Table A2) with the Formations representing a broad structural type, and the Divisions representing the regional geographical representations of these Formations. The WGT Divisions exhibit a wide range of long-term average levels of F_{PV}, F_{NPV}, and F_{BS} and substantial variation in the levels of the upper and lower hinges and fences (Figure 5) There is also major variation in the standard deviations of VFC among GLPS classes with a WGT Division. Across the WGT, F_{PV} exhibits greater variation in upper and lower hinge values than F_{NPV} and F_{BS}. The Divisions within the TLSGS and TMSGS Formations (lowland and montane savanna woodlands) exhibit relatively high average F_{PV} and low average F_{BS} except for the NSSDSG Division which covers the Sahelian region of northern Africa. Among the other Divisions, the MBDG, MSACSDSG, PCSDSG, TACSDSG, and AWSDSG within the CSDSG (semi-desert) and MSGFM (Mediterranean) Formations are notable for very low average F_{PV}, median levels of F_{NPV} close to 50%, and relatively high levels of F_{BS}. The WEGS Division in western Eurasia, and the montane AMGS, and the alpine CAASFMG Divisions exhibit wide variation in average F_{PV} but WEGS exhibits a lower median and much lower range between hinge values for standard deviation than AMGS and CAASFMG.

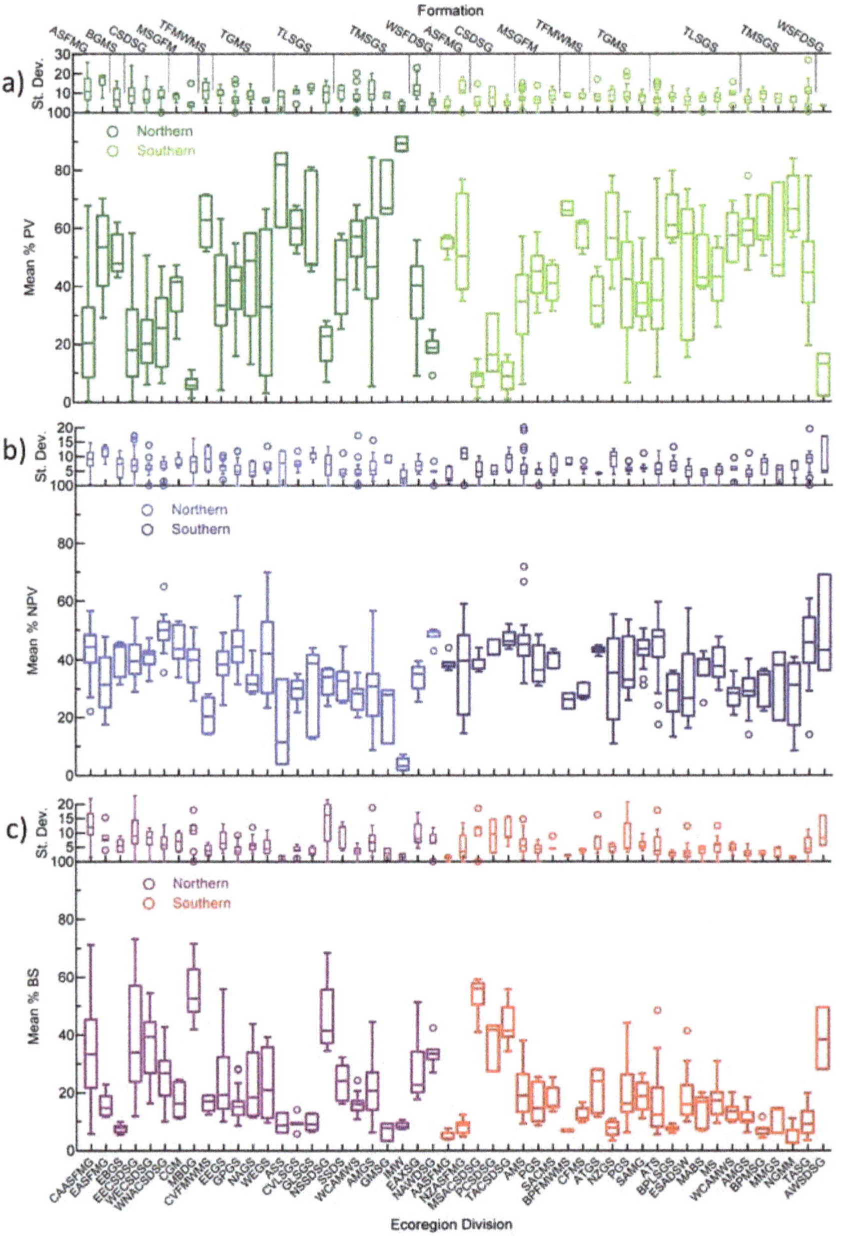

Figure 5. Fence box plots showing the medians, hinges, and inner and outer fences of average long-term vegetation fractional cover of ecoregions within World Grassland Type Divisions (listed in Table A2. (**a**) F_{PV}; (**b**) F_{NPV}; and (**c**) F_{BS}.

3.4. Trends in Vegetation Fractional Cover

3.4.1. World Grassland Types

The spatial distribution of pixels exhibiting significant trends at p < 0.1 is shown in Figure 6. Across the WGT Ecoregions, the extent of significant trends was greater for F_{BS} than for the vegetation fractions since vegetation cover trends were split between change in F_{PV} and change in F_{NPV} and these vary in different ways for different geographies. There are significant positive and negative long-term trends in F_{PV}, F_{NPV}, and across the WGT Ecoregions (Figure 7). The WGT Ecoregions are displayed over the global map of countries, showing that the significant trends affect many African and Eurasian countries whilst the USA, Brazil, Argentina, and Australia with their large areas of grassland and savanna experience change over very large areas. There are positive trends in F_{PV} in the northern great plains of North America, parts of China, and parts of eastern and southern Africa (Figure 7a). There are notable negative trends in F_{NPV} in Patagonia, across Sahelian and Sudanian Africa, in Mongolia and on the Tibetan Plateau, and in the Mitchell Grasslands of Northern Australia (Figure 7b). However, the largest areas of significant trends occur for F_{BS} with negative trends across the Great Plains of North America, across central China, across northern and southern Australia, and areas of positive trends in Sahelian, Sudanian and East Africa, western Mongolia, Ukraine and southern Russia, and the Mitchell Grasslands of Northern Australia (Figure 7c).

Examination of the variation due to GLPS in trends for WGT Divisions show that although the median values within Divisions seldom exceed ± 0.2%, the lower hinge values are as low as −0.5% and the upper hinge values are as high as 0.4% for F_{PV}, as low as −0.4%, as high as 0.4% for F_{NPV}, and as high as 0.4% for F_{BS} and as low as −0.5% for F_{BS} (Figure 8). It is particularly notable that certain Divisions (CAASMG, EECSDSG, WECSDSG, EEGS, NAGS, and WEGS) contain individual GLPS-Ecoregion combinations with positive and negative trends in F_{NPV} as great as ± 0.5% and in F_{BS} as low as −1.0% and as high as 0.5% that sit well beyond the upper and lower fences (outliers). These Divisions are in the northern hemisphere and are in the alpine ASFMG, semi-desert CSDSG, Mediterranean MSGFM, and temperate TGMS Formations suggesting major land use effects between GLPS types. Among the Divisions in the TLSGS and TMSGS Formations (savanna woodlands), most northern hemisphere Divisions show small increasing trends in F_{PV} and relatively insignificant changes in F_{NPV} and F_{BS}. However, in the southern hemisphere the lower hinge and fence values for F_{PV}, upper and lower hinge and fence values for F_{NPV}, and upper hinge and fence values for F_{BS} in the mopane and bushveld savannas (MS), and the montane WCAMWS, AMGS, and BPMSG identify GLPS classes with negative trends in F_{PV} and positive trends in F_{NPV} and F_{BS} (Figure 8).

The fence box analysis describes the range in variation in VFC trends across GLPS types within WGT Divisions, however the Divisions vary in size and this may mask localised strong trends in large Divisions (such as the Sahelian Acacia Savanna which is a single enormous Ecoregion) and emphasize trends in small Divisions with more uniform climate, terrain, and edaphic features (such as the Ethiopian Montane Grasslands and Woodlands). Hence, it is important to define the area and area percentages where significant positive and negative trends in VFC are occurring (Figure 9). The area analysis identifies eight Divisions where areas in excess of 200,000 km^2 show significant positive trends in F_{PV} (EBGS, EEGS, GPGS, WCAMWS, EAXSG, AMS, ATS, and BPLSGS) and seven Divisions where areas in excess of 100,000 km^2 show significant negative trends in F_{PV} (EBGS, EEGS, GPGS, EAXSG, PGS, BPLSGS, and AWSDSG; Figure 9a). Note that five Divisions have both large areas of positive and negative trends. By contrast, some of the smallest Divisions in area, have significant positive or negative trends in F_{PV} over a large proportion of their area: greater than 60% of IMW and AAFSMG and about 50% of NGMM have positive trends in F_{PV}, while greater than 20% of CGM, CVFMWMS, PGS, AMGS, and MMGS have significant negative trends in F_{PV}.

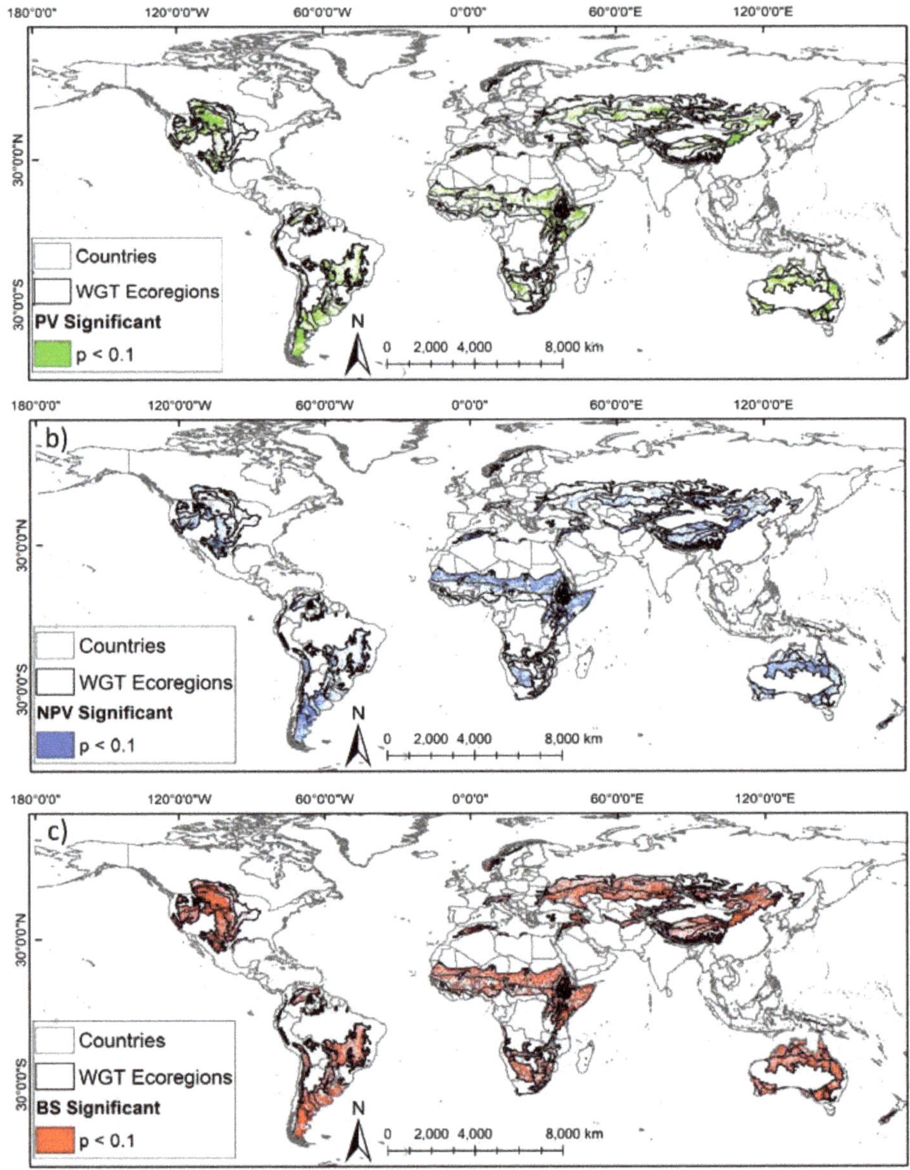

Figure 6. (**a**) The spatial distribution of pixels exhibiting significant long-term trends at p < 0.1 from 2001–2018 for F_{PV}; (**b**) The spatial distribution of pixels exhibiting significant long-term trends at p < 0.1 from 2001–2018 for F_{NPV}; and (**c**) The spatial distribution of pixels exhibiting significant long-term trends at p < 0.1 from 2001–2018 for F_{BS}.

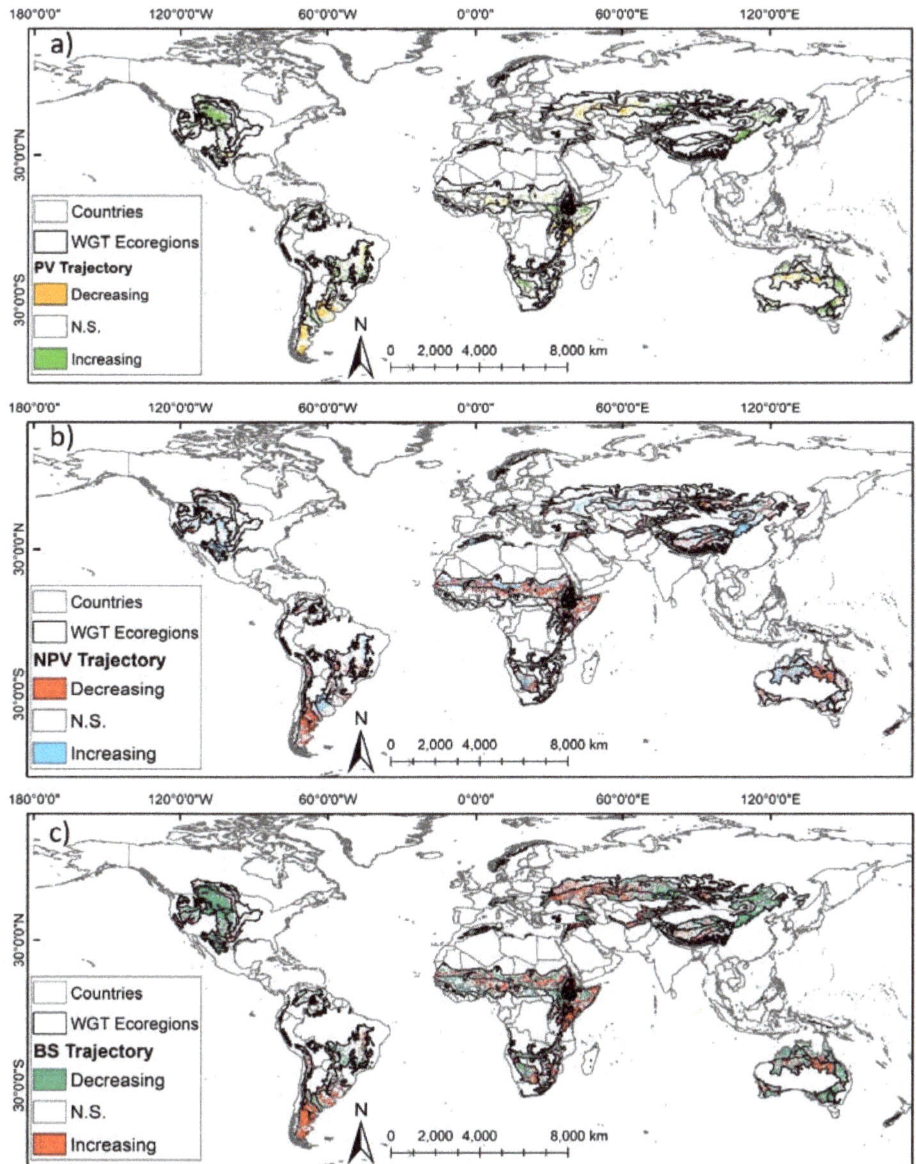

Figure 7. (**a**) Geographical pattern of significant ($p < 0.1$) positive or negative long-term trajectory of vegetation fractional cover for: F_{PV}; (**b**) Geographical pattern of significant ($p < 0.1$) positive or negative long-term trajectory of vegetation fractional cover for F_{NPV}; and (**c**) Geographical pattern of significant ($p < 0.1$) positive or negative long-term trajectory of vegetation fractional cover for F_{BS}.

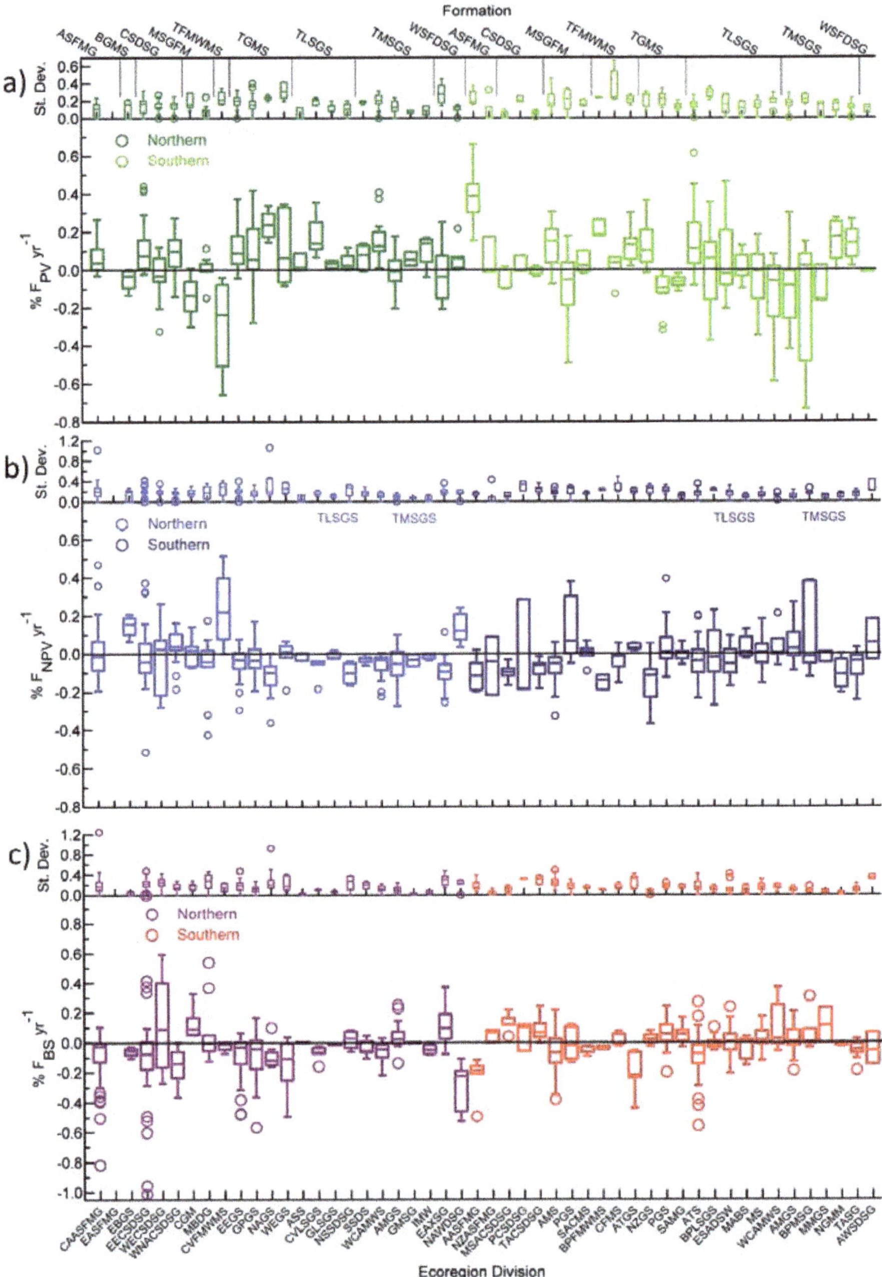

Figure 8. Fence box plots showing the median, upper and lower hinges and upper and lower fences of the average slope of the regression trend line from 2001–2018 for fractional cover within WGT vegetation divisions (listed in Table A2). (**a**) F_{PV}; (**b**) F_{NPV}; and (**c**) F_{BS}.

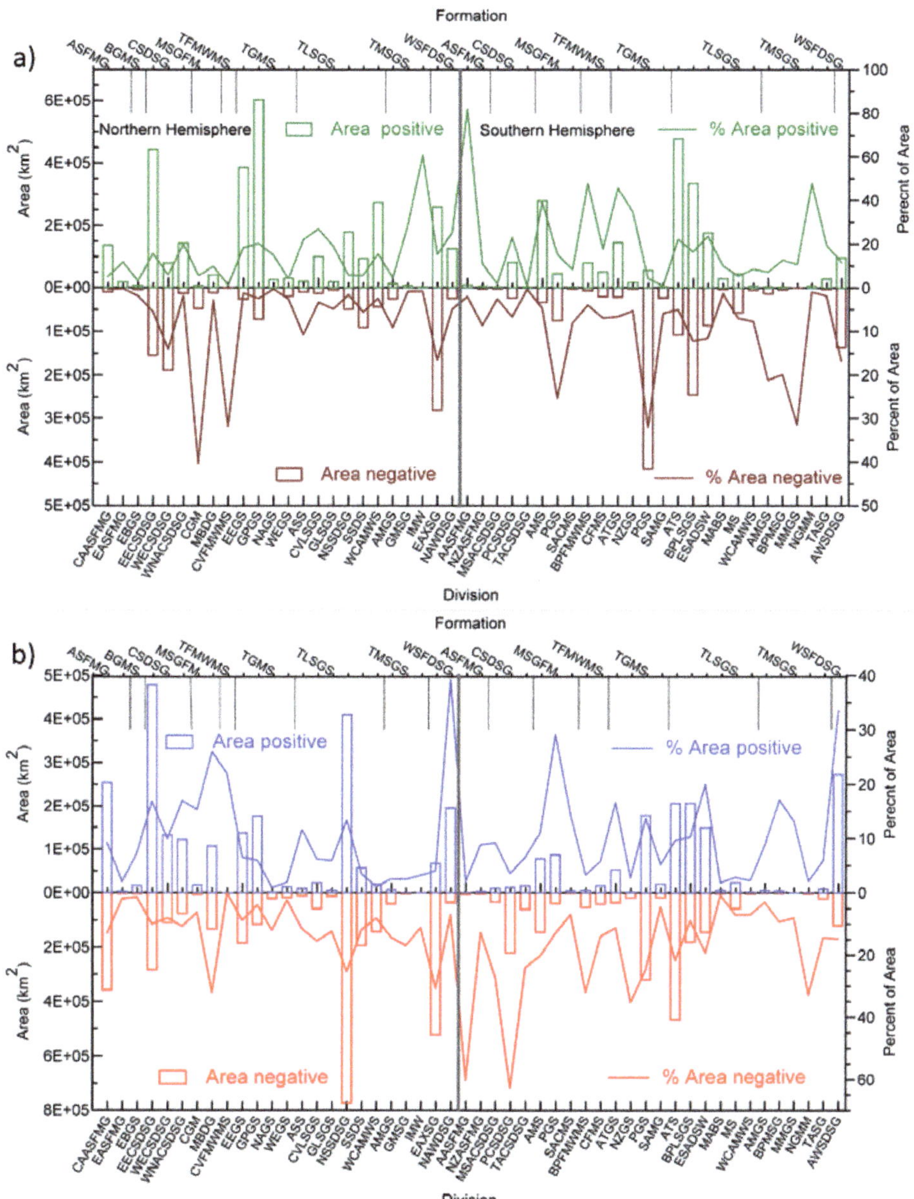

Figure 9. *Cont.*

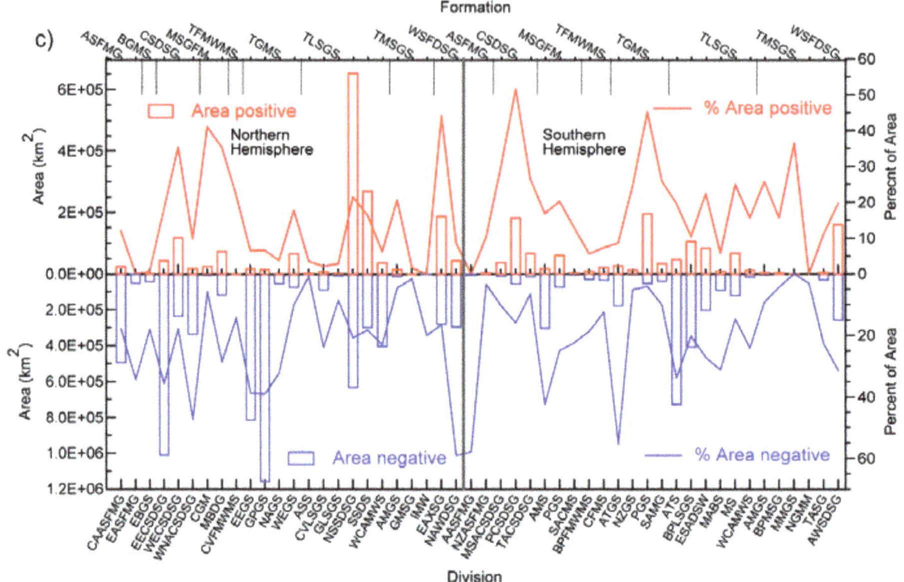

Figure 9. Area (km^2) and percentage of area of WGT Divisions exhibiting significant positive or negative trends in (**a**) F_{PV}, (**b**) F_{NPV}, and (**c**) F_{BS}. Key to Formation and Division acronyms in Table A2.

There are also large areas of significant change in F_{NPV} with 13 Divisions having areas in excess of 100,000 km^2 showing significant positive trends in F_{NPV} and seven Divisions having areas in excess of 200,000 km^2 show significant negative trends in F_{NPV} (Figure 9b). As with F_{PV}, there are several Divisions with large areas of positive and negative trends. In addition, greater than 20% of the area of MBDG, NAWDSG, PGS, and AWSDSG show positive trends and greater than 20% of the area of MBDG, NSSDSG, EAXSG, AASFMG, PCSDSG, BPFMWMS, NZGS, and NGMM show a negative trend in F_{NPV}. Percentage areas of significant trends tended to be greater for F_{BS} than for F_{NPV} and F_{PV}, since F_{BS} reflects combined changes in F_{PV} and F_{NPV} (Figure 9c). There were 15 Divisions with greater than 200,000 km^2 showing negative trends in F_{BS}.

Only five Divisions had areas greater than 200,000 km^2 showing positive trends in F_{BS}. However, 21 Divisions had greater than 20% of their area showing positive trends in F_{BS} (WECSDSG, CVFMWMS, CGM, MBDG, WEGS, NSSDSG, SSDS, EAXBG, MSACSDSG, PCSDSG, TACSGSG, AMS, NZGS, PGS, SAMG, ATS, ESADSW, MS, AMGS, BPMSG, AWSDSG, MMGS). Large areas of both positive and negative trends in F_{BS} only occurred in three Divisions (NSSDSG, SSDS, AWSDSG). There 19 Divisions with greater than 20% of their area showing negative trends in F_{BS}.

3.4.2. Savanna Woodland and Scrub Grasslands

Within the SWSG (see Table A3 for a list of ecoregion names and associated Divisions), individual ecoregions exhibit wide variation in the percentage of their area with significant positive and negative trends in F_{PV}, F_{NPV}, and F_{BS} (Figure 10). In the northern hemisphere parts of Africa, greater than 50% of the area of the Saharan flooded grasslands and the Kinabalu montane alpine meadows show significant positive trends in F_{PV} (Figure 10a). By contrast, more than 20% of the areas of the Masai Xeric Grasslands and Shrublands and the Northern and Southern Acacia-Commiphora Bushlands show significant negative trends in F_{PV}. In North and South America, the Chihuahan desert, Llanos, Guianian savanna, Rio Negro campinarana, and Pantepui have more than 20% of their areas exhibiting significant positive trends in F_{PV}. In the southern hemisphere, the Australian savannas and scrublands and South American lowland and montane savannas exhibit significant positive trends in F_{PV}, with 50%

or more of the area of the Brigalow Tropical Savanna, Central Range Sub-Alpine Grasslands, Cordillera de Merida píramo, and Northern Andean píramo exhibiting positive trends F_{PV}. However, in southern Africa, the Victoria Basin forest-savanna mosaic, Madagascan ericoid thicket, Angolan montane forest-grassland mosaic, and the Rwenzori-Virunga montane moorlands all have greater than 20% of their areas exhibiting negative trends in F_{PV}.

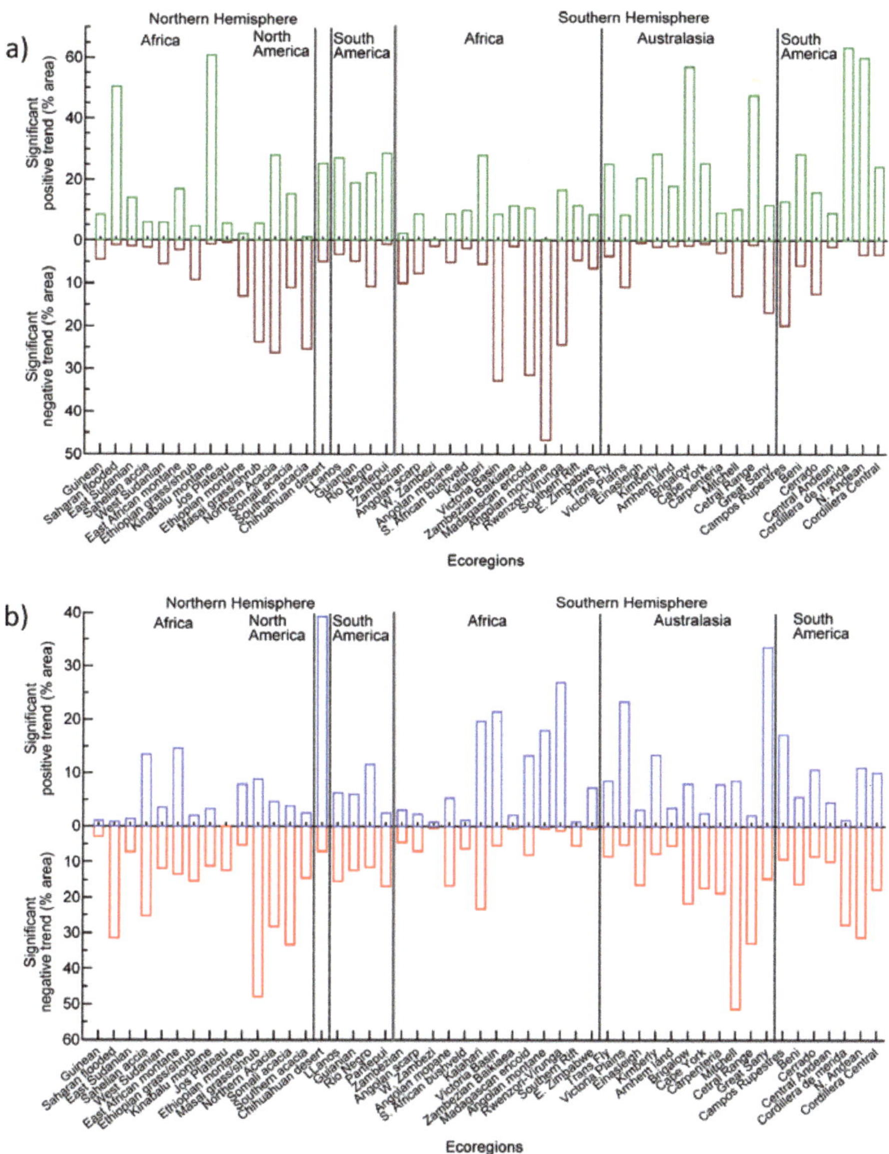

Figure 10. *Cont.*

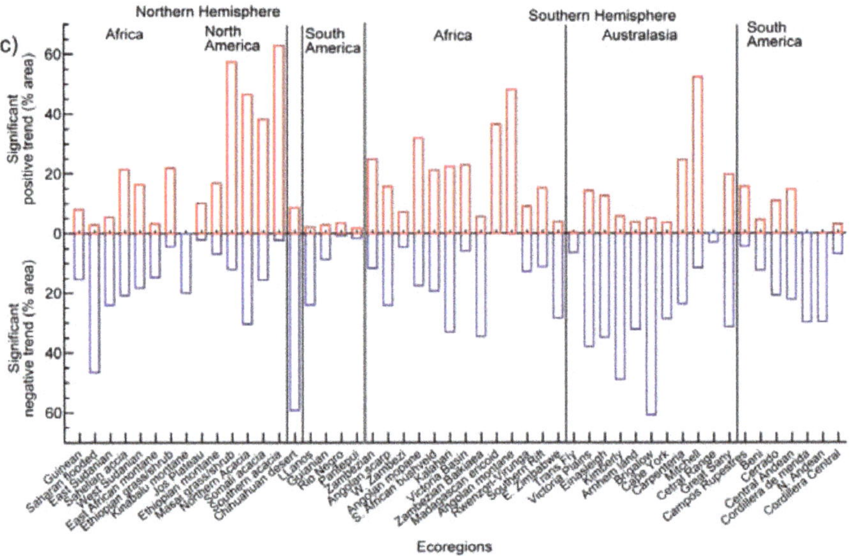

Figure 10. Area (%) of savanna, woodland and shrubland WGT Ecoregions showing significant monotonic trends in (**a**) F_{PV}; (**b**) F_{NPV}; and (**c**) F_{BS}.

The Masai Xeric Grasslands and Shrublands and the Northern Acacia-Commiphora Bushlands also had large areas of negative trends in F_{NPV}, while the Saharan Flooded Grasslands, Victoria Basin forest-savanna mosaic, Madagascan ericoid thicket, Angolan montane forest-grassland mosaic, and the Rwenzori-Virunga montane moorlands had large areas of positive trends in F_{NPV} (Figure 10b). However, there are several ecoregions showing large areas of significant trends in F_{NPV} not strongly associated with area patterns for F_{PV}. Large areas of positive trends in F_{NPV} occur in the Chihuahan desert or North America and the Great Sandy Desert of Australia, and large areas of negative trends in F_{NPV} occur in the Mitchell Grasslands and Central Range Sub-Alpine Grasslands of Australia (Figure 10b).

There are very large areas of significant positive and negative trends in F_{BS} in many savanna ecoregions (Figure 10c). Across much of the Australian and South America savannas much larger percentages (> 30%) of the ecoregion areas exhibit negative trends in F_{BS} than exhibit positive trends; the Brigalow savanna has a very large area of negative trends in F_{BS}. However, the Mitchell Grasslands are the exception with around 50% of the area of this very large ecoregion exhibiting a positive trend in F_{BS}. There are some sharp distinctions between African regions in the northern hemisphere, with Masai Xeric Grasslands and Shrublands and the Northern, Southern and Somali Acacia-Commiphora Bushlands having greater than 40% of their areas exhibiting significant positive trends in F_{BS} while West African and montane East African ecoregions show much smaller and balanced areas between positive and negative trends. About 60% of the Chihuahan desert ecoregion exhibits a negative trend in F_{BS} that corresponds to the areas of positive trends in F_{NPV} and F_{PV}. Across the SWSG Formations, 3.13 M km^2 exhibited significant positive trends in F_{BS} and 2.79 M km^2 exhibited negative trends in F_{NPV} (Table 2). These areas represented 18.1% and 16.8%, respectively of the total area of SWSG Formations (Table 3). However, 2.31 M km^2 exhibited significant positive trends in F_{PV}, 3.89 M km^2 exhibited negative trends in F_{BS}, and 1.69 M km^2 exhibited positive trends in F_{NPV} representing 13.8%, 10.1%, and 23.3% of the total area of SWSG Formations. The TLSGS Formation contained large areas where trends exceeded ± 0.1 FC units yr^{-1} but were not significant at p < 0.1.

Table 2. Area of Savanna Woodland and Scrub Grassland (SWSG) Divisions exhibiting significant trends in Vegetation Fractional Cover with magnitudes > 0.1 or < −0.1 % yr^{-1}.

Form	Area (M km^2)	Negative Trend F_{BS} (M km^2)	Positive Trend F_{BS} (M km^2)	Negative Trend F_{NPV} (M km^2)	Positive Trend F_{NPV} (M km^2)	Negative Trend F_{PV} (M km^2)	Positive Trend F_{PV} (M km^2)
				North			
TLSGS	7.002	1.435	1.045	1.186	0.527	0.211	0.688
TMSGS	0.307	0.014	0.058	0.044	0.008	0.026	0.022
WSDSG	2.194	0.574	0.791	0.557	0.264	0.306	0.386
				South			
TLSGS	6.095	1.569	1.034	0.855	0.595	0.506	1.073
TMSGS	0.254	0.042	0.038	0.030	0.019	0.022	0.044
WSDSG	0.816	0.256	0.161	0.120	0.274	0.137	0.096
Total	16.668	3.890	3.127	2.793	1.687	1.208	2.309

Table 3. Percentage of area of SWSG Divisions exhibiting significant trends in VFC > 0.1 or < −0.1 % yr^{-1}.

Form	WGT % Area	Negative Trend F_{BS} (%)	Positive Trend F_{BS} (%)	Negative Trend F_{NPV} (%)	Positive Trend F_{NPV} (%)	Negative Trend F_{PV} (%)	Positive Trend F_{PV} (%)
				North			
TLSGS	19.7	20.5	14.9	16.9	7.5	3.0	9.8
TMSGS	0.9	0.2	0.8	0.6	0.1	0.4	0.3
WSDSG	6.2	8.2	11.3	8.0	3.8	4.4	5.5
				South			
TLSGS	17.2	22.4	14.8	12.2	8.5	7.2	15.3
TMSGS	0.7	0.6	0.5	0.4	0.3	0.3	0.6
WSDSG	2.3	3.7	2.3	1.7	3.9	2.0	1.4
Total	46.9	23.3	18.8	16.8	10.1	7.2	13.8

3.5. Variation in Vegetation Fractional Cover in Example Ecoregions

While extensive examination of VFC behaviours within ecoregions is beyond the scope of this study, it is worthwhile to examine several notable regional trends in F_{PV}, F_{NPV}, and F_{BS} and compare these across GLPS types. The east African Acacia-Commiphora Bushlands and Masai Xeric Grassland and Shrublands contained large areas of significant trends, but relativities among GLPS types were different (Figure 11a). In the Northern and Somali ecoregions, the LGH and MRH systems showed positive trends in F_{PV}, while the MRA system showed a positive trend in F_{BS}. By contrast, all GLPS systems across the Southern Acacia-Commiphora Bushland exhibited positive trends in F_{BS} and negative trends in F_{PV} and F_{NPV}. In the Masai Xeric Grassland and Shrubland, the main features are a positive trend in F_{BS} for LGA and a negative trend in F_{PV} for the Other class.

The Chihuahan Desert exhibits small to moderate positive trends in F_{PV} and F_{NPV} across all GLPS types and large negative trends in F_{BS} especial for LGA, LGT, MRA, and MRT (Figure 11b). In the Mitchell Grasslands, LGA and MRA systems dominate and both show positive trends in F_{BS} and negative trends in F_{NPV}. In the Llanos, positive trends in F_{PV} across LGA, LGH, MRA, and MRH types are matched by small negative trends in both F_{NPV} and F_{BS}.

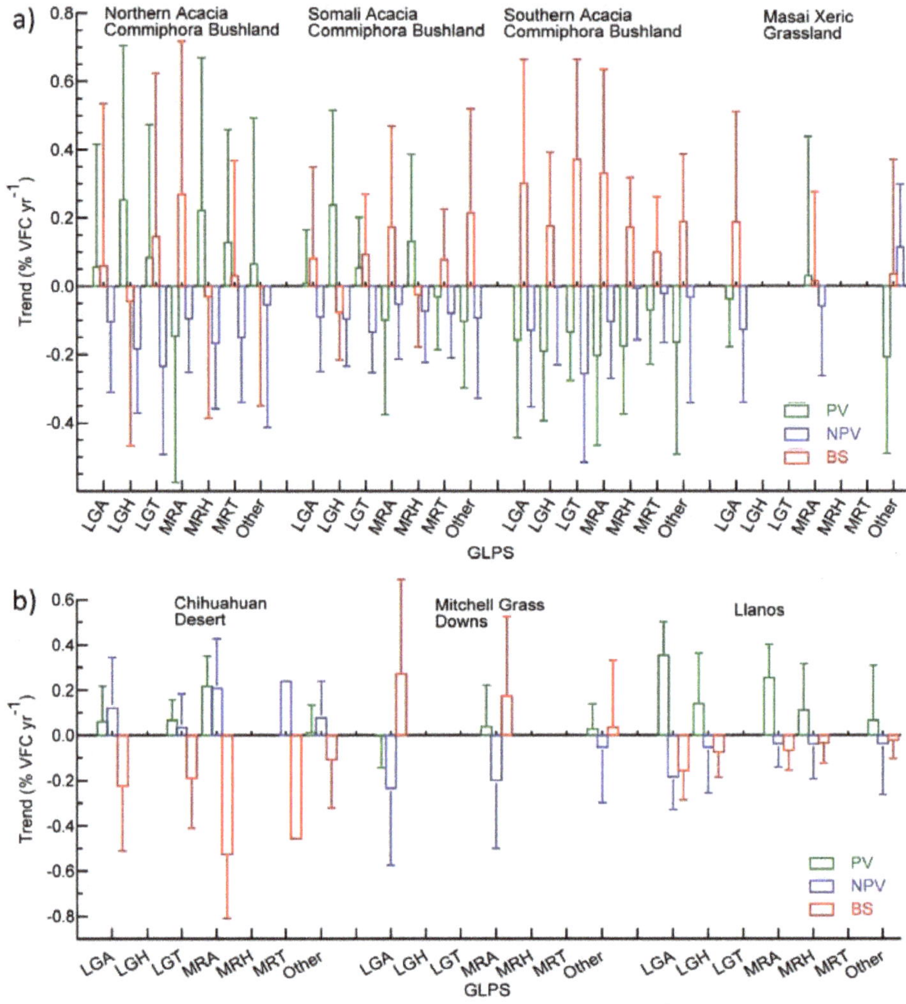

Figure 11. Trends in vegetation fractional cover among Global Livestock Production Systems (GLPS) areas within selected Ecoregions in (**a**) Africa; and (**b**) North America, South America, and Australia. The histogram indicates the average and the whiskers indicate the standard deviation among GLPS pixels within the Ecoregion.

4. Discussion

This study has introduced a unique remotely-sensed global vegetation product derived from MODIS reflectance data, the GVFCP, which is available as both an eight day and monthly 500 m resolution product. We have mapped global averages for F_{PV}, F_{NPV}, and F_{BS} for data derived from MODIS for 2001-2018. The study then explored the behaviour of F_{PV}, F_{NPV}, and F_{BS} across GLPS types within the WGT Divisions and Ecoregions including analysis of long-term trends in vegetation fractional cover with a focus on the SWSG Formations. The analysis illustrated the variation in average long-term levels and long-term trends of VFC associated with different GLPS types, with vegetation Formations and with geographically separate Divisions within the same Formation. The variation observed emphasized the strong interaction between land use management and vegetation characteristics. Although other products have previously provided global quantitative retrievals of F_{PV} and F_{BS}, the GVFCP is new and important since it provides an explicit, spectrally functional measure of F_{NPV}. This provides effective discrimination of both global regions with high average levels of F_{NPV} and high seasonal fluctuations in F_{NPV}, such as in the grassy understorey or tropical savannas, and the massive leaf litter associated with the deciduous forests of the eastern USA. It also highlights global regions where certain GLPS types or cropland expansion produce large amounts of stubble and crop residue such as southern and central China, south-eastern and south-western Australia, and the US corn belt.

The study has focused more detailed evaluation of levels and trends in F_{PV}, F_{NPV}, and F_{BS} on the grassy biomes of the world represented by WGT Ecoregions, and for this Special Issue on the SWSG Formations within the WGT. There are large differences between average levels and long-term trends in F_{PV}, F_{NPV}, and F_{BS} across the WGT. Fence box graphical analysis identified specific Divisions where average F_{BS} was low, where average F_{PV} was high, and where the upper and lower hinges, fences, and outlier points indicated massive variation in average levels within regional Formations (i.e., in different hemispheres and continents) indicative of possible climate or land use effects. Recent studies have identified global "greening" with an increase in the F_{PV} attributable to developments in India and China [20]. Our analysis agrees with [20] in terms of the "greening" of north-eastern China, the US Great Plains, and eastern Australia. It also shows that the SWSG Formations, much of which experience an LGH production system, and contain all the major grassy savanna, woodland and scrub ecoregions, tend to exhibit more areas of positive to neutral trends in F_{PV} than negative trends in F_{NPV} or positive trends in F_{BS}. Although it is difficult to make direct comparisons with the study of [19], there is broad agreement on the magnitude of areas (SWSG here, tropical dry forest and tropical shrubland in [19] experiencing negative trends in bare ground, and a positive trend in short vegetation (SV) in that study, most of the F_{PV} and F_{NPV} in SWSG in this study). These tropical humid systems either have reliable seasonal rainfall to support land cover changes (e.g., South American tropical savannas), or no conversion potential (e.g., Australian tropical savannas). However, our analysis also identifies specific Divisions and individual Ecoregions in the WGT where "browning" is occurring in Argentina, Australia, East Africa, Southern Africa, and Eurasia, some of which were also identified by [19,20]. Here, the GVFCP is able to associate the "browning" with explicit trends in both F_{BS} and F_{NPV} providing more insight into the changes that are occurring.

The analysis at WGT Divisional level identifies "hot spots" of change in VFC based on both area and percentage area of significant positive or negative trends. The total area is important because of the implications for the productivity of the ecosystem, the implications for livelihoods and food supplies, and the impacts on human and wildlife populations. The percentage areas are important because they identify levels of stress on regional and unique representatives of global vegetation Formations with consequences for biodiversity, endangered species, and extinction risks. When WGT Divisions are ranked by percentage of their area with significant positive trends in F_{BS}, the first five Divisions are the Patagonian Cool Semi-Desert Scrub and Grassland, the Pampean Grassland and Shrubland, the East African Xeric Scrub and Grassland, the Californian Grassland Meadow, and the Madagascan Montane Grassland and Shrubland with between 38% and 50% of their area affected. Potential for land degradation arising from increasing F_{BS} is occurring in large regions such as Patagonia, where aridity and grazing pressure are reducing the cover of palatable grasses [47] and the pampas of South America where conversion from woodland and pasture to cropland has occurred [48]. It can also arise in smaller regions such as the Central Valley Grasslands of California with recent land use change [49], and the very small Madagascan ericoid thicket ecoregion subject to the same pressures on land cover more widely evident across the island [50,51].

However, some of the largest areas of positive trends in F_{BS} and negative trends in F_{PV} and F_{NPV} occur in the largest WGT Divisions such as the North Sahel Semi-Desert Scrub and Grassland which is so large that it crosses many national boundaries and contains a patchwork of areas with both positive and negative trends in F_{PV}, F_{NPV}, and F_{BS}. Analysis found that although there was about 200,000 km^2 of positive trends in F_{PV}, there were massive areas of almost 800,000 km^2 showing a negative trend in F_{NPV} and about 600,000 km^2 showing a positive trend in F_{BS}. The drivers and manifestations of environmental change in the Sahel have been scientifically contested for some time [52]. The mixed trends in VFC across the NSSDSG tend to reinforce the contested debate and emphasize that both "greening" and "browning" are occurring at the same time in different locations [53,54]. For example, agricultural practices that reduce vegetation cover, i.e., produce a positive trend in F_{BS}, increase potential for wind erosion, dune migration, and soil loss [55]. Hence, negative trends in F_{NPV} may indicate removal of both dry grass from rangelands, and residue from croplands. However, relationships with precipitation support areas of greening (positive trends in F_{PV}) which may be due to agroforestry [56] and recovery of woody vegetation [54], and promotion of tree cover around Sahelian farms contrasts with reduced woody cover on farmlands in the sub-humid zone such as the WCAMWS [57].

For the SWSG Formations (TLSGS, TMSGS, and WSFDSG), the trend analysis revealed a mosaic of significant positive and negative trends in F_{BS}, F_{NPV}, and F_{PV} covering 42.1%, 26.9%, and 21% of the area, respectively. This represents a massive change in vegetation fractional cover in the grassy savanna, woodland, and scrub biomes of the world over the past 20 years. When the analysis is focused on the individual Ecoregions, trends in VFC can be associated with particular factors unique to specific Ecoregions. In East Africa, in the Acacia-Commiphora Woodlands (EAXSG), positive trends in F_{BS} may be associated with increased logging of woodlands for charcoal production [58–60], recent high frequency of droughts [61] due to a decline in long-season rains [62,63], and long-term impacts of expansion of populations, cropland and pastoralism ([64–66]. Land use changes from forest cover to cultivation, pastoralism, and plantations has led to increased surface run-off but variability exists due to site-specific factors [67]. In the Chihuahan Desert ecoregion, invasive grasses and shrubs may be changing the ecosystem dynamics leading to the observed negative trends in F_{BS} and positive trends in F_{NPV} with consequent increase in fire risk [68–70]. Since 2000, there has been a major expansion of cropping, exotic pastures, and oil palm plantations especially in the western Llanos of Colombia leading to the moderate positive trends in F_{PV} [71], with recent potential for acceleration post the Peace Agreement [72,73]. The Mitchell Grasslands of Northern Australia and especially the western Barkly Tableland exhibit a strong positive trend in F_{BS} that may relate to long term interactions between cycles of precipitation and grazing pressure [74].

The GVFCP has explicit quantitative uncertainty estimates for the eight-day 500 m product based on the multiple comprehensive cycles of calibration and validation undertaken for Australia [4,5]. In this study we used the aggregated monthly 5 km^2 resolution product which has values that represent the medoid of the eight-day values for each month averaged across a hundred 500 m pixels. The global scale for comparison of WGT Formations, Divisions, and Ecoregions and the size of even the small Ecoregions leads to the values presented in the fence box plots being derived from hundreds to hundreds of thousands of pixels with variation within and between Ecoregions well in excess of the published RMSE values of between 11% and 16% for F_{PV}, F_{NPV}, and F_{BS}. If the magnitude of trends, is examined for Ecoregions with trend values beyond the upper and lower fences, values of between ± 0.5% to ± 1.0% yr^{-1} result in changes in average VFC across Divisions and Ecoregions of between 9% and 18% over the 18 years of the time series. Upper fence values for standard deviations of pixel trend values within ecoregions were as high as 1.2% yr^{-1} and as low as 0.1% yr^{-1} indicating that there was major heterogeneity in responses within Ecoregions, related to their size, population distributions, land use and land cover, and between Ecoregions where vegetation Formation and geographical location, as well as anthropogenic factors drive variation.

There is extensive scope for much more detailed examination of the trends in the GVFCP across the globe, and especially within the WGT, and at finer scale for smaller grassy vegetation types. Specifically, there is a need to segment the time period from 2001–2018 and explore discontinuities and changes in trends that may be attributable to changes in climate cycles, land use, and production systems. In addition, there is a need to explore with spatial explicitness, the association between changes in F_{PV}, F_{NPV}, and F_{BS}; and examine the trends in total cover (F_{TC}) as an indicator of wind and water erosion potential [13]. There is also a need for the product to be widely used beyond current Australian applications and be subject to further field validation analysis especially in particular ecosystems with very different vegetation structures, arboreal vegetation phenology, and soil surface colours and conditions. Since the GVFCP is freely available on the GEOGLAM RAPP-MAP website, the authors are encouraging scientists and agencies to explore and utilize these data.

5. Conclusions

This study has presented the GVFCP and explored the levels and dynamics of F_{PV}, F_{NPV}, and F_{BS} across the grassy ecoregions of the world. It has documented benchmark long-term average levels, and areas and magnitudes of significant trends in VFC at Divisional level for WGT Ecoregions. Focused exploration of the savanna woodlands showed that:

(1) Ecoregions in both Africa and Australia are exhibiting concerning positive trends in F_{BS} probably associated with climate and land use interactions.

(2) Large areas of both positive and negative trends are occurring in individual ecoregions requiring more detailed examination of both fine scale spatial pattern and short-term trends.

(3) There is value in explicit measures of level and trend in F_{NPV} since change in dry intact herbaceous vegetation cover has huge implications for ecosystem function, livestock feed reserves, and carbon dynamics of grassy systems.

The GVFCP is made freely available to the global scientific community in the hope that they will use it and provide further validation studies. There is great potential for development of a Landsat/Sentinel 2 global product using the same methodology. This will become necessary upon the demise of the MODIS sensors since the Visible Infrared Imaging Radiometer Suite sensor does not carry the 2105–2155 nm short wave infrared channel and sensitivity to F_{NPV} and F_{BS} will likely be diminished.

Online Resources: Global and Australian products are freely available along with analytical tools at https://map.geo-rapp.org.

Author Contributions: Conceptualization, J.P.G. and M.J.H.; Methodology, J.P.G.; Validation, J.P.G. and M.J.H.; Formal analysis, M.J.H. and J.P.G.; Investigation, M.J.H.; Resources, J.P.G.; Data curation, J.P.G.; Writing—original draft preparation, M.J.H.; Writing—review and editing, M.J.H.; Visualization, J.P.G.; Project administration, J.P.G.; Funding acquisition, J.P.G. All authors have read and agreed to the published version of the manuscript.

Funding: This research spans 15 years. The initial stage was partially funded by the NASA Earth Science Enterprise Carbon Cycle Science research program (NRA04-OES-010). Development of the current Australian and Global products was funded by the Australian Government's National Landcare Program and CSIRO. GEOGLAM RAPP is an initiative of CSIRO, GEO, and GEOGLAM.

Acknowledgments: The GVFCP is maintained for GEOGLAM RAPP by Biswajit Bala.

Conflicts of Interest: The authors declare no conflict of interest. The funders had no role in the design of the study; in the collection, analyses, or interpretation of data; in the writing of the manuscript, or in the decision to publish the results.

Appendix A

Table A1. Acronyms and abbreviations used (in addition to WGT acronyms documented in Tables A2 and A3).

Acronyms	Full Name
AVFCP	Australian Vegetation Fractional Cover Product
AVHRR	Advanced Very High Resolution Radiometer
BRDF	Bi-directional Reflectance Distribution Function
BS	Bare Soil
CAI	Cellulose Absorption Index
FAPAR	Fraction of Absorbed Photosynthetically Active Radiation
GEOGLAM	Group on Earth Observations Global Agricultural Monitoring
GIMMS	Global Inventory Modeling and Mapping Studies
GLPS	Global Livestock Production Systems
GVFCP	Global Vegetation Fractional Cover Product
MODIS	Moderate Resolution Imaging Spectroradiometer

Table A1. *Cont.*

Acronyms	Full Name
NBAR	Nadir BRDF-Adjusted Reflectance
NDVI	Normalized Difference Vegetation Index
NPV	Non-Photosynthetic Vegetation
PV	Photosynthetic Vegetation
RAPP	Rangeland and Pasture Productivity
SWIR32	Short Wave Infrared Ratio
TEoW	Terrestrial Ecoregions of the World
VFC	Vegetation Fractional Cover
WGT	World Grassland Type

Table A2. World Grassland Type Formations and Divisions (Dixon et al., 2014).

Formation Name (Formation Code) Division Name	Division Code	Number of Ecoregions
NORTHERN HEMISPHERE		
Alpine Scrub, Forb Meadow and Grassland (ASFMG)		
Central Asian Alpine Scrub, Forb Meadow and Grassland	CAASFMG	14
European Alpine Scrub, Forb Meadow and Grassland	EASFMG	1
Boreal Grassland, Meadow and Shrubland (BGMS)		
Eurasian Boreal Grassland, Meadow and Shrubland	EBGS	2
Cool Semi-Desert Scrub and Grassland (CSDSG)		
Eastern Eurasian Cool Semi-Desert Scrub and Grassland	EECSDSG	12
Western Eurasian Cool Semi-Desert Scrub and Grassland	WECSDSG	4
Western North American Cool Semi-Desert Scrub and Grassland	WNACSDSG	4
Mediterranean Scrub, Grassland and Forb Meadow (MSGFM)		
California Grassland and Meadow	CGM	2
Mediterranean Basin Dry Grassland	MBDG	2
Temperate Grassland, Meadow and Shrubland (TGSM)		
Eastern Eurasian Grassland and Shrubland	EEGS	8
Great Plains Grassland and Shrubland	GPGS	15
Northeast Asia Grassland and Shrubland	NAGS	4
Western Eurasian Grassland and Shrubland	WEGS	2
Tropical Freshwater Marsh, Wet Meadow and Shrubland TFMWMS)		
Colombian-Venezuelan Freshwater Marsh, Wet Meadow and Shrubland	CVFMWMS	1
Tropical Lowland Shrubland, Grassland and Savanna (TLSGS)		
Amazonian Shrubland and Savanna	ASS	1
Colombian-Venezuelan Lowland Shrubland, Grassland and Savanna	CVLSGS	
Guianan Lowland Shrubland, Grassland and Savanna	GLSGS	1
North Sahel Semi-Desert Scrub and Grassland	NSSDSG	1
Sudano Sahelian Dry Savanna	SSDS	1
West-Central African Mesic Woodland and Savanna	WCAMWS	3
Tropical Montane Shrubland, Grassland and Savanna (TMSGS)		
African Montane Grassland and Shrubland	AMGS	4
Guianan Montane Shrubland and Grassland	GMSG	1
Indomalayan Montane Meadow	IMW	1
Warm Semi-Desert Scrub and Grassland (WSDSG)		
Eastern Africa Xeric Scrub and Grassland	EAXSG	4
North American Warm Desert Scrub and Grassland	NAWDSG	1
SOUTHERN HEMISPHERE		
Alpine Scrub, Forb Meadow and Grassland (ASFMG)		
Australian Alpine Scrub, Forb Meadow and Grassland	AASFMG	1
New Zealand Alpine Scrub, Forb Meadow and Grassland	NZASFMG	1
Cool Semi-Desert Scrub and Grassland (CSDSG)		
Mediterranean and Southern Andean Cool Semi-Desert Scrub and Grassland	MSACSDSG	1
Patagonian Cool Semi-Desert Scrub and Grassland	PCSDSG	1
Tropical Andean Cool Semi-Desert Scrub and Grassland	TACSDSG	1
Mediterranean Scrub, Grassland and Forb Meadow (MSGFM)		
Australian Mediterranean Scrub	AMS	7
Pampean Grassland and Shrubland (semi-arid Pampa)	PGS	4
South African Cape Mediterranean Scrub	SACMS	1

Table A2. *Cont.*

Formation Name (Formation Code) Division Name	Division Code	Number of Ecoregions
Temperate Grassland, Meadow and Shrubland (TGMS)		
Australian Temperate Grassland and Shrubland	ATGS	1
New Zealand Grassland and Shrubland	NZGS	1
Southern African Montane Grassland	SAMG	3
Tropical Freshwater Marsh, Wet Meadow and Shrubland (TFMWMS)		
Brazilian-Parana Freshwater Marsh, Wet Meadow and Shrubland	BPFMWMS	1
Chaco Freshwater Marsh and Shrubland	CFMS	1
Tropical Lowland Shrubland, Grassland and Savanna (TLSGS)		
Australian Tropical Savanna	ATS	9
Brazilian-Parana Lowland Shrubland, Grassland and Savanna	BPLSGS	2
Eastern and Southern African Dry Savanna and Woodland	ESADSW	2
Miombo and Associated Broadleaf Savanna	MABS	2
Mopane Savanna	MS	3
Tropical Montane Shrubland, Grassland and Savanna (TMSGS)		
Madagascan Montane Grassland and Shrubland	MMGS	1
African Montane Grassland and Shrubland	AMGS	4
Brazilian-Parana Montane Shrubland and Grassland	BPMSG	1
New Guinea Montane Meadow	NGMM	1
Tropical Andean Shrubland and Grassland	TASG	4
Warm Semi-Desert Scrub and Grassland (WSDSG)		
Australia Warm Semi-Desert Scrub and Grassland	AWSDSG	1

Table A3. Savanna, woodland and shrubland ecoregions with World Grassland Types (Dixon et al., 2014). (See Table A2 for description of Formations and Divisions; Continent codes as follows: AF—Africa; NA—North America; SA—South America; AU—Australasia (includes New Guinea and New Zealand)).

Formation Code	ECO_CODE	Division Code	Hemisphere	Continent	Ecoregion Name
TLSGS	AT0707	WCAMWS	N	AF	Guinean forest-savanna mosaic
TLSGS	AT0905	WCAMWS	N	AF	Saharan flooded grasslands
TLSGS	AT0705	WCAMWS	N	AF	East Sudanian savanna
TLSGS	AT0713	NSSDSG	N	AF	Sahelian Acacia savanna
TLSGS	AT0722	SSDS	N	AF	West Sudanian savanna
TMSGS	AT1005	AMGS	N	AF	East African montane moorlands
TMSGS	AT1007	AMGS	N	AF	Ethiopian montane grasslands and woodlands
TMSGS	IM1001	IMW	N	AF	Kinabalu montane alpine meadows
TMSGS	AT1010	AMGS	N	AF	Jos Plateau forest-grassland mosaic
TMSGS	AT1008	AMGS	N	AF	Ethiopian montane moorlands
WSDSG	AT1313	EAXSG	N	AF	Masai xeric grasslands and shrublands
WSDSG	AT0711	EAXSG	N	AF	Northern Acacia-Commiphora bushlands and thickets
WSDSG	AT0715	EAXSG	N	AF	Somali Acacia-Commiphora bushlands and thickets
WSDSG	AT0716	EAXSG	N	AF	Southern Acacia-Commiphora bushlands and thickets
WSDSG	NA1303	NAWDSG	N	NA	Chihuahuan desert
TLSGS	NT0709	CVLSGS	N	SA	Llanos
TLSGS	NT0707	GLSGS	N	SA	Guianan savanna
TLSGS	NT0158	ASS	N	SA	Rio Negro campinarana
TMSGS	NT0169	GMSG	N	SA	Pantepui

Table A3. *Cont.*

Formation Code	ECO_CODE	Division Code	Hemisphere	Continent	Ecoregion Name
TLSGS	AT0725	MS	S	AF	Zambezian and Mopane woodlands
TLSGS	AT1002	WCAMWS	S	AF	Angolan scarp savanna and woodlands
TLSGS	AT0724	MABS	S	AF	Western Zambezian grasslands
TLSGS	AT0702	MS	S	AF	Angolan Mopane woodlands
TLSGS	AT0717	MS	S	AF	Southern Africa bushveld
TLSGS	AT1309	ESADSW	S	AF	Kalahari xeric savanna
TLSGS	AT0721	ESADSW	S	AF	Victoria Basin forest-savanna mosaic
TLSGS	AT0726	MABS	S	AF	Zambezian Baikiaea woodlands
TMSGS	AT1011	MMGS	S	AF	Madagascar ericoid thickets
TMSGS	AT1001	AMGS	S	AF	Angolan montane forest-grassland mosaic
TMSGS	AT1013	AMGS	S	AF	Rwenzori-Virunga montane moorlands
TMSGS	AT1015	AMGS	S	AF	Southern Rift montane forest-grassland mosaic
TMSGS	AT1006	AMGS	S	AF	Eastern Zimbabwe montane forest-grassland mosaic
TLSGS	AA0708	ATS	S	AU	Trans Fly savanna and grasslands
TLSGS	AA0709	ATS	S	AU	Victoria Plains tropical savanna
TLSGS	AA0705	ATS	S	AU	Einasleigh upland savanna
TLSGS	AA0706	ATS	S	AU	Kimberly tropical savanna
TLSGS	AA0701	ATS	S	AU	Arnhem Land tropical savanna
TLSGS	AA0702	ATS	S	AU	Brigalow tropical savanna
TLSGS	AA0703	ATS	S	AU	Cape York Peninsula tropical savanna
TLSGS	AA0704	ATS	S	AU	Carpentaria tropical savanna
TLSGS	AA0707	ATS	S	AU	Mitchell grass downs
TMSGS	AA1002	NGMM	S	AU	Central Range sub-alpine grasslands
WSDSG	AA1304	AWSDSG	S	AU	Great Sandy-Tanami desert
TMSGS	NT0703	BPMSG	S	NA	Campos Rupestres montane savanna
TLSGS	NT0702	BPLSGS	S	SA	Beni savanna
TLSGS	NT0704	BPLSGS	S	SA	Cerrado
TMSGS	NT1003	TASG	S	SA	Central Andean wet puna
TMSGS	NT1005	TASG	S	SA	Cordillera de Merida píramo
TMSGS	NT1006	TASG	S	SA	Northern Andean píramo
TMSGS	NT1004	TASG	S	SA	Cordillera Central píramo

References

1. Ustin, S.L.; Gamon, J.A. Remote sensing of plant functional types. *New Phytol.* **2006**, *186*, 795–816. [CrossRef] [PubMed]
2. Guan, K.; Wood, E.F.; Caylor, K.K. Multi-sensor derivation of regional vegetation fractional cover in Africa. *Remote Sens. Environ.* **2012**, *124*, 653–665. [CrossRef]
3. Guerschman, J.P.; Hill, M.J.; Renzullo, L.J.; Barrett, D.J.; Marks, A.S.; Botha, E.J. Estimating fractional cover of photosynthetic vegetation, non-photosynthetic vegetation and bare soil in the Australian tropical savanna region upscaling the EO-1 Hyperion and MODIS sensors. *Remote Sens. Environ.* **2009**, *113*, 928–945. [CrossRef]
4. Guerschman, J.P.; Scarth, P.F.; McVicar, T.R.; Renzullo, L.J.; Malthus, T.J.; Stewart, J.B.; Rickards, J.E.; Trevithick, R. Assessing the effects of site heterogeneity and soil properties when unmixing photosynthetic vegetation, non-photosynthetic vegetation and bare soil fractions from Landsat and MODIS data. *Remote Sens. Environ.* **2015**, *161*, 12–26. [CrossRef]
5. Guerschman, J.P.; Hill, M.J. Calibration and validation of the Australian fractional cover product for MODIS collection 6. *Remote Sens. Lett.* **2018**, *9*, 696–705. [CrossRef]
6. Rickards, J.; Stewart, J.; McPhee, R.; Randall, L. Australian ground cover reference sites database 2014: User guide for PostGIS. *Victoria* **2014**, *119*, 74. Available online: https://data.gov.au/data/dataset/68963ddf-fa83-43fe-86bd-2f27ec0284f4 (accessed on 25 January 2020).

7. Nagler, P.L.; Daughtry, C.S.T.; Goward, S.N. Plant litter and soil reflectance. *Remote Sens. Environ.* **2000**, *71*, 207–215. [CrossRef]

8. Nagler, P.L.; Inoue, Y.; Glenn, E.P.; Russ, A.L.; Daughtry, C.S.T. Cellulose absorption index (CAI) to quantify mixed soil-plant litter scenes. *Remote Sens. Environ.* **2003**, *87*, 310–325. [CrossRef]

9. Daughtry, C.S.T. Discriminating crop residues from soil by shortwave infrared reflectance. *Agron. J.* **2001**, *93*, 125–131. [CrossRef]

10. Daughtry, C.S.T.; Hunt, E.R.; Doraiswamy, P.C.; McMurtrey, J.E. Remote sensing the spatial distribution of crop residues. *Agron. J.* **2005**, *97*, 864–871. [CrossRef]

11. Daughtry, C.S.T.; Doraiswamy, P.C.; Hunt, E.R.; Stern, A.J.; McMurtrey, J.E.; Prueger, J.H. Remote sensing of crop residue cover and soil tillage intensity. *Soil Tillage Res.* **2006**, *91*, 101–108. [CrossRef]

12. Guerschman, J.P.; Held, A.A.; Donohue, R.J.; Renzullo, L.J.; Sims, N.; Kerblat, F.; Grundy, M. The GEOGLAM Rangelands and Pasture Productivity Activity: Recent Progress and Future Directions. *AGU Fall Meeting Abstracts* **2015**. Available online: http://adsabs.harvard.edu/abs/2015AGUFM.B43A0531G (accessed on 25 January 2020).

13. Guerschman, J.P.; Leys, J.; Rozas Larraondo, P.; Henrikson, M.; Paget, M.; Barson, M. *Monitoring Groundcover: An Online Tool for Australian Regions*; Technical Report; CSIRO: Canberra, Australia, 21 November 2018; 58p. [CrossRef]

14. Guerschman, J.P.; Hill, M.J.; Leys, J.; Heidenreich, S. Vegetation cover dependence on accumulated antecedent precipitation in Australia: Relationships with photosynthetic and non-photosynthetic vegetation fractions. *Remote Sens. Environ.* **2020**, in press.

15. Hansen, M.C.; Potapov, P.V.; Moore, R.; Hancher, M.; Turubanova, S.A.A.; Tyukavina, A.; Thau, D.; Stehman, S.V.; Goetz, S.J.; Loveland, T.R.; et al. High-resolution global maps of 21st-century forest cover change. *Science* **2013**, *342*, 850–853. [CrossRef]

16. Zomer, R.J.; Neufeldt, H.; Xu, J.; Ahrends, A.; Bossio, D.; Trabucco, A.; Van Noordwijk, M.; Wang, M. Global Tree Cover and Biomass Carbon on Agricultural Land: The contribution of agroforestry to global and national carbon budgets. *Sci. Rep.* **2016**, *6*, 29987. [CrossRef]

17. Bastin, J.-F.; Finegold, Y.; Garcia, C.; Mollicone, D.; Rezende, M.; Routh, D.; Zohner, C.M.; Crowther, T.W. The global tree restoration potential. *Science* **2019**, *365*, 76–79. [CrossRef]

18. Mousivand, A.; Arsanjani, J.J. Insights on the historical and emerging global land cover changes: The case of ESA-CCI-LC datasets. *Appl. Geogr.* **2019**, *106*, 82–92. [CrossRef]

19. Song, X.P.; Hansen, M.C.; Stehman, S.V.; Potapov, P.V.; Tyukavina, A.; Vermote, E.F.; Townshend, J.R. Global land change from 1982 to 2016. *Nature* **2018**, *560*, 639–643. [CrossRef]

20. Chen, C.; Park, T.; Wang, X.; Piao, S.; Xu, B.; Chaturvedi, R.K.; Fuchs, R.; Brovkin, V.; Ciais, P.; Fensholt, R.; et al. China and India lead in greening of the world through land-use management. *Nat. Sustain.* **2019**, *2*, 122–129. [CrossRef]

21. Yang, W.; Tan, B.; Huang, D.; Rautiainen, M.; Shabanov, N.; Wang, Y.; Privette, J.; Huemmrich, K.; Fensholt, R.; Sandholt, I.; et al. MODIS leaf area index products: from validation to algorithm improvement. *IEEE Trans. Geosci. Remote Sens.* **2006**, *44*, 1885–1898. [CrossRef]

22. Okin, G.S.; Clarke, K.D.; Lewis, M.M. Comparison of methods for estimation of absolute vegetation and soil fractional cover using MODIS normalized BRDF-adjusted reflectance data. *Remote Sens. Environ.* **2013**, *130*, 266–279. [CrossRef]

23. Ying, Q.; Hansen, M.C.; Potapov, P.V.; Tyukavina, A.; Wang, L.; Stehman, S.V.; Moore, R.; Hancher, M. Global bare ground gain from 2000 to 2012 using Landsat imagery. *Remote Sens. Environ.* **2017**, *194*, 161–176. [CrossRef]

24. Hansen, M.C.; DeFries, R.S.; Townshend, J.R.G.; Carroll, M.; Dimiceli, C.; Sohlberg, R.A. Global percent tree cover at a spatial resolution of 500 meters: First results of the MODIS vegetation continuous fields algorithm. *Earth Interact.* **2003**, *7*, 1–15. [CrossRef]

25. Sexton, J.O.; Song, X.-P.; Feng, M.; Noojipady, P.; Anand, A.; Huang, C.; Kim, -H.; Collins, K.M.; Channan, S.; DiMiceli, C.; et al. Global, 30-m resolution continuous fields of tree cover: Landsat-based rescaling of MODIS vegetation continuous fields with lidar-based estimates of error. *Int. J. Digit. Earth* **2013**, *6*, 427–448. [CrossRef]

26. McCallum, I.; Wagner, W.; Schmullius, C.; Shvidenko, A.; Obersteiner, M.; Fritz, S.; Nilsson, S. Comparison of four global FAPAR datasets over Northern Eurasia for the year 2000. *Remote Sens. Environ.* **2010**, *114*, 941–949. [CrossRef]

27. Pickett-Heaps, C.A.; Canadell, J.; Briggs, P.R.; Gobron, N.; Haverd, V.; Paget, M.J.; Pinty, B.; Raupach, M.R. Evaluation of six satellite-derived Fraction of Absorbed Photosynthetic Active Radiation (FAPAR) products across the Australian continent. *Remote Sens. Environ.* **2014**, *140*, 241–256. [CrossRef]

28. Myneni, R.; Hoffman, S.; Knyazikhin, Y.; Privette, J.; Glassy, J.; Tian, Y.; Wang, Y.; Song, X.; Zhang, Y.; Smith, G.; et al. Global products of vegetation leaf area and fraction absorbed PAR from year one of MODIS data. *Remote Sens. Environ.* **2002**, *83*, 214–231. [CrossRef]

29. Fang, H.; Wei, S.; Jiang, C.; Scipal, K. Theoretical uncertainty analysis of global MODIS, CYCLOPES, and GLOBCARBON LAI products using a triple collocation method. *Remote Sens. Environ.* **2012**, *124*, 610–621. [CrossRef]

30. Zhu, Z.; Bi, J.; Pan, Y.; Ganguly, S.; Anav, A.; Xu, L.; Samanta, A.; Piao, S.; Nemani, R.; Myneni, R. Global data sets of vegetation leaf area index (LAI) 3g and fraction of photosynthetically active radiation (FPAR) 3g derived from global inventory modeling and mapping studies (GIMMS) normalized difference vegetation index (NDVI3g) for the period 1981 to 2011. *Remote Sens.* **2013**, *5*, 927–948. [CrossRef]

31. Bucini, G.; Hanan, N.P. A continental-scale analysis of tree cover in African savannas. *Glob. Ecol. Biogeogr.* **2007**, *16*, 593–605. [CrossRef]

32. Hill, M.J.; Román, M.O.; Schaaf, C.B.; Hutley, L.; Brannstrom, C.; Etter, A.; Hanan, N.P. Characterizing vegetation cover in global savannas with an annual foliage clumping index derived from the MODIS BRDF product. *Remote Sens. Environ.* **2011**, *115*, 2008–2024. [CrossRef]

33. Southworth, J.; Zhu, L.; Bunting, E.; Ryan, S.J.; Herrero, H.; Waylen, P.R.; Hill, M.J. Changes in vegetation persistence across global savanna landscapes, 1982–2010. *J. Land Use Sci.* **2016**, *11*, 7–32. [CrossRef]

34. Staver, A.C.; Archibald, S.; Levin, S.A. The Global Extent and Determinants of Savanna and Forest as Alternative Biome States. *Science* **2011**, *334*, 230–232. [CrossRef] [PubMed]

35. Hanan, N.P.; Tredennick, A.T.; Prihodko, L.; Bucini, G.; Dohn, J. Analysis of stable states in global savannas: is the CART pulling the horse? *Glob. Ecol. Biogeogr.* **2014**, *23*, 259–263. [CrossRef] [PubMed]

36. Staver, A.C.; Hansen, M.C. Analysis of stable states in global savannas: is the CART pulling the horse? - a comment. *Glob. Ecol. Biogeogr.* **2015**, *24*, 985–987. [CrossRef]

37. Hanan, N.P.; Tredennick, A.T.; Prihodko, L.; Bucini, G.; Dohn, J. Analysis of stable states in global savannas - a response to Staver and Hansen. *Glob. Ecol. Biogeogr.* **2015**, *24*, 988–989. [CrossRef]

38. Olson, D.M.; Dinerstein, E.; Wikramanayake, E.D.; Burgess, N.D.; Powell, G.V.N.; Underwood, E.C.; D'Amico, J.A.; Itoua, I.; Strand, H.E.; Morrison, J.C.; et al. Terrestrial Ecoregions of the World: A New Map of Life on Earth. *Bioscience* **2001**, *51*, 933–938. [CrossRef]

39. Dixon, A.P.; Faber-Langendoen, D.; Josse, C.; Morrison, J.; Loucks, C.J. Distribution mapping of world grassland types. *J. Biogeogr.* **2014**, *41*, 2003–2019. [CrossRef]

40. Robinson, T.P.; Thornton, P.K.; Franceschini, G.; Kruska, R.L.; Chiozza, F.; Notenbaert, A.; Cecchi, G.; Herrero, M.; Epprecht, M.; Fritz, S.; et al. *Global livestock production systems*; Food and Agriculture Organization of the United Nations (FAO) and International Livestock Research Institute (ILRI): Rome, Italy, 2011; 52p, Available online: http://www.fao.org/3/i2414e/i2414e.pdf (accessed on 25 January 2020).

41. Schaaf, C.B.; Wang, Z. MCD43A4 MODIS/Terra+Aqua BRDF/Albedo Nadir BRDF Adjusted Reflectance Daily L3 Global - 500m V006. *NASA EOSDIS Land Processes DAAC* **2015**. [CrossRef]

42. Flood, N. Seasonal Composite Landsat TM/ETM+ Images Using the Medoid (a Multi-Dimensional Median). *Remote Sens.* **2013**, *5*, 6481–6500. [CrossRef]

43. Faber-Langendoen, D.; Keeler-Wolf, T.; Meidinger, D.; Tart, D.; Hoagland, B.; Josse, C.; Navarro, G.; Ponomarenko, S.; Saucier, J.-P.; Weakley, A.; et al. EcoVeg: a new approach to vegetation description and classification. *Ecol. Monogr.* **2014**, *84*, 533–561. [CrossRef]

44. Mann, H.B. Nonparametric Tests Against Trend. *Econometrica* **1945**, *13*, 245. [CrossRef]

45. Kendall, M.G. *Rank correlation methods*; Oxford University Press: New York, NY, USA, 1962. Available online: https://trove.nla.gov.au/version/264239415 (accessed on 26 January 2020).

46. Meals, D.W.; Spooner, J.; Dressing, S.A.; Harcum, J.B. *Statistical Analysis for Monotonic Trends*; Tech Notes 6; Developed for U.S. Environmental Protection Agency by Tetra Tech, Inc.: Fairfax, VA, USA, November 2011; 23p. Available online: https://www.epa.gov/sites/production/files/2016-05/documents/tech_notes_6_dec2013_trend.pdf (accessed on 25 January 2020).

47. Gaitán, J.J.; Bran, D.E.; Oliva, G.E.; Aguiar, M.R.; Buono, G.G.; Ferrante, D.; Nakamatsu, V.; Ciari, G.; Salomone, J.M.; Massara, V.; et al. Aridity and overgrazing have convergent effects on ecosystem structure and functioning in Patagonian rangelands. *Land Degrad. Dev.* **2018**, *29*, 210–218. [CrossRef]

48. Piquer-Rodríguez, M.; Butsic, V.; Gärtner, P.; Macchi, L.; Baumann, M.; Pizarro, G.G.; Volante, J.; Gasparri, I.; Kuemmerle, T. Drivers of agricultural land-use change in the Argentine Pampas and Chaco regions. *Appl. Geogr.* **2018**, *91*, 111–122. [CrossRef]

49. Soulard, C.E.; Wilson, T.S. Recent land-use/land-cover change in the Central California Valley. *J. Land Use Sci.* **2015**, *10*, 59–80. [CrossRef]

50. Brown, K.A.; Parks, K.E.; Bethell, C.A.; Johnson, S.E.; Mulligan, M. Predicting Plant Diversity Patterns in Madagascar: Understanding the Effects of Climate and Land Cover Change in a Biodiversity Hotspot. *PLoS ONE* **2015**, *10*, 0122721. [CrossRef] [PubMed]

51. Vieilledent, G.; Grinand, C.; Rakotomalala, F.A.; Ranaivosoa, R.; Rakotoarijaona, J.-R.; Allnutt, T.F.; Achard, F. Combining global tree cover loss data with historical national forest cover maps to look at six decades of deforestation and forest fragmentation in Madagascar. *Boil. Conserv.* **2018**, *222*, 189–197. [CrossRef]

52. Rasmussen, K.; D'haen, S.; Fensholt, R.; Fog, B.; Horion, S.; Nielsen, J.O.; Rasmussen, L.V.; Reenberg, A. Environmental change in the Sahel: reconciling contrasting evidence and interpretations. *Reg. Environ. Chang.* **2016**, *16*, 673–680. [CrossRef]

53. Aleman, J.C.; Blarquez, O.; Staver, C.A.; Staver, A.C. Land-use change outweighs projected effects of changing rainfall on tree cover in sub-Saharan Africa. *Glob. Chang. Boil.* **2016**, *22*, 3013–3025. [CrossRef]

54. Anchang, J.Y.; Prihodko, L.; Kaptué, A.T.; Ross, C.W.; Ji, W.; Kumar, S.S.; Lind, B.; Sarr, M.A.; Diouf, A.A.; Hanan, N.P. Trends in Woody and Herbaceous Vegetation in the Savannas of West Africa. *Remote Sens.* **2019**, *11*, 576. [CrossRef]

55. Touré, A.A.; Tidjani, A.; Rajot, J.; Marticorena, B.; Bergametti, G.; Bouet, C.; Ambouta, K.; Garba, Z. Dynamics of wind erosion and impact of vegetation cover and land use in the Sahel: A case study on sandy dunes in southeastern Niger. *Catena* **2019**, *177*, 272–285. [CrossRef]

56. Hanan, N.P. Agroforestry in the Sahel. *Nat. Geosci.* **2018**, *11*, 296–297. [CrossRef]

57. Brandt, M.; Rasmussen, K.; Hiernaux, P.; Herrmann, S.; Tucker, C.J.; Tong, X.; Tian, F.; Mertz, O.; Kergoat, L.; Mbow, C.; et al. Reduction of tree cover in West African woodlands and promotion in semi-arid farmlands. *Nat. Geosci.* **2018**, *11*, 328–333. [CrossRef]

58. Bolognesi, M.; Vrieling, A.; Rembold, F.; Gadain, H. Rapid mapping and impact estimation of illegal charcoal production in southern Somalia based on WorldView-1 imagery. *Energy Sustain. Dev.* **2015**, *25*, 40–49. [CrossRef]

59. Kiruki, H.M.; van der Zanden, E.H.; Malek, Ž.; Verburg, P.H. Land cover change and woodland degradation in a charcoal producing semi-arid area in Kenya. *Land Degrad. Dev.* **2017**, *28*, 472–481. [CrossRef]

60. Ndegwa, G.M.; Nehren, U.; Anhuf, D.; Iiyama, M. Estimating sustainable biomass harvesting level for charcoal production to promote degraded woodlands recovery: A case study from Mutomo District, Kenya. *Land Degrad. Dev.* **2018**, *29*, 1521–1529. [CrossRef]

61. Rowell, D.P.; Booth, B.B.B.; Nicholson, S.E.; Good, P. Reconciling Past and Future Rainfall Trends over East Africa. *J. Clim.* **2015**, *28*, 9768–9788. [CrossRef]

62. Lyon, B.; DeWitt, D.G. A recent and abrupt decline in the East African long rains. *Geophys. Res. Lett.* **2012**, *39*. [CrossRef]

63. Lyon, B. Seasonal Drought in the Greater Horn of Africa and Its Recent Increase during the March–May Long Rains. *J. Clim.* **2014**, *27*, 7953–7975. [CrossRef]

64. Holechek, J.L.; Cibils, A.F.; Bengaly, K.; Kinyamario, J.I. Human Population Growth, African Pastoralism, and Rangelands: A Perspective. *Rangel. Ecol. Manag.* **2017**, *70*, 273–280. [CrossRef]

65. Lankester, F.; Davis, A. Pastoralism and wildlife: historical and current perspectives in the East African rangelands of Kenya and Tanzania. *Rev. Sci. et Tech. de l'OIE* **2016**, *35*, 473–484. [CrossRef] [PubMed]

66. Rukundo, E.; Liu, S.; Dong, Y.; Rutebuka, E.; Asamoah, E.F.; Xu, J.; Wu, X. Spatio-temporal dynamics of critical ecosystem services in response to agricultural expansion in Rwanda, East Africa. *Ecol. Indic.* **2018**, *89*, 696–705. [CrossRef]
67. Guzha, A.; Rufino, M.; Okoth, S.; Jacobs, S.; Nóbrega, R. Impacts of land use and land cover change on surface runoff, discharge and low flows: Evidence from East Africa. *J. Hydrol. Reg. Stud.* **2018**, *15*, 49–67. [CrossRef]
68. Bradley, B.A.; Houghton, R.A.; Mustard, J.F.; Hamburg, S.P. Invasive grass reduces aboveground carbon stocks in shrublands of the Western US. *Glob. Chang. Boil.* **2006**, *12*, 1815–1822. [CrossRef]
69. Mau-Crimmins, T.M.; Schussman, H.R.; Geiger, E.L. Can the invaded range of a species be predicted sufficiently using only native-range data? Lehmann lovegrass (Eragrostis lehmanniana) in the southwestern United States. *Ecol. Model.* **2006**, *193*, 736–746. [CrossRef]
70. Abatzoglou, J.T.; Kolden, C.A. Climate Change in Western US Deserts: Potential for Increased Wildfire and Invasive Annual Grasses. *Rangel. Ecol. Manag.* **2011**, *64*, 471–478. [CrossRef]
71. Romero-Ruiz, M.; Flantua, S.; Tansey, K.; Berrío, J. Landscape transformations in savannas of northern South America: Land use/cover changes since 1987 in the Llanos Orientales of Colombia. *Appl. Geogr.* **2012**, *32*, 766–776. [CrossRef]
72. Ocampo-Peñuela, N.; Garcia-Ulloa, J.; Ghazoul, J.; Etter, A. Quantifying impacts of oil palm expansion on Colombia's threatened biodiversity. *Boil. Conserv.* **2018**, *224*, 117–121. [CrossRef]
73. Ordoñez, D.A.R.; Leal, M.R.L.V.; Bonomi, A.; Cortez, L.A.B. Expansion assessment of the sugarcane and ethanol production in the Llanos Orientales region in Colombia. *Biofuels, Bioprod. Biorefining* **2018**, *12*, 857–872. [CrossRef]
74. McKeon, G.M.; Stone, G.S.; Syktus, J.I.; Carter, J.O.; Flood, N.R.; Ahrens, D.G.; Bruget, D.N.; Chilcott, C.R.; Cobon, D.H.; Cowley, R.A.; et al. Climate change impacts on northern Australian rangeland livestock carrying capacity: a review of issues. *Rangel. J.* **2009**, *31*, 1–29. [CrossRef]

remote sensing

MDPI

Review

Terrestrial Laser Scanning for Vegetation Analyses with a Special Focus on Savannas

Tasiyiwa Priscilla Muumbe [1,*], Jussi Baade [2], Jenia Singh [3], Christiane Schmullius [1] and Christian Thau [1,4]

[1] Department for Earth Observation, Friedrich Schiller University Jena, 07743 Jena, Germany; c.schmullius@uni-jena.de (C.S.); christian.thau@uni-jena.de (C.T.)
[2] Department of Physical Geography, Friedrich Schiller University Jena, 07743 Jena, Germany; jussi.baade@uni-jena.de
[3] Department of Organismic and Evolutionary Biology, Harvard University, Cambridge, MA 02138, USA; jeniasingh@fas.harvard.edu
[4] Department for Urban Development and Environment, Jena City Administration, 07743 Jena, Germany
* Correspondence: tasiyiwa.muumbe@uni-jena.de

Abstract: Savannas are heterogeneous ecosystems, composed of varied spatial combinations and proportions of woody and herbaceous vegetation. Most field-based inventory and remote sensing methods fail to account for the lower stratum vegetation (i.e., shrubs and grasses), and are thus underrepresenting the carbon storage potential of savanna ecosystems. For detailed analyses at the local scale, Terrestrial Laser Scanning (TLS) has proven to be a promising remote sensing technology over the past decade. Accordingly, several review articles already exist on the use of TLS for characterizing 3D vegetation structure. However, a gap exists on the spatial concentrations of TLS studies according to biome for accurate vegetation structure estimation. A comprehensive review was conducted through a meta-analysis of 113 relevant research articles using 18 attributes. The review covered a range of aspects, including the global distribution of TLS studies, parameters retrieved from TLS point clouds and retrieval methods. The review also examined the relationship between the TLS retrieval method and the overall accuracy in parameter extraction. To date, TLS has mainly been used to characterize vegetation in temperate, boreal/taiga and tropical forests, with only little emphasis on savannas. TLS studies in the savanna focused on the extraction of very few vegetation parameters (e.g., DBH and height) and did not consider the shrub contribution to the overall Above Ground Biomass (AGB). Future work should therefore focus on developing new and adjusting existing algorithms for vegetation parameter extraction in the savanna biome, improving predictive AGB models through 3D reconstructions of savanna trees and shrubs as well as quantifying AGB change through the application of multi-temporal TLS. The integration of data from various sources and platforms e.g., TLS with airborne LiDAR is recommended for improved vegetation parameter extraction (including AGB) at larger spatial scales. The review highlights the huge potential of TLS for accurate savanna vegetation extraction by discussing TLS opportunities, challenges and potential future research in the savanna biome.

Keywords: terrestrial laser scanning (TLS); savanna; Above Ground Biomass (AGB); 3D point cloud; vegetation structure

check for
updates

Citation: Muumbe, T.P.; Baade, J.; Singh, J.; Schmullius, C.; Thau, C. Terrestrial Laser Scanning for Vegetation Analyses with a Special Focus on Savannas. *Remote Sens.* **2021**, *13*, 507. https://doi.org/10.3390/rs13030507

Academic Editor: Michael J. Hill
Received: 28 December 2020
Accepted: 28 January 2021
Published: 31 January 2021

Publisher's Note: MDPI stays neutral with regard to jurisdictional claims in published maps and institutional affiliations.

1. Introduction

1.1. The Role and Importance of Savannas

Savannas are heterogeneous ecosystems characterized by mixed physiognomies of woody and herbaceous layer [1]. Savanna ecosystems span one-fifth of the global land [2,3] which makes up approximately 20% of the terrestrial vegetation and contribute 30% of all terrestrial ecosystem gross primary productivity [4,5]. Savanna ecosystems also contribute significantly to the global carbon cycle with a net primary productivity

Remote Sens. **2021**, *13*, 507. https://doi.org/10.3390/rs13030507

https://www.mdpi.com/journal/remotesensing

of 1 to 12 t C ha^{-1} year^{-1} and a carbon sequestration rate averaging 0.14 t C ha^{-1} to 0.39 Gt C year^{-1} [4].

Savannas are vital biomes that play a major role in the provision of ecosystem services. Moreover, they contribute significantly to the socio-economic needs of most rural households, especially in Africa, where many are dependent on natural resources for their livelihood [6,7]. Savannas support approximately 20% of the world population and account on average 28% of household income in rural areas of developing countries [8,9]. In west and central Africa, income from bushmeat is equivalent to approximately $1000 per year [7].

Since savannas are home to 20% of the world's population, humans have caused changes in these ecosystems through land clearing for agriculture and livestock production, infrastructure, fire lighting or suppression and fuelwood collection [9,10]. The alteration of savanna structure and function has direct effects on woody vegetation and results in woody encroachment as a result of the interaction of drivers mainly herbivory, fire and atmospheric carbon dioxide [11,12]. Anthropogenic land use and land cover change affects vast areas of the savanna ecosystem and the consequences can be substantial for the regional climate cycle [9]. Against this background, it is important to regularly map and monitor savannas and to quantify the carbon pools and dynamics of these ecosystems because they contribute significantly to global productivity [13].

1.2. Monitoring of Savannas Using Remote Sensing

Much of the current understanding on savanna vegetation structure is derived from traditional field-based methods, such as forest inventories, permanent sample plots and transects [14,15]. Field-based estimates are essential in the validation of remote sensed data [16], though they are limited in that they are time consuming, expensive, labour intensive and often require destructive sampling [17,18]. Remote sensing techniques are more appropriate for monitoring when compared to traditional methods [19]. Especially for consistent analyses at different spatial and temporal scales and when considering inaccessible areas, remote sensing with limited fieldwork has been proposed as an efficient method [17,19].

The main types of Earth observation data used for vegetation studies are based on optical (multispectral, hyperspectral), Radio Detection and Ranging (RADAR) and Light Detection and Ranging (LiDAR) sensors [20].

Multispectral data can be acquired from spaceborne (satellites) or airborne (aircrafts, Unmanned Aerial Vehicles (UAVs) platforms and from the ground imagery (spectrometers). Multispectral data has the advantage of higher spatial resolution [21] and is publicly available for most regions in the world e.g., Landsat data. It is limited in terms of spectral resolution [21] and spaceborne optical data often suffer from cloud cover and mixed pixels [20,22].

Hyperspectral data can be acquired from spaceborne and airborne sensors and hyperspectral cameras [23]. Hyperspectral data is acquired in hundreds of narrow bands, and thus it has a higher spectral resolution enabling distinguishing of vegetation types [24]. It is limited though in terms of spatial resolution [21] and suffers from band redundancy where bands are strongly correlated [24].

RADAR data have the advantage of being largely independent of atmospheric and illumination conditions. However, they are limited because data-specific issues need to be compensated for or at least considered when using RADAR for vegetation monitoring [25]. Among others, these issues comprise the geometry of Synthetic Aperture RADAR (SAR) imagery, the dielectric properties of the surveyed surfaces, RADAR speckle complications due to the separation between signal and noise in the data, RADAR shadow and saturation problems [25].

To date, LiDAR systems are considered to be the most accurate and recommended remote sensing technology for vegetation parameter retrieval by producing high-resolution 3D data called point clouds [26,27]. They are available as space and airborne, mobile and

UAV as well as terrestrial LiDAR sensors [27]. Spaceborne LiDAR provides the opportunity to make large-scale assessments of vegetation structure for global to regional biomass estimation [28]. Airborne LiDAR is also able to cover relatively large areas and measures vegetation structure with high accuracy [29]. Both spaceborne and airborne LiDAR systems are limited in detecting smaller trees and shrubs. The latter are key attributes in savanna ecosystems [30]. By using airborne or spaceborne LiDAR these attributes and are restricted to landscape level estimates. Mobile and UAV LiDAR sensors offer a faster approach for acquiring 3D point clouds. However, mobile LiDARs are constrained in extracting canopy and height attributes [31], whereas UAV LiDARs have difficulties in capturing tree trunks [32] and are also limited in flight time due to batteries [33]. An overview of the main advantages and disadvantages of the described monitoring technologies i.e., (field surveys, multi-spectral, hyperspectral, RADAR and LiDAR) platforms are also discussed in Béland et al. [27], Eisfelder et al. [34] and Galidaki et al. [35].

1.3. Terrestrial Laser Scanning (TLS)

Terrestrial Laser Scanning (TLS) has grown in interest because of its ability to produce 3D point cloud data for characterizing vegetation structure [36]. Accordingly, over the past decade, a considerable amount of research has been directed towards the 3D quantification of vegetation structure using TLS [37,38]. Vegetation structural parameters are difficult and laborious to measure in the field, especially in complex ecosystems like the savanna. However, TLS has proven efficient in deriving these parameters because it produces consistent and objective measurements with high precision. Furthermore, TLS measurements are relatively easy to interpret and parameter extraction can be automated [39,40].

A terrestrial laser scanner or ground-based LiDAR is a LiDAR system from which detailed 3D information called point clouds can be collected [41]. Using a laser beam, terrestrial laser scanners provide a high-resolution 3D view of the objects [36]. Depending on their range measurement principle, terrestrial laser scanners are classified into two types: phase shift or time of flight scanners [36,42]. Phase shift scanners measure distances by analysing the shift between the emitted and the received laser wave. In contrast, time of flight scanners compute distances based on the difference between the emission and reception of the laser pulse [42].

TLS measurements can be collected based on single or multiple scans. Using single scans, the scanner is placed at a single location within the study area whilst multiple scans are obtained when the scanner is placed at several positions within the test site [43]. For rapid and efficient extraction of vegetation parameters with reduced fieldwork and faster post-processing, single scans are preferred [44]. However, this approach is prone to occlusions (e.g., a tree being in the scan shadow of another tree or any other object) reducing the accuracy of vegetation structure retrieval and the ability to get a full 3D representation of an object [44]. For the best plot coverage and a complete 3D description of the objects under consideration, a multi-scan TLS setup is necessary [31,36].

Multi-scan TLS data allow for reliably retrieving common tree attributes such as Diameter at Breast Height (DBH), tree height, canopy structure and vertical height profiles [45,46]. In addition to these parameters, more detailed and complex vegetation attributes such as leaf water content [47], basal area and stem density [48], Above Ground Biomass (AGB) change [49], tree crown change [50] stem curve profiles [51], stem shrinkage [52], have been derived from TLS point clouds.

TLS provides unmatched 3D information on vegetation structure right down to branch and leaf scales, making it far superior to data obtained through traditional field work [38,53]. Using the intensity of the returned laser pulses, TLS can be used to map pigment concentration, photosynthetic capacity and water content of the vegetation [27]. It is also possible to separate wood from leaf material [54,55]. TLS offers a non-destructive approach to quantify canopy and stem volume with high accuracies, thereby enabling the estimation of AGB with reduced uncertainties as well as the development and improvement of reliable allometric equations [56,57]. TLS also enables the estimation of individual tree components [58] and it can also be employed as a calibration and validation dataset because of its ability to acquire data from the ground that is similar to traditional measures of vegetation structure [59,60].

The inherent heterogeneity of savannas presents a challenge regarding accurate vegetation characterization. To overcome this challenge, 3D data on the horizontal and vertical structure of the savanna are needed. TLS is able to account for savanna heterogeneity (Figure 1) by providing detailed information on all savanna vegetation types and layers [61,62]. It therefore holds much potential for making holistic AGB assessments, including long and short stature vegetation [63–65].

Figure 1. 3D point cloud showing the inherent heterogeneity of savanna ecosystems. The point cloud was captured with a time of flight Riegl VZ 1000 scanner in the vicinity of Skukuza Flux Tower in Kruger National Park, South Africa. One colour pass equals 5 m in height.

Different drivers can result in the loss of large trees in the savanna landscape [66] and changes in vegetation structure. The non-destructive nature of TLS makes it more so suitable for quantification of change in vegetation structure and reduction of uncertainties when estimating AGB [50,67]. The use of repeated TLS scans allows for monitoring these and other losses (and potentially gains) in savanna AGB over time. Multi-temporal TLS therefore holds the potential to support carbon accounting programmes such as REDD+.

While TLS is able to capture the 3D configuration of vegetation at high spatial resolution, data can only be acquired for a limited areal extent and data acquisition usually requires extensive field work [68]. This problem is solved by the integration of TLS with other remote sensing data to ensure the coverage at landscape scales. For example, landscape studies were conducted in the savanna by the integration of TLS with airborne LiDAR [69], TLS with MODIS satellite data [70] and TLS with ALOS PALSAR L band data [62]. Moreover, TLS scans frequently feature occlusions ("shadows") caused by objects close to the scanner which, in turn, block the view of other objects further away. Although occlusions are not a major problem in open savannas, they can be minimized by the use of multiple TLS scans [30,64]. Another shortcoming of TLS systems is that the point density decreases with distance from the scanner, an effect called beam divergence [68,71] which causes the point cloud to be unreliable at a certain distance.

1.4. Overall Goal and Objectives of the Review

This paper aims to review the scientific literature on the use of TLS for vegetation analyses. It puts a special but not exclusive geographic focus on savannas in order to establish the current status quo in savanna vegetation parameter extraction. In contrast to existing reviews in the TLS literature [72,73], this study presents the current TLS methods in practice and identifies those suitable for accurate biomass assessments in the savanna biome. To address this gap, our meta-analysis takes a detailed look at the geographic distribution of TLS studies, the parameters retrieved from TLS point clouds, known retrieval methods as well as the main opportunities and challenges arising from the studied literature. The remainder of the paper is structured as follows: Section 2 presents the approach and Section 3 the main findings of the review. The main discussion points and conclusions derived from the review are presented in Sections 4 and 5, respectively.

2. Materials and Methods

A systematic literature review was conducted in Web of Science [74]. Access to its catalogue was granted through the Thüringer Universitäts- und Landesbibliothek Jena (ThULB). The advanced search functionality in Web of Science was used to develop and apply a set of predefined queries. The search included all journals in the Web of Science Core Collection™ database. Peer-reviewed conference papers were also included. Searching for keywords in the title (TI) and in some instances in all fields (ALL) (when a query returned fewer than ten articles) was performed with a time frame up to May 2020. The time span for the search was all years that is from 1945 to 2020. In the event that a search query returned more than 100 articles, the search was refined by the research areas forestry or remote sensing to narrow down on specific studies employing TLS for vegetation parameter extraction. Each query consisted of different keywords and related synonyms (Table 1) which were combined through logical operators (i.e., "AND", "OR").

Table 1. Search terms used to build the search queries for this review.

Search Aspect	Search Terms
Vegetation type	forest *; grassland *; savanna *; shrub *; woodland *
TLS	ground-based lidar; lidar; terrestrial laser scann *; terrestrial lidar; TLS
Extraction of vegetation parameters	3D reconstruction; reconstruction; stem; tree parameter; tree structur *;
TLS experimental setup	full waveform; intensity; multi-temporal; reflection
TLS as auxiliary data	calibration; validation

An example of one of the search queries developed is: TI = ((TLS OR Terrestrial LiDAR OR Terrestrial Laser Scann *) AND (Tree Structur *)). This search query would return all journal articles that have either the word or terrestrial LiDAR or terrestrial laser scanner or terrestrial laser scanning and the word tree structure or tree structural in their title.

Following the above search scheme, a total of 158 research articles were initially retrieved from Web of Science. In a second step, certain papers were retained for the review according to a prescribed set of criteria. The criteria used to refine the initial search results were as follows:

➤ The study under consideration focused on the extraction of one or more vegetation parameters from TLS point clouds. For example, a study was only considered if parameters like the DBH or AGB were derived on the basis of TLS point clouds [75];
➤ The study under consideration used TLS point clouds or products as a calibration and/or validation data set for other remote sensing technologies (e.g., UAV LiDAR [32] airborne LiDAR [76] and ALOS PALSAR [62];
➤ The study under consideration reviewed TLS as a remote sensing technology [39];
➤ The study under consideration conducted technical experiments with TLS data to investigate, e.g., how specular reflectors affect distance in TLS measurements [77] or which scan setup (single vs. multiple scans) in the field is best suited for a particular application [51].

This resulted in a final number of 113 reviewed publications. To qualitatively and quantitatively assess the final selection of papers, a comprehensive meta-analysis was conducted based on 18 predefined study attributes. Besides author(s), year of publication and publishing journal, the list of reviewed meta-parameters included: study area(s) (country, continent, biome); type of laser scanner used (TLS model/device); range finder of the scanner (phase shift or time of flight); type of TLS data used (xyz and intensity); scanner setup (TLS settings for scanning); field setup (use of reflectors, GPS/dGNSS); number of scans (single vs. multiple scans); parameters retrieved (e.g., DBH, vegetation height, crown diameter); retrieval method (algorithm used); number of studied trees and shrubs (total number of trees/shrubs measured in the research); performance metrics such as coefficient of determination (R^2), Root Mean Square Error (RMSE) and bias; main discussion points (e.g., advantages and limitations of a used approach); and lastly research gaps and potential avenues for future research. The above search efforts were undertaken to ensure that all relevant TLS literature was collected. The full list of papers considered and finally selected in this review is available in Table S1 (supplementary material).

3. Results

3.1. General Overview

Of the 113 publications used for the detailed review, 80% of the publications focused on the extraction of vegetation parameters from TLS point clouds, 12% used TLS as auxiliary data (e.g., TLS data used to validate airborne LiDAR) [78–80] and 8% applied TLS for technical experiments (e.g., reflectance modelling using TLS intensity data) [81–83]. Among the 113 papers, 97 were publications from peer-reviewed journals whilst 16 were peer-reviewed proceedings. The most popular journals were Remote Sensing (20 studies), Remote Sensing of Environment (9 studies), Agricultural and Forest Meteorology (8 studies) and ISPRS Journal of Photogrammetry and Remote Sensing (6 studies). Most proceedings papers stem from the International Geoscience and Remote Sensing Symposium (IGARSS), with 9 studies.

3.2. Temporal and Spatial Patterns

Figure 2 shows the trend observed in the number of TLS studies. Beginning in the early 2000s, TLS technology evolved slowly. New systems with a range of capabilities and applications emerged and this explains the rise in the number of studies from the early 2000s to date. TLS technology had a real breakthrough in vegetation studies around 2013 and the upward trend seems to continue in 2019. A further significant increase in peer-reviewed studies can be observed for 2018 and 2019 (25 and 22 articles, respectively). During this time, the number of published papers per year has more than doubled when compared to the period after the breakthrough (on average 10.8 studies per year for 2013–2017).

The geographic distribution of the reviewed studies by country is shown in Figure 3. To date, most studies were conducted in the United States, China and Finland. From a continental perspective, Europe has the highest number of studies (38%) [84–86] followed by North America (24%) [46,59,87] and Asia (19%) [88–90]. Africa (6%), Australia (5%) and South America (7%), with more than half of their landmass belonging to the savanna biome [1,2], had few TLS studies [64,91,92].

The geographic distribution of studies by biome (after Olson et al. [93]) and the percentage of reviewed studies per biome are shown in Figures 3 and 4. Studies that were conducted in more than one biome were counted multiple times and four laboratory studies were excluded from this analysis, resulting in a sample size of n = 114. The reviewed papers covered a total of 11 biomes. While most studies were conducted in temperate forests (47%) [75,94,95], much less papers were published in boreal (16%) [96–98], tropical and subtropical forests (10%) [99–101] as well as grasslands and savannas (9%) [70,91,92]. The cumulative total for the savanna biome was 13%, including 4% of savanna research conducted in temperate regions ("Temperate Grasslands Savannas and Shrubland").

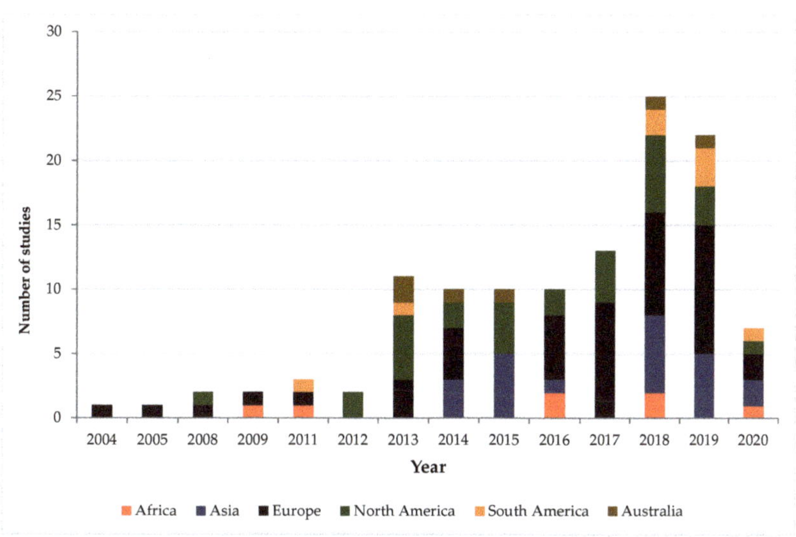

Figure 2. Number of published studies per year and continent. Over the past 15 years, research on TLS for vegetation assessments has increased significantly and knowledge has been generated especially for the European, North American and Asian region. Please note that the low number of studies in 2020 is an artefact of the review period ending on 31 May 2020.

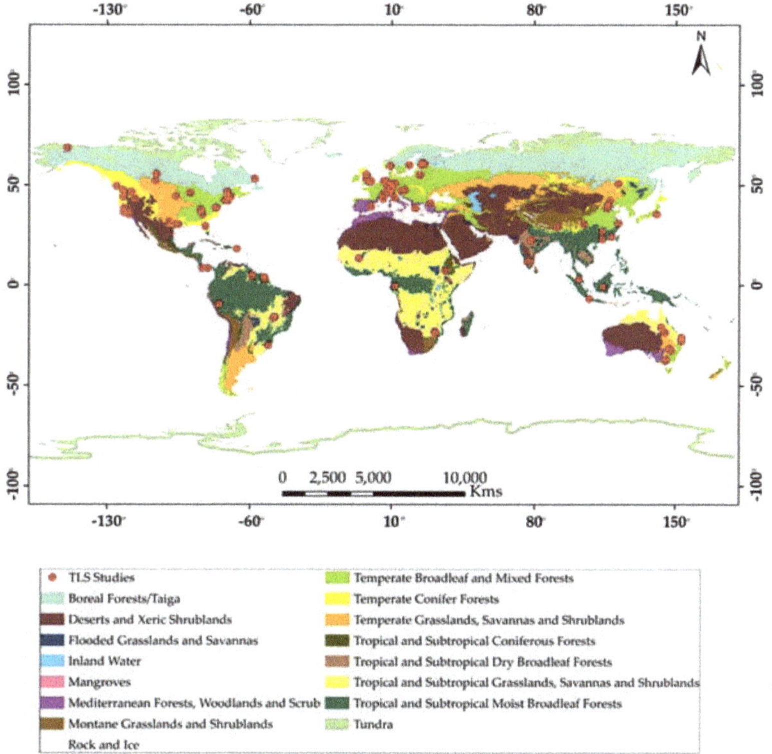

Figure 3. Geographic distribution of reviewed studies. Biome classification according to Olson et al. [93].

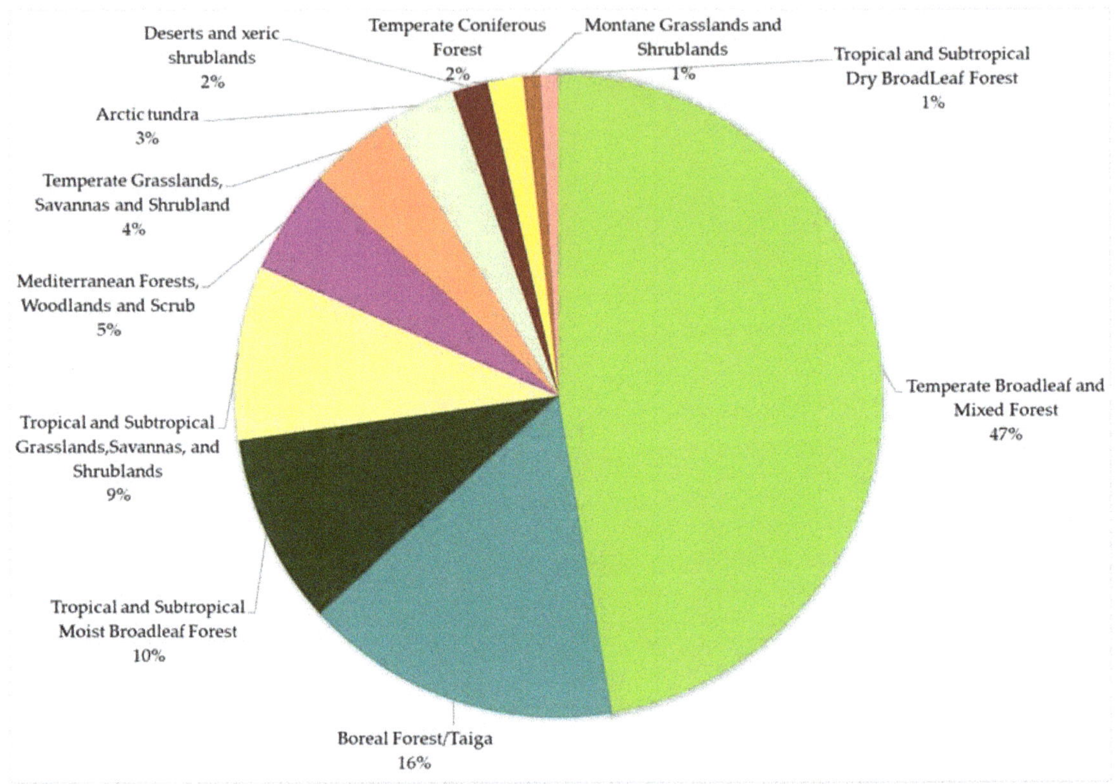

Figure 4. Percentage of reviewed studies per biome. Biome classification according to Olson et al. [93].

3.3. TLS Instruments and Data Used

A variety of TLS instruments can be purchased nowadays. They are offered by different manufacturers, with the main companies being RIEGL Laser Measurement Systems GmbH [102], FARO Technologies Inc [103] and Leica Geosystems [104]. Among others, TLS instruments differ in their, measurement range (the maximum distance that can be scanned e.g., 600 m), sample rate (pulses per second e.g., 122,000), wavelength (e.g. 1550 nm (near-infrared)), angular sampling (angle measurement resolution e.g., 0.02°), waveform (discrete or full-waveform), range measurement principle (time of flight or phase shift), size, mass and power efficiency. Based on our review, Table 2 provides an overview of frequently used TLS instrument models. Among the most popular scanners were Riegl VZ 400 (26 studies) [90,105,106], Leica HDS6100 (10 studies) [107–109], Riegl VZ 1000 (9 studies) [110–112], Faro Focus 3D (7 studies) [50,67,113] and Faro Focus 3D X 330 (also 7 studies) [83,114,115].

Figure 5 shows the number of scan positions and the area covered in the reviewed studies. It becomes clear that a larger area requires a higher number of scan positions, the higher the number of scan positions the fewer the studies and most studies use up to 10 scan positions and cover an area of less 5000 m². The number of scan positions and area covered also depend on the vegetation type and the level of detail required for the specific research. For high detail data, some studies employed a high number of scan positions on a relatively small area. Li et al. [63] used 18 scan positions on a 400 m² plot on shrubs and Lau et al. [99] used 16 scan positions on each selected tree in a 1200 m² plot in a tropical forest. Most inventories follow a field setup with multiple scans, with five

(28 studies) [121–123], four (23 studies) [124–126] and three (13 studies) [113,127,128] scan positions being the most popular. However, a single scan experimental design is also very common (28 studies [83,89,129]).

Table 2. Commonly used terrestrial laser scanners.

	Riegl VZ 400	Leica HDS6100	Riegl VZ 1000	Faro Focus 3D	Faro Focus 3D X 330
Range finder	Time of flight	Phase shift	Time of flight	Phase shift	Phase shift
Range of Measurement [m]	1.5–600	79	2.5–1400	0.6–120	0.6–330
Scan angle range [°]	100/360	360/310	100/360	305/360	300/360
Accuracy [mm]	5	±2 at 25 m	8	±2	±2
Beam divergence [mrad]	0.35	0.22	0.3	0.19	0.19
Laser wavelength [nm]	1550	690	1550	905	1550
Effective measurement rate [points/sec]	42,000–122,000	508,000	42,000–122,000	122,000–976,000	122,000–976,000
Angel measurement resolution [°]	0.0005° (1.8arc sec)	0.009° Hor × 0.009° Ver	0.0005° (1.8arc sec)	0.011°	0.011°
Weight [kg]	9.6	14	9.8	5.0	5.2
Reference	[116]	[117]	[118]	[119]	[120]

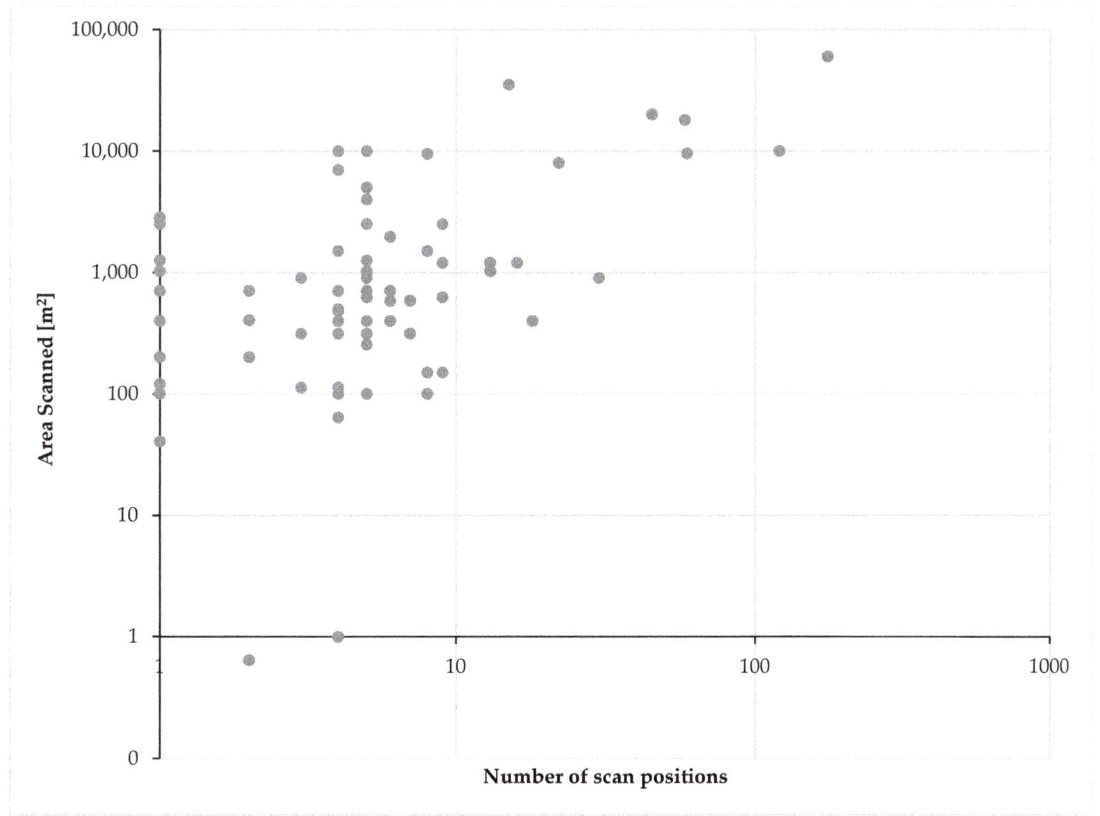

Figure 5. Number of scan positions and area covered (m^2) in the reviewed studies.

TLS instruments provide two types of data that can be used for forestry and vegetation studies. These are 3D coordinates (XYZ data, point clouds) and the intensity of the reflected signal. The signal can either be discrete or full-waveform. Discrete data provide single or multiple returns for each laser pulse [76,130]. Full -waveform contain measurement of the reflected laser beam as a function of range [76,130]. These two data types can be integrated with auxiliary data from the field, including RGB images from digital cameras. Most studies used XYZ data (85 studies) [131–133] whereas intensity information was considered much less (8 studies) [129,134,135]. A triple combination of both data types and RGB images (2 studies) [121,136] as well as a combination of XYZ and RGB (3 studies) [75,111,137] was not commonly observed. Instead, a number of studies took advantage of XYZ and intensity data (15 studies) [127,138,139]. No paper presented results based on a combination of TLS intensity data and RGB images.

3.4. TLS Point Cloud Pre-Processing

The scanner in use usually comes with software for pre-processing the point cloud; for example, Riegl scanners come with Riscan Pro, Faro with FARO Scene and Leica with Leica Cyclone. The first stage in prepossessing if a multi scan setup was used is to conduct coarse registration and georeferencing to the local coordinate system using the retroreflectors as tie points [43,96]. Fine registration can also be achieved using multi-station adjustment through an iterative closest point algorithm in Riscan Pro software [43,140]. Multi-station adjustment requires a minimum of 3 TLS setups with a minimum overlap of 30%. The algorithm uses surface and point normals to compute the adjustment [141]. The registered point cloud is then clipped to the radius of the plot in the case of circular plots or to the extent of the plots or to remove low density points at the plot edges [140,142].

The final pre-processing stage involves filtering the point cloud to remove noise from the data [43,121,142]. Noise and outlier removal is done to increase accuracy and reduce errors when extracting information from the point cloud data [115]. The noise can be classified into internal and external noise, internal noise being the inherent noise from the scanning device and external noise from the reflective nature of the scene [143,144]. The common noise points from vegetation are leaves, extremely high or low points, and from occlusions [115,144]. The filtering is performed using various approaches such as statistical, neighbourhood, projection, signal and partial differential based filtering techniques [144]. For example, in the case of statistical outlier removal, the filtering reduces the error on the point cloud by computing the average distance to its neighbours and rejects points that are further away from the average distance [115,144]. Point cloud filtering can be performed in software packages such as Cloud Compare, LAStools, MeshLab and programming libraries such as Point Cloud and Open Graphics Library [143]. After pre-processing various algorithms are then applied to extract the vegetation parameters.

3.5. Methods Used with TLS Point Clouds

Of the 113 papers reviewed, 87 papers focused on the extraction of vegetation parameters from TLS point clouds. Of the 87, only 15 studies worked in the savanna biome [39,145,146]. The most frequently employed methods to derive vegetation attributes are the RANdom SAmple Consensus (RANSAC) algorithm for extracting DBH, stem curve profiles and the detection of stems [97,147], the use of the highest z coordinate of the point cloud for estimating heights [61,148], voxel-based and radiative transfer models for assessing LAI [149,150], Canopy Height Models (CHMs) for delineating crown attributes [91,92], and Quantitative Structure Models (QSMs) for computing tree volume and branch parameters [151,152]. Popular software packages to implement and work with these methods are Lastools [153], Matlab [154], R [155] and Python Programming [156], Cyclone [157], FARO Scene [158], RiscanPro [159], CompuTree [160] and Cloud Compare [161]. For evaluating the above extraction techniques, performance metrics such as R^2, RMSE and bias are typically used [51].

3.6. Vegetation Parameters Extracted from TLS Data

The vegetation parameters that can be retrieved from TLS data can be broadly classified into primary and secondary parameters [22]. Primary parameters are those directly derived from the TLS point cloud, such as DBH, height, volume and tree canopy characteristics [22]. Secondary parameters can only be estimated indirectly and are derivatives of primary parameters. They are usually obtained through modelling [22]. Examples of secondary parameters are AGB, basal area, stem density and LAI [22]. Besides the common vegetation parameters e.g., DBH and height, some studies focused on extracting unique and complex vegetation parameters such as tree planar projection [162], the elevation of angle and azimuth of stem inclination [52], clumping index [163], ground cover [92] and structural loss [164].As shown in Figure 6, most of the reviewed studies retrieved primary parameters, including DBH (41), height (32), tree canopy features (e.g., crown diameter, canopy cover) (20), trees or tree stems (20) and tree volume (15). With only five and six articles, respectively, branch parameters and stem curve profiles gained the least attention. Of the secondary parameters, LAI was the most extracted parameter (21), followed by AGB (11), basal area (5) and stem density (3). Details regarding specific parameters and the methods to retrieve them are provided in the following subsections. Particular emphasis is put on savanna ecosystems.

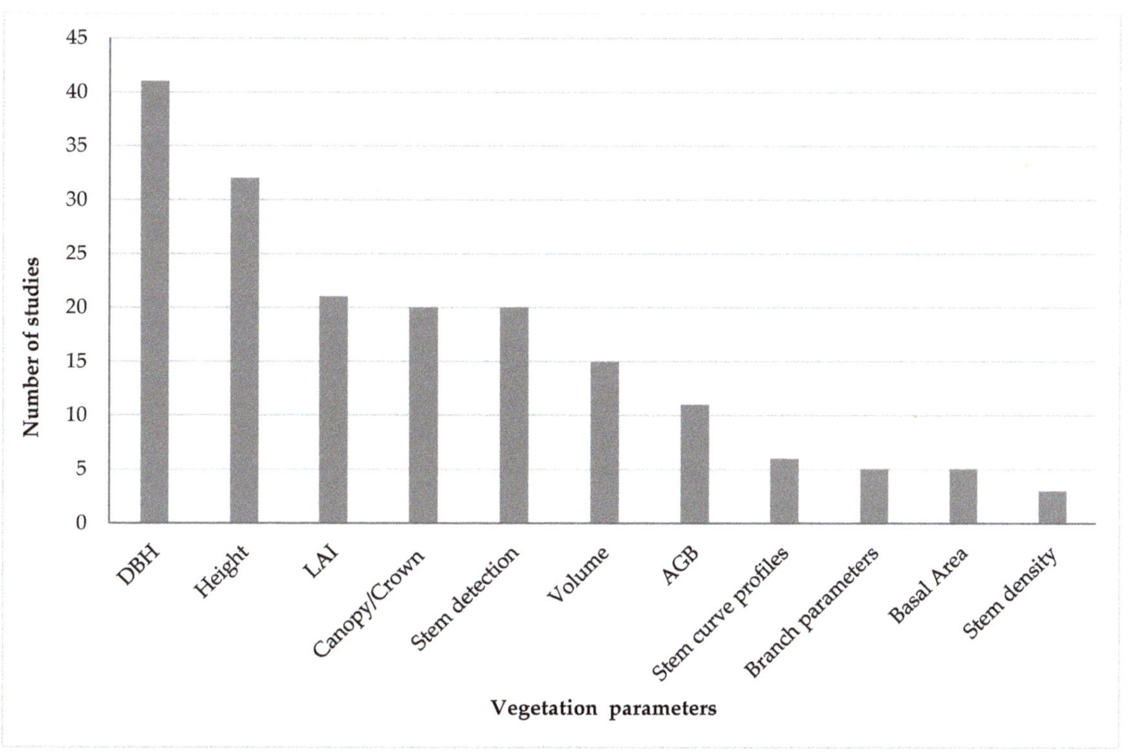

Figure 6. Primary and secondary vegetation parameters retrieved in the reviewed TLS studies. Studies were counted multiple times if two or more parameters were retrieved.

3.7. Primary Vegetation Attributes

3.7.1. DBH

From the reviewed papers, only one study derived DBH in the savanna [92]. Muir et al. [92] employed circle fitting to single-scan TLS data at heights between 1.28 m and 1.32 m above the Digital Elevation Model (DEM) in an open woodland site in Central Western Queensland Australia. They compared the resulting stem diameter measurements with data collected in the field and obtained an R^2 value of 0.77 [92]. Slightly better performances are often observed for DBH extraction in temperate, tropical and boreal forests, with R^2 values ranging from 0.81 to 0.99 [94,165,166] when compared to field reference data. For a pine woodland test site in France, least-squares circle fitting was reported to be the most accurate method (RMSE = 0.019 m) for estimating DBH when compared to least squares cylinder fitting and circular Hough transformation [85]. Similarly, Calders et al. [40] and Chen et al. [167] obtained R^2 values of 0.97 when using least squares circle fitting for DBH estimation with a multi-scan TLS scan with 5 positions in a eucalypt open forest and boreal forest respectively. Yurtseven et al. [148] obtained an R^2 value of 0.99 with a multi-scan with 8 positions in a Mediterranean forest employing randomised hough transformation with least square regression. In general, multiple TLS scans and a lower stem density yielded the best results [114,168,169].

3.7.2. Vegetation Height

Tree height estimation in the savanna was reported in five studies [30,62,91,92,145]. Odipo et al. [62] and Muir et al. [92] employed the use of CHMs to estimate tree height from multiple and single TLS scans (30 positions and single scan) and obtained R^2 values of 0.64 and 0.98 respectively when compared to ground referenced data. Singh et al. [30] observed an RMSE of 0.25 to 1.45 m based on multi-scan Long Range -TLS (LR-TLS) with 4 positions. They computed heights by measuring the vertical distance between treetops and the stem base. Kirton et al. [145] employed cylinder fitting to extract tree height and obtained a standard error of ±0.002 m using a TLS single scan. It was observed in this study that it is more challenging to derive the height of savanna shrubs than it is for savanna trees. Muir et al. [92] also reported that shrub heights are difficult to calculate from TLS data, with R^2 values of 0.381 and 0.08 for maximum and average shrub height, respectively. In contrast to this observation, the extraction of height from short stature vegetation was achieved with reasonable accuracy in montane areas as well as desert and xeric shrubland biomes [170]. Vrlingie et al. [170] applied a 2-D spatial wavelet analysis on a TLS derived CHM to extract shrub height and achieved an R^2 of 0.94 when compared to field measured shrub height.

3.7.3. Stem Detection

The detection of stems from TLS point clouds is an important first step in processing, to enable extraction of other vegetation parameters [171]. Only one study [147] from the review was conducted in the savanna biome. Burt et al. [147] employed treeseg algorithm (which extracts tree point clouds from larger point clouds) on a tropical and on an open vegetation cloud using TLS multi scans with 5 positions. They achieved 70% and 96% stem detection accuracy, respectively [147]. The detection of stems was achieved with high accuracies in temperate and boreal forests [138,171]. Kelbe et al. [138] used voxel stem segment modelling on a single TLS scan dataset acquired using low resolution TLS in various forest types in New England. Stems were detected with accuracies of (R^2 = 0.99; RMSE = 0.16 m) [138]. In a boreal forest Zhang et al. [171] employed connected component segmentation using both single and multi- scan with 5 positions. They achieved stem detection accuracies of 93% for single scan and 99% for multi-scan TLS setup when compared to reference data [171]. To date, most of the commonly used algorithms (e.g., density-based clustering [172], Hough transformation [85] and RANSAC cylinder fitting [48,173] used for stem detection achieved stem detection accuracies of 70% to 99% when applied in temperate, tropical and boreal forests.

3.7.4. Crown Attributes

Among the tree crown attributes dealt with in the scientific literature are canopy cover, canopy height/length, crown width/diameter and tree crown change. In the savanna biome, extraction accuracies for trees were better as compared to shrubs. Singh et al. [30], Odipo et al. [62] and Muir et al. [92] employed CHMs to determine canopy cover. Singh et al. [30] achieved an RMSE of 5.7–15.9% at distances up to 500 m with 4 scan positions, whilst Odipo et al. [62] and Muir et al. [92] achieved R^2 of 0.56 and 0.93 as compared to the ground referenced data with 30 and single scan positions respectively. Lower accuracies were obtained when estimating shrub canopy cover, with Muir et al. [92] reporting an R^2 of 0.23 when compared to field measurements. Yet, the application of 2D spatial wavelet analysis achieved an R^2 of 0.51 for 36 shrubs in a desert and xeric shrubland environment [170]. In this study, shrub crown area was calculated using spatial wavelet analysis using the detected radius while the actual shrub crown area was determined by measuring the minor and major axis and then calculating the crown as an ellipse [170]. The use of different algorithms for accurate extraction of canopy cover was also investigated by Yurtseven et al. [148] in a mediterranean woodland and scrub. They employed convex and concave hulls and they concluded that the concave planar underestimates canopy cover area as compared to the convex hull. A method to quantify canopy cover change was considered by Olivier et al. [50] in a sugar maple and balsam fir forest in Canada. They used bi-temporal multi-scan TLS data to identify vegetation boundaries, extract new material formed or displaced between the two-time period [56].

3.7.5. Tree Volume and Branch Parameters

Volume of standing trees or above ground volume is total stem wood and branches [174]. Branch parameters are parameters like branch total length, branch diameter, branch insertion angle, branch depth into the crown and branch radius. Over the years, a range of algorithms has been developed to compute tree volume and branch parameters from TLS point clouds [151,175]. However, none of the corresponding studies was conducted in the savanna biome. QSMs are the most common algorithms used in literature [96,151,175]. They have the ability to provide reliable volumes estimates of stem and branches of trees, but also have limitations in reconstructing leaves [176]. Figure 7 shows an example of a 3D point cloud and the resulting QSM. Branching structure and branch lengths were determined in boreal and tropical forests using QSM [151,175]. Kaasalainen et al. [175] used a multi scan with 2 positions to reconstruct branching structure and concluded that branching structure can be reproduced with ±10% accuracy when compared to laboratory reference measurements. Lau et al. [151] used a multi-scan with 13 positions and successfully reconstructed 87% of the branch lengths. They compared the reconstructed branch lengths to manually measured branches from destructively harvested trees [151]. For the estimation of shrub total and green biomass, Olsoy et.al [54] employed voxel-based and 3D convex hull algorithms to first determine shrub volume in a xeric sagebrush ecosystem in Southern Idaho, USA using multi scan with 2 positions. They then modelled shrub total and green biomass by subsetting the point cloud into green and non-green woody components based on the difference in their reflectivity. Their findings showed that convex hull algorithms ($R^2 = 0.92$, $R^2 = 0.83$) performed slightly better than the voxel-based method ($R^2 = 0.86$, $R^2 = 0.73$) in estimating total and green biomass respectively [54]. The TLS derived shrub biomass was compared to the destructive samples of shrubs [54]. These and other volume-related algorithms (e.g., bounding cuboids, outer hull models, square-based columns) have also been compared for test sites in the temperate and tropical regions [101]. It was found that convex hull techniques tend to overestimate while the square-based column method underestimates volume. Moreover, voxel-based estimates are mainly affected by the point density and level of occlusions in the point cloud [101].

Figure 7. (**a**) TLS point cloud coloured by height (lowest values in blue and highest values in red). (**b**) Resulting QSM model from the algorithm by Raumonen et al. [177]. Data Source: (Madhibha, 2016) [178].

3.7.6. Stem Curve Profiles

Stem profiles or stem taper is the "stem diameter measured as a function of height" [179]; that is, how the diameter of the tree changes at different heights along the stem. Six studies used TLS point clouds to determine stem curve profiles, five of the studies were conducted in boreal forests [48,51,52,162,166] and one was conducted in temperate broadleaf and mixed forest [58]. The methods used to extract stem curve profiles from TLS point clouds are either circle or cylinder fitting at different height bins [48,51,52,58,162,166]. Maas et al. [58] extracted stem curve profiles using circle fitting at multiple heights and obtained an RMSE of 4.7 cm when compared to a harvested stem. Liang et al. [51] observed that most algorithms used for the extraction of stem curve profiles results in an RMSE of 1.3 and 6.0 cm for single scan data and 0.9 and 5.0 cm from multi scan data for three different stem densities (i.e. low, medium and high stem densities) in a boreal forest. The RMSE values obtained are a function of both the stem density and the number of scan positions employed [51].

3.8. Secondary Vegetation Attributes

3.8.1. AGB

AGB is defined as "the total amount of biological material (oven-dried) present above the surface in an area" [180]. AGB is the most important secondary attribute derived from TLS point clouds. The accurate extraction of all other parameters is a necessity in improving tree biomass estimation. A total of five studies estimated AGB in the savanna using TLS [62,64,70,90,126]. Zimbres et al. [64] predicted plot AGB of three vegetation types in the Brazilian cerrado. The TLS derived AGB was validated against the AGB values obtained from allometric equations. The R^2 values obtained were 0.92 and 0.88 for the woodland savanna dry and rainy season respectively and 0.58 for the forested savanna. Due to occlusions, no conclusive results were obtained for the gallery forest [70]. Cuni-Sanchez et al. [70] estimated AGB change in five different vegetation types using vertical plant profiles extracted from a multi-scan TLS over a period of 20 years in Gabon. However, the savanna plots were not assessed because all the trees had a diameter of less than 10 cm [70]. Odipo et al. [62] also estimated AGB change in a south african savanna using TLS derived canopy cover and height metrics for plot level AGB and then further extrapolating to a wider scale using multi-temporal L band SAR. The TLS derived metrics were validated with field measured parameters. Significant relationships were obtained between the TLS derived predictors (canopy cover & height) with field biomass (R^2 = 0.93). Seven variables extracted from TLS data were used to build regression models with field measured AGB to estimate grass biomass in China [90]. The regression methods used were Simple Regression (SR, Stepwise Multiple Regression (SMR), Random Forest (RF) and Artificial Neural Network (ANN). Results obtained showed that the highest prediction of biomass was obtained using SMR model (R^2 = 0.84), followed by SR (R^2 = 0.80), RF (R^2 = 0.78) and lastly ANN (R^2 = 0.73) [90]. Cooper et al. [126] investigated the application of Structure from Motion (SfM) photogrammetry and TLS for the estimation of grass biomass and compared this to destructive harvest grass biomass. The R^2 values for the TLS derived grass biomass was 0.46 and total biomass was 0.57, whilst for the SfM it was 0.54 for the grass biomass and 0.72 for the total biomass [126]. TLS data have also been used to derive tree and grass AGB change [49,115]. Srinivasan et al. [49] extracted various tree parameters (including AGB) from loblolly pines in Texas, USA by the use of multi-temporal TLS data. The best results (R^2 = 0.50, RMSE = 10.09 kg) were obtained by directly estimating AGB change from the point cloud when compared to field measured AGB change. Guimarães-Steinicke et al. [115] also investigated grass biomass temporal change in Thuringia, Germany. They used 25 TLS derived metrics to model biomass [115].

While most of the reviewed studies focused on the estimation of tree biomass, only few papers dealt with shrub biomass. Shrub biomass was investigated by Greaves et al. [59] in the arctic tundra. They employed a leave one out cross-validation technique of the canopy volume against the harvested shrub biomass to establish predictive relationships between the TLS canopy volume and harvested shrub biomass and between Airborne Laser Scanning (ALS) canopy volume and TLS derived shrub biomass estimates. The TLS produced more accurate predictions of shrub biomass as compared to the ALS (TLS R^2 = 0.78 and ALS R^2 = 0.62). Li et al. [181] merged TLS and ALS to estimate the biomass of sagebrush in southwest Idaho, USA by building predictive models. TLS derived biomass was used for calibration of a biomass predictive model with an R^2 of 0.87. The TLS derived shrub volume was compared to the destructive shrub biomass measurements (R^2 = 0.88) [181].

3.8.2. LAI

LAI is defined "as the ratio of leaf area to per unit ground surface area" [39,130]. A study was conducted in the savanna biome by Béland et al. [39]. They applied a voxel-based approach making use of the contact frequency method on six trees. To this end, they employed two multi-scan setups with 2 and 3 positions and compared the TLS derived estimates against direct measurements obtained from leaf harvesting. The leaf area ranged from 30 to 530 m^2 whilst crown LAI was 0.8 and 7.2 with a general bias of 14%. To ensure an accurate extraction of LAI, the green and non-green components were distinguished in the TLS data based on a separation threshold applied to the return intensities [39]. Olsoy et al. [182] estimated the LAI of sagebrush from TLS data with 2 scan positions and found that the estimated LAI was highly correlated with their reference field data (R^2 = 0.78).

3.8.3. Basal Area

Basal area is defined as a "cross-sectional area of all stems in a plot measured at breast height and usually expressed per unit of the land" [183]. Five studies estimated basal area from TLS point clouds, with four being conducted in a temperate broadleaf and mixed forest [75,78,85,184] and one from the boreal zone [48]. No study has been published yet for the savanna biome. Yrttimaa et al. [48] and Tansey et al. [85] derived basal area from TLS point clouds by first calculation basal area for a single stem, then scaling it to one hectare. The TLS derived basal area was compared to the field derived basal area which was derived by considering the cross-sectional area of a tree to be circular [48]. The R^2 observed were 0.87, 0.86 and 0.63 for a plot radius of 6, 11 and 16 m respectively [48]. Côté et al. [184] used an architectural model called LiDAR to Tree Architecture (L-Architect) by reconstructing forest stands with scans of individual trees. The L-Architect derived basal area was compared to the measured values of basal area to assess accuracy. A relative difference of -0.01% was obtained between the L-Architect and field measured basal area [184]. A methodology based on voxel data structure was applied to derive the basal area of individual trees in a heterogeneous vegetation structure in Washington Park Arboretum [75]. The TLS derived basal area was derived by fitting a 3D cylinder primitive to the trunk assuming that the trunk was a geometric circle. The TLS derived basal area was compared to the manually measured basal area. The TLS derived basal area was underestimated compared to the field-based methods for conifers, deciduous and mixed stands (Conifer: Field = 12.3116 m^2, TLS = 6.9217 m^2; Deciduous: Field = 1.4275 m^2, TLS = 1.0172 m^2 & Mixed: Field = 1.8431 m^2, TLS = 1.575 m^2) [75]. Aijazi et al. [78] used a super voxel segmentation method to estimate basal area and concluded that parameter estimation error ranged from 1.6 to 9% from TLS point clouds.

3.8.4. Stem Density

Stem density is defined as "the number of trees per hectare" [185]. Three of the reviewed studies estimated stem density, with one being performed in a savanna ecosystem [92]. Stem density is derived from the number of stems identified in the scanned area and comparing it to the field derived stem density [48,85,92]. Yrttimaa et al. [48] varied the plot size in order to investigate its effects on stem density. The smaller the plot size, the higher the correlation of the TLS-derived stem density with field measurements (r = 6 m I R^2 = 0.77; r = 11 m I R^2 = 0.67; r = 16 m I R^2 = 0.38). Muir et al. [92] determined stem density from TLS point clouds as well. They achieved an R^2 of 0.1 and reported a large RMSE of 130.9 trees/hectare that was attributed to plot complexity.

4. Discussion

4.1. Geographical Trends in the TLS Literature

From a global perspective, most TLS studies have been conducted in Europe, North America and China (Figure 3). Despite their large geographical extent in Africa, South America and Australia, savannas have rarely been investigated in these areas. In the early 2000s, TLS-based vegetation studies were mainly conducted in Europe and North America (Figure 2). From 2009 onwards, TLS studies expanded into the rest of the world, including Africa, Asia, Australia and South America. An increasing collaboration among TLS research groups from different parts of the world can explain this observation [57,100,147,171]. To further advance TLS research in savannas and to learn more about their ecology, building scientific capacity in financially constrained countries is key.

Savannas cover ca. 50% of Africa's landmass. Yet, only 6% of the reviewed studies were conducted in Africa. Of those studies, the majority focused on South Africa. Only a few articles presented results for other African countries, including Mali [39], Ethiopia [186] and Gabon [70]. One reason for this observation is that TLS research in Africa is mainly constrained by financial limitations. The acquisition and processing of TLS data usually require expensive equipment and software [22]. Moreover, it necessitates trained personnel, which is limited in most third-world countries [176]. With the advent of low-cost terrestrial laser scanners [65,187,188], more TLS research can be expected in the future for Africa.

Since savannas cover large geographical areas, TLS data can usually not be used by itself to study an entire region due to its limited footprint. Instead, an integration of TLS point clouds with other geospatial data is necessary for large-scale assessments. In the past, the creation of wall to wall maps of savannas has been achieved by combining TLS with L-band SAR [62] and airborne LiDAR [69]. Currently, the existence of freely available satellite data from Sentinel-1/2 and Landsat 5–8, as well as coarser resolution imagery from MODIS [70] can resolve upscaling issues from fine spatial resolution from TLS to medium resolution from Landsat and Sentinel then to coarse resolution using MODIS. The planned launch of BIOMASS by the European space agency and Multi-footprint Observation LiDAR and Imager (MOLI) by JAXA after Global Ecosystems Dynamics Investigation (GEDI) and Ice Cloud & Elevation Satellite-2 (ICESat-2) will present an opportunity for landscape-level studies of the savanna [60]. Based on these and similar data sets, TLS point clouds can be used in wide-area mapping for calibration and validation purposes [60].

4.2. Challenges & Opportunities

Single and multiple scans (five, four, three and two positions) were found to be the most commonly used TLS setups in the reviewed literature. Single scan approaches have been used successfully to estimate DBH [138,166], height [92] and to detect stems [189]. Single scan inventories received a lot of attention because they allow for fast and less labour intensive data acquisitions and do not require the registration of multiple point clouds [109,112]. However, single scan approaches are also more prone to occlusions [43]. Data from multi-scan TLS campaigns were shown to yield higher accuracies and reduced error margins [45,139]. Therefore, multiple TLS scans with larger plot sizes (>15 m in radius) provide the only opportunity to adequately capture the heterogeneous nature of savannas and to characterize the full spectrum of savanna vegetation structure [92]. Depending on the parameters to be extracted, the requirements to be met and the available resources, the choice between a single- or multi-scan TLS setup should be made carefully. For example, long range single scans give reliable data in the dry season (leaf-off) and not in the wet season (leaf-on) [30].

DBH is one of the most reliable proxies for biomass [190]. This is why it is also the most frequently measured quantity not only by foresters, but also with regard to the reviewed literature. It has been shown that DBH can be delineated automatically based on multiple TLS scans [40]. However, these techniques have not yet been adjusted or tested for more complex vegetation structures [88]. To date, most algorithms to retrieve DBH (and also stems and stem curve profiles) assume a circular or cylindrical stem geometry. However, these rather simple geometries are not well-suited to describe the short stature vegetation and deformed trees that are typically found in savannas [40,191]. While a number of DBH extraction techniques (e.g., RANSAC algorithm [186], point cloud slicing at 1.3 m [75], least-squares circle fitting [52] and Hough transformation [86] exist), they have only been developed and demonstrated in temperate and boreal forests. For savannas, only the circle fitting method has been explored yet [92].

Height was extracted from TLS point clouds in the savanna, mostly through the employment of CHMs to achieve the goal. Muir et al. [92] obtained an R^2 of 0.97 and 0.38 for TLS-based estimates of tree and shrub heights, respectively. Small stature woody vegetation is often not well captured by CHMs, which leads to an underestimation of its height when interpolation is used to create the canopy surface [170]. This issue has not yet been adequately addressed, and more accurate height assessment techniques still need to be developed [170]. Potential ways to achieve this would be to elevate the scanner above the vegetation layer or place the scanner on a suitable vantage position [30,92], to capture TLS data of shrubs with higher point density (e.g., 16 pts per m^2, [170]) and to distinguish between trees and shrubs in the TLS point cloud and to create two corresponding tree- and shrub-related CHMs that can be analysed separately [92].

Other methods for vegetation height extraction are available, including the combined analysis of TLS and airborne (also UAV) LiDAR data [140,192], the highest z-coordinate (height is derived by the highest value of the return) [51], the vertical distance between the highest point and stem base at the ground (the vertical distance is measured from the base of the stem to the highest point on a segmented point cloud) [30] and cylinder fitting (a cylinder is fitted on a segmented point cloud and the height is denoted as the highest point in a vertical cylinder) [58,145]. While they are also not without errors, these techniques have neither been developed, adjusted nor evaluated for savannas. Known problems, e.g., uncertainties from treetop shading [86], are especially common in the riparian zones of savanna ecosystems.

The extraction of LAI using TLS point clouds is not very common across all biomes [22] including the savanna. Calders et al. [79] argue that caution is needed when using LAI estimates and the variable that is measured from ground-based sensors is effective Plant Area Index (ePAI) and effective Wood Area Index (eWAI). They recommend that TLS measurements are ideal for estimation of gap fraction, ePAI and eWAI as compared to instruments that are dependent on certain illumination conditions [79]. Gap fraction is an indirect way of estimating LAI and TLS is recommended for the estimation of gap fraction to derive ePAI and eWAI [79]. It has been used successfully in the past to estimate LAI [193]. TLS intensity data can also be used to compute LAI. They allow for differentiating between photosynthetically active (green) from non-active (woody) vegetation parts [39,54]. However, they can also be compromised by instrumental, atmospheric and object properties as well as the scanning geometry [82,129]. In order to reliably make use of TLS intensity data for vegetation applications, conversion of instrumental intensity into the proportional or equal target reflectance is required [129,135].

While tree crown attributes have been successfully derived from TLS point clouds, more research needs to be directed towards crown change and shrub parameters. With regard to the former, the method by Olivier et al. [50] needs to be tested in the savanna biome. Vierling et al. [170] suggested the use of vantage points, high cliffs or tall tripods in order to achieve steep viewing zenith angles. These enable the acquisition of enough returns from both vegetation and ground. Following this procedure, reliable construction of DEMs and characterization of vegetation structure is ensured. Other ways to capture more accurate canopy details are the acquisition of full-waveform TLS data [170], the separation of trees and shrubs on the basis of height thresholds and noise filtering [92,148,170] and also separation through the application of reflectance thresholds and LiDAR return deviations [69].

Various reconstruction algorithms are available to extract plant volume from TLS data [48,54,96,101]. A limitation common to many of these techniques is that they rely on simple geometries (e.g., cylinders) to model vegetation components. This usually leads to an overestimation of volume due to the tapering of branches and the complexity of the tree architecture [176]. Another challenge lies in the exact evaluation of TLS-based volume estimates. True validation requires destructive sampling which is seldomly possible and desired [176]. Despite these limitations, volume reconstructions from TLS point clouds are key to improving tree allometries, non-destructively estimating and monitoring biomass change as well as assessing structural gains and losses in savanna ecosystems [30,175].

AGB is a key parameter in the study of vegetation structure and crucial indicator of terrestrial carbon pools and productivity [140,190]. Accordingly, it is also a popular secondary vegetation attribute in the TLS literature. Volume estimates form the basis of AGB assessments with TLS data, which are less prone to errors when compared to the simple use of allometric equations [40,176]. QSMs are most frequently employed for estimating AGB from TLS point clouds. However, reconstruction quality still suffers from certain errors and uncertainties. These are related to the modelling of complex tree crowns and smaller branches, the optimal use of wood density values (which vary both within and between species and regions), the use of simple reconstruction geometries such as cylinders, the separation of leaf and woody components as well as the validation of AGB values [71,176]. Despite these limitations, volume-based AGB estimates do not require information on DBH and show better agreement with reference data. As a non-destructive proxy for estimation of AGB, they should be further explored in the savanna context [40,176].

Using TLS point clouds, branch level parameters from a much larger percentage of a whole tree can be extracted [194]. Combined with other tree attributes, they hold the potential to calculate whole tree carbon and water use [194]. Especially reconstruction algorithms like QSMs [177] have enabled their retrieval. To date, there are very few studies on this topic and none of them focused on savannas. QSMs have been reported to yield accurate results when modelling thick branches [195]. Accuracies usually drop for smaller branches, particularly in cases where leaf-on scans are employed [195]. In order to improve QSMs with regard to branch parameter retrieval, they should be able to correctly separate the point cloud into leaf and woody components [71].

Shrub volume and AGB has also been estimated using TLS-based reconstruction algorithms. Greaves et al. [142] employed volumetric surface differencing and voxel counting in the arctic tundra whereas Olsoy et al. [54] used voxel-based and 3D convex hull techniques to assess the volume of shrubs. The existing methods hold the potential to be successfully transferred and adjusted to savanna ecosystems. This would allow for quantifying shrub layer biomass and help making more accurate AGB assessments overall. Moreover, these integrated biomass estimates could be used as calibration and validation reference data for mapping and monitoring savannas at larger spatial scales [142].

4.3. Future Outlook

The accurate extraction of vegetation parameters from TLS data is a challenging task. This review has shown that there are still a lot of research gaps that need to be addressed. This is especially true for savannas, which are in many ways much different from those biomes that have so far gained more attention in the literature. Savannas are heterogeneous and dynamic ecosystems that feature unique and complex mixtures of trees, shrubs and grasses. This requires novel developments, (re-) adjustments as well as extensive tests and experiments when it comes to TLS algorithms and applications.

Savannas are regularly subject to disturbances from fire, herbivory, droughts and humans. These disturbances lead to vegetation structural changes (including the loss of large trees [66,196] that can affect savanna ecological functioning [196]. As a future avenue for mapping and monitoring savanna vegetation, the acquisition and analysis of multi-temporal TLS point clouds should be considered. While research on this subject is still in its infancy (e.g., [49,50,115]), there is no doubt that it holds a large potential for characterizing and understanding savanna dynamics [70], including phenological processes (i.e., leaf-on vs. leaf-off conditions) [54]. Among others, this is because of the non-destructive nature of TLS [49,50] and its ability to repeatedly capture 3D vegetation information with great detail [58]. Levick and Asner et al. [196] quantified the loss of large savanna trees and observed declines of more than 30% for some sites. However, since they were using ALS, below-canopy treefall and canopy-related losses (e.g., branch breakages) could not detected [196]. Such limitations can be addressed by acquiring and investigating time series of TLS scans. By developing suitable change detection algorithms, multi-temporal TLS will enable accurate and regular accounting of carbon losses and gains in savannas.

To date, woody AGB mapping in the savanna has mainly been focusing on tree biomass. Despite being a key component of savannas, shrubs have not been considered in this regard. Future research should therefore focus on an accurate and more holistic assessment of woody AGB in savannas, including trees, small understory trees and shrubs [65]. One pathway to achieve this would be a combination of TLS and airborne LiDAR data. TLS is well-suited to characterize understory vegetation structure in terms of both trees and shrubs [64,91]. Especially regarding the latter, existing algorithms that have initially been developed for sagebrush and arctic shrubs [54,110,142] could be tested and refined. Airborne (including UAV) LiDAR offers 3D point clouds with a bird's eye perspective of the savanna. While it has been found to be of limited use in shrub-dominated ecosystems [170], it is superior for tree height and canopy structure mapping [27]. By combining TLS and airborne LiDAR data, synergies between both technologies are exploited while their individual limitations can be overcome. To increase spatial coverage and to create wall-to-wall AGB maps, optical (e.g., Landsat and Sentinel-2) and SAR (e.g., Sentinel-1, ALOS PALSAR-1/2, BIOMASS) satellite time series should be integrated into such a monitoring scheme.

From a technical perspective, TLS-based assessments of volume and AGB have not adequately taken into account the different types of woody vegetation in the savanna. An opportunity to improve them would be the separation of TLS point clouds into individual trees and shrubs. Afterwards, specific 3D models tailored to each savanna life form as well as dedicated noise filtering algorithms could be applied to yield more accurate reconstruction results [64,92,170]. For instance, QSMs could be employed to extract volume information from tree point clouds while convex hulls or voxels could be used in the case of shrub point clouds. To this end, methods exist that have been tested on trees in temperate [37], boreal [96,175] and tropical forests [101,151] as well as shrubs in the temperate regions [54,181]. Future studies in the savanna need to transfer, train and validate these techniques.

TLS intensity data have already been demonstrated to hold potential for vegetation analyses in a few studies. However, only limited overall evidence can be found in the literature so that a number of unresolved scientific questions still remain. In particular, more research is needed on the improved retrieval of the LAI with TLS intensity [80,150]. Among others, this could be achieved by a better separation of leaf and wood material [176]. Other challenging tasks based on TLS reflectance data comprise the study of seasonal changes (i.e., leaf-on vs. leaf-off vegetation properties) [54] and leaf biochemical parameters to understand the photosynthetic capacity of savanna vegetation [27].

5. Conclusions

3D data from TLS can make a vital contribution to mapping the current state of savanna vegetation, monitoring and quantifying its dynamics over space and time and, in this way, provide a better understanding of the underlying ecological factors and processes as the very basis for sustainable savanna management and long-term conservation. This study reviewed the scientific literature on the use of TLS for extracting vegetation parameters with a special geographic focus on savannas. The current state of the art was presented and methods relevant for TLS research in the savanna biome were identified and discussed. To achieve this, a comprehensive quantitative and qualitative meta-analysis of 113 research articles was conducted using 18 attributes. It was found that only few studies have dealt with TLS data for vegetation analyses in the savanna. Among others, research gaps exist with respect to a better separation of leaf and woody components in TLS point clouds, the accurate 3D reconstruction of savanna vegetation volume and biomass, more holistic assessments of woody AGB by considering both trees and shrubs as well as the application of TLS time series data for multi-temporal assessments of vegetation structural change, including studies quantifying seasonal differences in vegetation dry and green biomass. Future research should thus concentrate on extending previous experiments to savannas as well as developing novel information extraction routines for these.

Supplementary Materials: The following are available online at https://www.mdpi.com/2072-429 2/13/3/507/s1. Table S1: Full list of reviewed research articles.

Author Contributions: Conceptualization and methodology by T.P.M. and C.T. Meta-analysis by T.P.M. with the help of C.T. Draft writing by T.P.M. and C.T. Reviewing and editing by C.T., J.B., J.S., C.S. and T.P.M. Supervision by C.T., C.S. and J.B. TLS data used in Figure 1 was acquired and processed by J.B. All authors have read and agreed to the published version of the manuscript.

Funding: This study was made possible by funding from the Deutscher Akademischer Austauschdienst: DAAD Ref No. SPACES II.2 CaBuDe 57531823, Bundesministerium für Bildung und Forschung (BMBF): South African Land Degradation Monitor: Grant No. 01LL1701A, Bundesministerium für Bildung und Forschung (BMBF): Ecosystem Management Support for Climate Change in Southern Africa (EMSAfrica) Project: Grant No. 01LL1801D.

Institutional Review Board Statement: Not applicable.

Informed Consent Statement: Not applicable.

Data Availability Statement: Data is contained within the article or supplementary material. The data presented in this study are available in Supplementary Material Table S1.

Conflicts of Interest: The authors declare no conflict of interest.

Abbreviations

ALS	Airborne Laser Scanning
3D	Three Dimensional
AGB	Above Ground Biomass
CHM	Canopy Height Model
DBH	Diameter at Breast Height
DEM	Digital Elevation Model
ePAI	effective Plant Area Index
eWAI	effective Wood Area Index
GPS	Global Positioning System
GNSS	Global Navigation Satellite System
IGARSS	International Geoscience and Remote Sensing Symposium
LAI	Leaf Area Index
LANDSAT	Land Remote-Sensing Satellite
L-Architect	LiDAR to tree Architecture
LiDAR	Light Detection And Ranging
MODIS	Moderate Resolution Imaging Spectroradiometer
QSM	Quantitative Structure Models
R^2	Coefficient of determination
RADAR	Radio Detection and Ranging
RANSAC	RANdom SAmple Consensus
RGB	Red, Green, Blue
RMSE	Root Mean Squared Error
TLS	Terrestrial Laser Scanner
UAV	Unmanned Aerial Vehicle

References

1. Scholes, R.J.; Archer, S.R. Tree-grass interactions in savannas. *Annu. Rev. Ecol. Syst.* **1997**, *28*, 517–544. [CrossRef]
2. Ma, X.; Huete, A.; Moore, C.E.; Cleverly, J.; Hutley, L.B.; Beringer, J.; Leng, S.; Xie, Z.; Yu, Q.; Eamus, D. Spatiotemporal partitioning of savanna plant functional type productivity along NATT. *Remote Sens. Environ.* **2020**, *246*, 111855. [CrossRef]
3. Sankaran, M.; Hanan, N.P.; Scholes, R.J.; Ratnam, J.; Augustine, D.J.; Cade, B.S.; Gignoux, J.; Higgins, S.I.; Le Roux, X.; Ludwig, F.; et al. Determinants of woody cover in African savannas. *Nature* **2005**, *438*, 846–849. [CrossRef] [PubMed]
4. Grace, J.; Jose, J.S.; Meir, P.; Miranda, H.S.; Montes, R.A. Productivity and carbon fluxes of tropical savannas. *J. Biogeogr.* **2006**, *33*, 387–400. [CrossRef]
5. Hutley, L.B.; Setterfield, S.A. Savanna. In *Encyclopedia of Ecology*, 2nd ed.; Elsevier Inc.: Amsterdam, The Netherlands, 2018. [CrossRef]
6. Galvin, K.A.; Reid, R.S. People in savanna ecosystems: Land use, change, and sustainability. In *Ecosystem Function in Savannas*; CRC Press: Boca Raton, FL, USA, 2010; pp. 521–536.
7. Egoh, B.N.; O'Farrell, P.J.; Charef, A.; Gurney, L.J.; Koellner, T.; Abi, H.N.; Egoh, M.; Willemen, L. An African account of ecosystem service provision: Use, threats and policy options for sustainable livelihoods. *Ecosyst. Serv.* **2012**, *2*, 71–81. [CrossRef]
8. Pritchard, R.; Grundy, I.M.; Van Der Horst, D.; Ryan, C.M. Environmental incomes sustained as provisioning ecosystem service availability declines along a woodland resource gradient in Zimbabwe. *World Dev.* **2019**, *122*, 325–338. [CrossRef]
9. Stevens, N.; Lehmann, C.E.R.; Murphy, B.P.; Durigan, G. Savanna woody encroachment is widespread across three continents. *Glob. Chang. Biol.* **2017**, *23*, 235–244. [CrossRef]
10. Guuroh, R.T.; Ruppert, J.C.; Ferner, J.; Čanak, K.; Schmidtlein, S.; Linstädter, A. Drivers of forage provision and erosion control in West African savannas—A macroecological perspective. *Agric. Ecosyst. Environ.* **2018**, *251*, 257–267. [CrossRef]
11. Mograbi, P.J.; Asner, G.P.; Witkowski, E.T.; Erasmus, B.F.; Wessels, K.J.; Mathieu, R.; Vaughn, N.R. Humans and elephants as treefall drivers in African savannas. *Ecography* **2017**, *40*, 1274–1284. [CrossRef]
12. Conradi, T. Woody encroachment in African savannas: Towards attribution to multiple drivers and a mechanistic model. *J. Biogeogr.* **2018**, *45*, 1231–1233. [CrossRef]

13. Ciais, P.; Bombelli, A.; Williams, M.; Piao, S.L.; Chave, J.; Ryan, C.M.; Henry, M.; Brender, P.; Valentini, R. The carbon balance of Africa: Synthesis of recent research studies. *Philos. Trans. R. Soc. A Math. Phys. Eng. Sci.* **2011**, *369*, 2038–2057. [CrossRef] [PubMed]

14. Cook, G.D.; Liedloff, A.C.; Eager, R.W.; Chen, X.; Williams, R.J.; O'Grady, A.P.; Hutley, L.B. The estimation of carbon budgets of frequently burnt tree stands in savannas of northern Australia, using allometric analysis and isotopic discrimination. *Aust. J. Bot.* **2005**, *53*, 621–630. [CrossRef]

15. Williams, R.J.; Carter, J.; Duff, G.A.; Woinarski, J.C.Z.; Cook, G.D.; Farrer, S.L. Carbon accounting, land management, science and policy uncertainty in Australian savanna landscapes: Introduction and overview. *Aust. J. Bot.* **2005**, *53*, 583–588. [CrossRef]

16. Tsalyuk, M.; Kelly, M.; Getz, W.M. Improving the prediction of African savanna vegetation variables using time series of MODIS products. *ISPRS J. Photogramm. Remote Sens.* **2017**, *131*, 77–91. [CrossRef] [PubMed]

17. Patenaude, G.; Milne, R.; Dawson, T.P. Synthesis of remote sensing approaches for forest carbon estimation: Reporting to the Kyoto Protocol. *Environ. Sci. Policy* **2005**, *8*, 161–178. [CrossRef]

18. Asner, G.P.; Levick, S.R.; Smit, I.P.J. Remote sensing of fractional cover and biochemistry in Savannas. In *Ecosystem Function in Savannas: Measurement and Modeling at Landscape to Global Scales*; CRC Press: Boca Raton, FL, USA, 2010; pp. 195–217. [CrossRef]

19. Viergever, K.M.; Woodhouse, I.H.; Stuart, N. Monitoring the world's savanna biomass by earth observation. *Scott. Geogr. J.* **2008**, *124*, 218–225. [CrossRef]

20. Hirata, Y.; Takao, G.; Sato, T.; Toriyama, J. (Eds.) *REDD-Plus Cookbook*; REDD Research and Development Centre, Forestry and Forest Products Research Institute: Tsukuba, Japan, 2012; 156p, ISBN 9784905304135.

21. Feng, X.; He, L.; Cheng, Q.; Long, X.; Yuan, Y. Hyperspectral and Multispectral Remote Sensing Image Fusion Based on Endmember Spatial Information. *Remote Sens.* **2020**, *12*, 1009. [CrossRef]

22. Gwenzi, D. Lidar remote sensing of savanna biophysical attributes: Opportunities, progress, and challenges. *Int. J. Remote Sens.* **2016**, *38*, 235–257. [CrossRef]

23. Bioucas-Dias, J.M.; Plaza, A.; Camps-Valls, G.; Scheunders, P.; Nasrabadi, N.; Chanussot, J. Hyperspectral remote sensing data analysis and future challenges. *IEEE Geosci. Remote Sens. Mag.* **2013**, *1*, 6–36. [CrossRef]

24. Cho, M.A.; Mathieu, R.; Asner, G.P.; Naidoo, L.; Van Aardt, J.A.N.; Ramoelo, A.; Debba, P.; Wessels, K.; Main, R.; Smit, I.P.; et al. Mapping tree species composition in South African savannas using an integrated airborne spectral and LiDAR system. *Remote Sens. Environ.* **2012**, *125*, 214–226. [CrossRef]

25. Sinha, S.; Jeganathan, C.; Sharma, L.K.; Nathawat, M.S. A review of radar remote sensing for biomass estimation. *Int. J. Environ. Sci. Technol.* **2015**, *12*, 1779–1792. [CrossRef]

26. Lim, K.; Treitz, P.; Wulder, M.; St-Onge, B.; Flood, M. LiDAR remote sensing of forest structure. *Prog. Phys. Geogr. Earth Environ.* **2003**, *27*, 88–106. [CrossRef]

27. Beland, M.; Parker, G.; Sparrow, B.; Harding, D.; Chasmer, L.; Phinn, S.; Antonarakis, A.; Strahler, A. On promoting the use of lidar systems in forest ecosystem research. *For. Ecol. Manag.* **2019**, *450*, 450. [CrossRef]

28. Gwenzi, D.; Lefsky, M.A. Plot-level aboveground woody biomass modeling using canopy height and auxiliary remote sensing data in a heterogeneous savanna. *J. Appl. Remote Sens.* **2016**, *10*, 16001. [CrossRef]

29. Hill, R.; Broughton, R. Mapping the understorey of deciduous woodland from leaf-on and leaf-off airborne LiDAR data: A case study in lowland Britain. *ISPRS J. Photogramm. Remote Sens.* **2009**, *64*, 223–233. [CrossRef]

30. Singh, J.; Levick, S.R.; Guderle, M.; Schmullius, C. Moving from plot-based to hillslope-scale assessments of savanna vegetation structure with long-range terrestrial laser scanning (LR-TLS). *Int. J. Appl. Earth Obs. Geoinf.* **2020**, *90*, 102070. [CrossRef]

31. Bauwens, S.; Bartholomeus, H.M.; Calders, K.; Lejeune, P. Forest Inventory with Terrestrial LiDAR: A Comparison of Static and Hand-Held Mobile Laser Scanning. *Forests* **2016**, *7*, 127. [CrossRef]

32. Brede, B.; Calders, K.; Lau, A.; Raumonen, P.; Bartholomeus, H.M.; Herold, M.; Kooistra, L. Non-destructive tree volume estimation through quantitative structure modelling: Comparing UAV laser scanning with terrestrial LIDAR. *Remote Sens. Environ.* **2019**, *233*, 111355. [CrossRef]

33. Boukoberine, M.N.; Zhou, Z.; Benbouzid, M. A critical review on unmanned aerial vehicles power supply and energy management: Solutions, strategies, and prospects. *Appl. Energy* **2019**, *255*, 113823. [CrossRef]

34. Eisfelder, C.; Kuenzer, C.; Dech, S. Derivation of biomass information for semi-arid areas using remote-sensing data. *Int. J. Remote Sens.* **2011**, *33*, 2937–2984. [CrossRef]

35. Galidaki, G.; Zianis, D.; Gitas, I.; Radoglou, K.; Karathanassi, V.; Tsakiri–Strati, M.; Woodhouse, I.; Mallinis, G. Vegetation biomass estimation with remote sensing: Focus on forest and other wooded land over the Mediterranean ecosystem. *Int. J. Remote Sens.* **2016**, *38*, 1940–1966. [CrossRef]

36. Dassot, M.; Constant, T.; Fournier, M. The use of terrestrial LiDAR technology in forest science: Application fields, benefits and challenges. *Ann. For. Sci.* **2011**, *68*, 959–974. [CrossRef]

37. Burt, A.; Disney, M.; Raumonen, P.; Armston, J.; Calders, K.; Lewis, P. Rapid characterisation of forest structure from TLS and 3D modelling. In Proceedings of the 2013 IEEE International Geoscience and Remote Sensing Symposium—IGARSS, Melbourne, VIC, Australia, 21–26 July 2013; pp. 3387–3390. Available online: http://128.197.168.195/wp-content/uploads/2013/08/Burt-Disney-IGARSS.pdf (accessed on 12 October 2020).

38. Newnham, G.J.; Armston, J.D.; Calders, K.; Disney, M.I.; Lovell, J.L.; Schaaf, C.B.; Strahler, A.H.; Danson, F.M. Terrestrial laser scanning for plot-scale forest measurement. *Curr. For. Rep.* **2015**, *1*, 239–251. [CrossRef]

39. Béland, M.; Widlowski, J.-L.; Fournier, R.A.; Côté, J.-F.; Verstraete, M.M. Estimating leaf area distribution in savanna trees from terrestrial LiDAR measurements. *Agric. For. Meteorol.* **2011**, *151*, 1252–1266. [CrossRef]
40. Calders, K.; Newnham, G.; Burt, A.; Murphy, S.; Raumonen, P.; Herold, M.; Culvenor, D.S.; Avitabile, V.; Disney, M.; Armston, J.; et al. Nondestructive estimates of above-ground biomass using terrestrial laser scanning. *Methods Ecol. Evol.* **2014**, *6*, 198–208. [CrossRef]
41. Wang, Y.; Spiecker, H.; Calders, K.; Disney, M.I.; Raumonen, P. SimpleTree—An efficient open source tool to build tree models from TLS clouds. *Forests* **2015**, *6*, 4245–4294. [CrossRef]
42. Lemmens, M. *Geo-Information: Technologies, Applications and the Environment*; Springer Science & Business Media: Dordrecht, The Netherlands, 2011; ISBN 9400716672. [CrossRef]
43. Wilkes, P.; Lau, A.; Disney, M.; Calders, K.; Burt, A.; De Tanago, J.G.; Bartholomeus, H.; Brede, B.; Herold, M. Data acquisition considerations for Terrestrial Laser Scanning of forest plots. *Remote Sens. Environ.* **2017**, *196*, 140–153. [CrossRef]
44. Liang, X.; Kankare, V.; Hyyppä, J.; Wang, Y.; Kukko, A.; Haggrén, H.; Yu, X.; Kaartinen, H.; Jaakkola, A.; Guan, F.; et al. Terrestrial laser scanning in forest inventories. *ISPRS J. Photogramm. Remote Sens.* **2016**, *115*, 63–77. [CrossRef]
45. Fan, G.; Nan, L.; Chen, F.; Dong, Y.; Wang, Z.; Li, H.; Chen, D. A new quantitative approach to tree attributes estimation based on LIDAR point clouds. *Remote Sens.* **2020**, *12*, 1779. [CrossRef]
46. Fang, R.; Strimbu, B.M. Comparison of mature douglas-firs' crown structures developed with two quantitative structural models using TLS point clouds for neighboring trees in a natural regime stand. *Remote Sens.* **2019**, *11*, 1661. [CrossRef]
47. Zhu, X.; Wang, T.; Darvishzadeh, R.; Skidmore, A.K.; Niemann, K.O. 3D leaf water content mapping using terrestrial laser scanner backscatter intensity with radiometric correction. *ISPRS J. Photogramm. Remote Sens.* **2015**, *110*, 14–23. [CrossRef]
48. Yrttimaa, T.; Saarinen, N.; Kankare, V.; Liang, X.; Hyyppä, J.; Holopainen, M.; Vastaranta, M. Investigating the feasibility of multi-scan terrestrial laser scanning to characterize tree communities in southern boreal forests. *Remote Sens.* **2019**, *11*, 1423. [CrossRef]
49. Srinivasan, S.; Popescu, S.C.; Eriksson, M.; Sheridan, R.D.; Ku, N.-W. Multi-temporal terrestrial laser scanning for modeling tree biomass change. *For. Ecol. Manag.* **2014**, *318*, 304–317. [CrossRef]
50. Olivier, M.-D.; Robert, S.; Fournier, R.A. A method to quantify canopy changes using multi-temporal terrestrial lidar data: Tree response to surrounding gaps. *Agric. For. Meteorol.* **2017**, *237*, 184–195. [CrossRef]
51. Liang, X.; Hyyppä, J.; Kaartinen, H.; Lehtomäki, M.; Pyörälä, J.; Pfeifer, N.; Holopainen, M.; Brolly, G.; Francesco, P.; Trochta, J.; et al. International benchmarking of terrestrial laser scanning approaches for forest inventories. *ISPRS J. Photogramm. Remote Sens.* **2018**, *144*, 137–179. [CrossRef]
52. Tian, W.; Lin, Y.; Liu, Y.; Niu, Z. Derivation of tree stem structural parameters from static terrestrial laser scanning data. *Lidar Remote Sens. Environ. Monit. Xiv* **2014**, *9262*, 92620Z. [CrossRef]
53. Dassot, M.; Colin, A.; Santenoise, P.; Fournier, M.; Constant, T. Terrestrial laser scanning for measuring the solid wood volume, including branches, of adult standing trees in the forest environment. *Comput. Electron. Agric.* **2012**, *89*, 86–93. [CrossRef]
54. Olsoy, P.J.; Glenn, N.F.; Clark, P.E.; Derryberry, D.R. Aboveground total and green biomass of dryland shrub derived from terrestrial laser scanning. *ISPRS J. Photogramm. Remote Sens.* **2014**, *88*, 166–173. [CrossRef]
55. Danson, F.M.; Sasse, F.; Schofield, L.A. Spectral and spatial information from a novel dual-wavelength full-waveform terrestrial laser scanner for forest ecology. *Interface Focus* **2018**, *8*, 20170049. [CrossRef]
56. Calders, K.; Origo, N.; Burt, A.; Disney, M.I.; Nightingale, J.; Raumonen, P.; Åkerblom, M.; Malhi, Y.; Lewis, P. Realistic forest stand reconstruction from terrestrial LIDAR for radiative transfer modelling. *Remote Sens.* **2018**, *10*, 933. [CrossRef]
57. De Tanago, J.G.; Lau, A.; Bartholomeus, H.; Herold, M.; Avitabile, V.; Raumonen, P.; Martius, C.; Goodman, R.C.; Disney, M.I.; Manuri, S.; et al. Estimation of above-ground biomass of large tropical trees with terrestrial LiDAR. *Methods Ecol. Evol.* **2018**, *9*, 223–234. [CrossRef]
58. Maas, H.-G.; Bienert, A.; Scheller, S.; Keane, E. Automatic forest inventory parameter determination from terrestrial laser scanner data. *Int. J. Remote Sens.* **2008**, *29*, 1579–1593. [CrossRef]
59. Greaves, H.E.; Vierling, L.A.; Eitel, J.U.H.; Boelman, N.T.; Magney, T.S.; Prager, C.M.; Griffin, K.L. Applying terrestrial lidar for evaluation and calibration of airborne lidar-derived shrub biomass estimates in Arctic tundra. *Remote Sens. Lett.* **2017**, *8*, 175–184. [CrossRef]
60. Levick, S.R.; Whiteside, T.; Loewensteiner, D.A.; Rudge, M.; Bartolo, R. Leveraging TLS as a calibration and validation tool for MLS and ULS mapping of savanna structure and biomass at landscape-scales. *Remote Sens.* **2021**, *13*, 257. [CrossRef]
61. Indirabai, I.; Nair, M.H.; Jaishanker, R.N.; Nidamanuri, R.R. Terrestrial laser scanner based 3D reconstruction of trees and retrieval of leaf area index in a forest environment. *Ecol. Inform.* **2019**, *53*, 100986. [CrossRef]
62. Odipo, V.O.; Nickless, A.; Berger, C.; Baade, J.; Urbazaev, M.; Walther, C.; Schmullius, C. Assessment of aboveground woody biomass dynamics using terrestrial laser scanner and L-band ALOS PALSAR Data in South African Savanna. *Forests* **2016**, *7*, 294. [CrossRef]
63. Li, H.; Gao, J.; Hu, Q.; Li, Y.; Tian, J.; Liao, C.; Ma, W.; Xu, Y. Assessing revegetation effectiveness on an extremely degraded grassland, southern Qinghai-Tibetan Plateau, using terrestrial LiDAR and field data. *Agric. Ecosyst. Environ.* **2019**, *282*, 13–22. [CrossRef]

64. Zimbres, B.; Shimbo, J.Z.; Bustamante, M.M.D.C.; Levick, S.R.; De Miranda, S.D.C.; Roitman, I.; Silvério, D.V.; Gomes, L.; Fagg, C.W.; Alencar, A. Savanna vegetation structure in the Brazilian Cerrado allows for the accurate estimation of aboveground biomass using terrestrial laser scanning. *For. Ecol. Manag.* **2020**, *458*, 117798. [CrossRef]
65. Luck, L.; Hutley, L.B.; Calders, K.; Levick, S.R. Exploring the variability of tropical savanna tree structural allometry with terrestrial laser scanning. *Remote Sens.* **2020**, *12*, 3893. [CrossRef]
66. Lindenmayer, D.B.; Laurance, W.F.; Franklin, J.F. Global Decline in Large Old Trees. *Science* **2012**, *338*, 1305–1306. [CrossRef]
67. Estornell, J.; Velázquez-Martí, A.; Fernández-Sarría, A.; Cortés, I.L.; Martí-Gavilá, J.; Salazar, D. Estimation of structural attributes of walnut trees based on terrestrial laser scanning. *Rev. Teledetec.* **2017**, *2017*, 67–76. [CrossRef]
68. Richardson, J.J.; Moskal, L.M.; Bakker, J.D. Terrestrial laser scanning for vegetation sampling. *Sensors* **2014**, *14*, 20304–20319. [CrossRef] [PubMed]
69. Marselis, S.M.; Tang, H.; Armston, J.D.; Calders, K.; Labrière, N.; Dubayah, R. Distinguishing vegetation types with airborne waveform lidar data in a tropical forest-savanna mosaic: A case study in Lopé National Park, Gabon. *Remote Sens. Environ.* **2018**, *216*, 626–634. [CrossRef]
70. Cuni-Sanchez, A.; White, L.J.T.; Calders, K.; Jeffery, K.J.; Abernethy, K.; Burt, A.; Disney, M.; Gilpin, M.; Gomez-Dans, J.L.; Lewis, S.L. African savanna-forest boundary dynamics: A 20-year study. *PLoS ONE* **2016**, *11*, e0156934. [CrossRef] [PubMed]
71. Calders, K.; Adams, J.; Armston, J.; Bartholomeus, H.; Bauwens, S.; Bentley, L.P.; Chave, J.; Danson, F.M.; Demol, M.; Disney, M.; et al. Terrestrial laser scanning in forest ecology: Expanding the horizon. *Remote Sens. Environ.* **2020**, *251*, 112102. [CrossRef]
72. Disney, M.I.; Burt, A.; Calders, K.; Schaaf, C.; Stovall, A. Innovations in Ground and Airborne Technologies as Reference and for Training and Validation: Terrestrial Laser Scanning (TLS). *Surv. Geophys.* **2019**, *40*, 937–958. [CrossRef]
73. Calders, K.; Disney, M.I.; Armston, J.; Burt, A.; Brede, B.; Origo, N.; Muir, J.; Nightingale, J. Evaluation of the range accuracy and the radiometric calibration of multiple terrestrial laser scanning instruments for data interoperability. *IEEE Trans. Geosci. Remote Sens.* **2017**, *55*, 2716–2724. [CrossRef]
74. Clarivate Analytics. Web of Science Core Collection. *Web Sci.* 2020. Available online: https://apps.webofknowledge.com/WOS_GeneralSearch_input.do?product=WOS&search_mode=GeneralSearch&SID=F2RTSaN12sn1d9ppMkH&preferencesSaved= (accessed on 23 September 2020).
75. Moskal, L.M.; Zheng, G. Retrieving forest inventory variables with terrestrial laser scanning (TLS) in urban heterogeneous forest. *Remote Sens.* **2011**, *4*, 1–20. [CrossRef]
76. Hancock, S.; Anderson, K.; Disney, M.; Gaston, K.J. Measurement of fine-spatial-resolution 3D vegetation structure with airborne waveform lidar: Calibration and validation with voxelised terrestrial lidar. *Remote Sens. Environ.* **2017**, *188*, 37–50. [CrossRef]
77. Tan, K.; Zhang, W.; Shen, F.; Cheng, X. Investigation of TLS intensity data and distance measurement errors from target specular reflections. *Remote Sens.* **2018**, *10*, 1077. [CrossRef]
78. Aijazi, A.K.; Checchin, P.; Malaterre, L.; Trassoudaine, L. Automatic detection and parameter estimation of trees for forest inventory applications using 3D terrestrial LiDAR. *Remote Sens.* **2017**, *9*, 946. [CrossRef]
79. Calders, K.; Origo, N.; Disney, M.I.; Nightingale, J.; Woodgate, W.; Armston, J.; Lewis, P. Variability and bias in active and passive ground-based measurements of effective plant, wood and leaf area index. *Agric. For. Meteorol.* **2018**, *252*, 231–240. [CrossRef]
80. Zheng, G.; Moskal, L.M.; Kim, S.-H. Retrieval of effective leaf area index in heterogeneous forests with terrestrial laser scanning. *IEEE Trans. Geosci. Remote Sens.* **2013**, *51*, 777–786. [CrossRef]
81. Bordin, F.; Teixeira, E.C.; Rolim, S.B.A.; Tognoli, F.M.W.; Souza, C.N.; Veronez, M.R. Analysis of the Influence of Distance on Data Acquisition Intensity Forestry Targets by a LIDAR Technique with Terrestrial Laser Scanner. *ISPRS—Int. Arch. Photogramm. Remote Sens. Spat. Inf. Sci.* **2013**, *XL-2/W1*, 99–103. [CrossRef]
82. Heinzel, J.; Huber, M.O. TLS field data based intensity correction for forest environments. *Int. Arch. Photogramm. Remote Sens. Spat. Inf. Sci.* **2016**, *41*, 643–649. [CrossRef]
83. Tan, K.; Cheng, X. Intensity data correction based on incidence angle and distance for terrestrial laser scanner. *J. Appl. Remote Sens.* **2015**, *9*, 94094. [CrossRef]
84. Vaccari, S.; Van Leeuwen, M.; Calders, K.; Coops, N.C.; Herold, M. Bias in lidar-based canopy gap fraction estimates. *Remote Sens. Lett.* **2013**, *4*, 391–399. [CrossRef]
85. Tansey, K.; Selmes, N.; Anstee, A.; Tate, N.J.; Denniss, A. Estimating tree and stand variables in a Corsican Pine woodland from terrestrial laser scanner data. *Int. J. Remote Sens.* **2009**, *30*, 5195–5209. [CrossRef]
86. Olofsson, K.; Holmgren, J.; Olsson, H. Tree stem and height measurements using terrestrial laser scanning and the RANSAC algorithm. *Remote Sens.* **2014**, *6*, 4323–4344. [CrossRef]
87. Magney, T.S.; Eitel, J.U.; Griffin, K.L.; Boelman, N.T.; Greaves, H.E.; Prager, C.M.; Logan, B.A.; Zheng, G.; Ma, L.; Fortin, E.A.; et al. LiDAR canopy radiation model reveals patterns of photosynthetic partitioning in an Arctic shrub. *Agric. For. Meteorol.* **2016**, *221*, 78–93. [CrossRef]
88. Sun, H.; Wang, G.; Lin, H.; Li, J.; Zhang, H.; Ju, H. Retrieval and Accuracy Assessment of Tree and Stand Parameters for Chinese Fir Plantation Using Terrestrial Laser Scanning. *IEEE Geosci. Remote Sens. Lett.* **2015**, *12*, 1993–1997. [CrossRef]
89. Wang, Z.; Zhang, L.; Fang, T.; Mathiopoulos, P.T.; Qu, H.; Chen, D.; Wang, Y. A structure-aware global optimization method for reconstructing 3-D tree models from terrestrial laser scanning data. *IEEE Trans. Geosci. Remote Sens.* **2014**, *52*, 5653–5669. [CrossRef]

90. Xu, K.; Su, Y.; Liu, J.; Hu, T.; Jin, S.; Ma, Q.; Zhai, Q.; Wang, R.; Zhang, J.; Li, Y.; et al. Estimation of degraded grassland aboveground biomass using machine learning methods from terrestrial laser scanning data. *Ecol. Indic.* **2020**, *108*, 105747. [CrossRef]

91. Singh, J.; Levick, S.R.; Guderle, M.; Schmullius, C.; Trumbore, S.E. Variability in fire-induced change to vegetation physiognomy and biomass in semi-arid savanna. *Ecosphere* **2018**, *9*, e02514. [CrossRef]

92. Muir, J.; Phinn, S.; Eyre, T.; Scarth, P. Measuring plot scale woodland structure using terrestrial laser scanning. *Remote Sens. Ecol. Conserv.* **2018**, *4*, 320–338. [CrossRef]

93. Olson, D.M.; Dinerstein, E.; Wikramanayake, E.D.; Burgess, N.D.; Powell, G.V.N.; Underwood, E.C.; D'Amico, J.A.; Itoua, I.; Strand, H.E.; Morrison, J.C.; et al. Terrestrial ecoregions of the worlds: A new map of life on Earth. *Bioscience* **2001**, *51*, 933–938. [CrossRef]

94. Srinivasan, S.; Popescu, S.; Eriksson, M.; Sheridan, R.D.; Ku, N.-W. Terrestrial laser scanning as an effective tool to retrieve tree level height, crown width, and stem diameter. *Remote Sens.* **2015**, *7*, 1877–1896. [CrossRef]

95. Watt, P.J.; Donoghue, D.N.M. Measuring forest structure with terrestrial laser scanning. *Int. J. Remote Sens.* **2005**, *26*, 1437–1446. [CrossRef]

96. Krooks, A.; Kaasalainen, S.; Kankare, V.; Joensuu, M.; Raumonen, P.; Kaasalainen, M. Tree structure vs. height from terrestrial laser scanning and quantitative structure models. *Silva Fenn.* **2014**, *48*, 48. [CrossRef]

97. Pitkänen, T.P.; Raumonen, P.; Kangas, A. Measuring stem diameters with TLS in boreal forests by complementary fitting procedure. *ISPRS J. Photogramm. Remote Sens.* **2019**, *147*, 294–306. [CrossRef]

98. Yu, X.; Liang, X.; Hyyppä, J.; Kankare, V.; Vastaranta, M.; Holopainen, M. Stem biomass estimation based on stem reconstruction from terrestrial laser scanning point clouds. *Remote Sens. Lett.* **2013**, *4*, 344–353. [CrossRef]

99. Lau, A.; Martius, C.; Bartholomeus, H.; Shenkin, A.; Jackson, T.D.; Malhi, Y.; Herold, M.; Bentley, L.P. Estimating architecture-based metabolic scaling exponents of tropical trees using terrestrial LiDAR and 3D modelling. *For. Ecol. Manag.* **2019**, *439*, 132–145. [CrossRef]

100. Moorthy, S.M.K.; Bao, Y.; Calders, K.; Schnitzer, S.A.; Verbeeck, H. Semi-automatic extraction of liana stems from terrestrial LiDAR point clouds of tropical rainforests. *ISPRS J. Photogramm. Remote Sens.* **2019**, *154*, 114–126. [CrossRef] [PubMed]

101. Paynter, I.; Genest, D.; Peri, F.; Schaaf, C. Bounding uncertainty in volumetric geometric models for terrestrial lidar observations of ecosystems. *Interface Focus* **2018**, *8*, 20170043. [CrossRef]

102. RIEGL Laser Measurement Systems GmbH. RIEGL—About RIEGL. 2020. Available online: http://www.riegl.com/company/about-riegl/ (accessed on 30 September 2020).

103. FARO Technologies Inc. High-Precision 3D Acquisition, Measurement and Analysis. 2020. Available online: https://www.faro.com/de-de/faro-im-uberblick/ (accessed on 30 September 2020).

104. Leica Geosystems. When it has to be Right | Leica Geosystems. 2020. Available online: https://leica-geosystems.com/ (accessed on 30 September 2020).

105. Kim, A.M.; Olsen, R.C.; Béland, M. Simulated full-waveform lidar compared to Riegl VZ-400 terrestrial laser scans. *Laser Radar Technol. Appl. XXI* **2016**, *9832*, 98320. [CrossRef]

106. Wu, D.; Phinn, S.; Johansen, K.; Robson, A.; Muir, J.; Searle, C. Estimating Changes in Leaf Area, Leaf Area Density, and Vertical Leaf Area Profile for Mango, Avocado, and Macadamia Tree Crowns Using Terrestrial Laser Scanning. *Remote Sens.* **2018**, *10*, 1750. [CrossRef]

107. Ferrara, R.; Virdis, S.G.; Ventura, A.; Ghisu, T.; Duce, P.; Pellizzaro, G. An automated approach for wood-leaf separation from terrestrial LIDAR point clouds using the density based clustering algorithm DBSCAN. *Agric. For. Meteorol.* **2018**, *262*, 434–444. [CrossRef]

108. Åkerblom, M.; Raumonen, P.; Casella, E.; Disney, M.I.; Danson, F.M.; Gaulton, R.; Schofield, L.A.; Kaasalainen, M. Non-intersecting leaf insertion algorithm for tree structure models. *Interface Focus* **2018**, *8*, 20170045. [CrossRef]

109. Wan, P.; Wang, T.; Zhang, W.; Liang, X.; Skidmore, A.K.; Yan, G. Quantification of occlusions influencing the tree stem curve retrieving from single-scan terrestrial laser scanning data. *For. Ecosyst.* **2019**, *6*, 1–13. [CrossRef]

110. Anderson, K.E.; Glenn, N.F.; Spaete, L.P.; Shinneman, D.J.; Pilliod, D.S.; Arkle, R.S.; McIlroy, S.K.; Derryberry, D.R. Estimating vegetation biomass and cover across large plots in shrub and grass dominated drylands using terrestrial lidar and machine learning. *Ecol. Indic.* **2018**, *84*, 793–802. [CrossRef]

111. Bailey, B.N.; Ochoa, M.H. Semi-direct tree reconstruction using terrestrial LiDAR point cloud data. *Remote Sens. Environ.* **2018**, *208*, 133–144. [CrossRef]

112. Reddy, R.S.; Rakesh, A.; Jha, C.S.; Rajan, K.S. Automatic estimation of tree stem attributes using terrestrial laser scanning in central Indian dry deciduous forests. *Curr. Sci.* **2018**, *114*, 201–206. [CrossRef]

113. Vaaja, M.T.; Virtanen, J.-P.; Kurkela, M.; Lehtola, V.; Hyyppä, J.; Hyyppä, H. The Effect of Wind on Tree Stem Parameter Estimation Using Terrestrial Laser Scanning. *ISPRS Ann. Photogramm. Remote Sens. Spat. Inf. Sci.* **2016**, *III-8*, 117–122. [CrossRef]

114. Ghimire, S.; Xystrakis, F.; Koutsias, N. Using terrestrial laser scanning to measure forest inventory parameters in a mediterranean coniferous stand of western Greece. *PFG-J. Photogramm. Remote Sens. Geoinf. Sci.* **2017**, *85*, 213–225. [CrossRef]

115. Guimarães-Steinicke, C.; Weigelt, A.; Ebeling, A.; Eisenhauer, N.; Duque-Lazo, J.; Reu, B.; Roscher, C.; Schumacher, J.; Wagg, C.; Wirth, C. Terrestrial laser scanning reveals temporal changes in biodiversity mechanisms driving grassland productivity. *Adv. Ecol. Res.* **2019**, *61*, 133–161. [CrossRef]

116. RIEGL Laser Measurement Systems GmbH. *RIEGL VZ-400*; RIEGL Laser Measurement Systems GmbH: Vienna, Austria, 2017.
117. Leica Geosystems. Leica HDS6100 Latest Generation of Ultra-High Speed Laser Scanner. 2009. Available online: https://w3.leica-geosystems.com/downloads123/hds/hds/HDS6100/brochures/Leica_HDS6100_brochure_us.pdf (accessed on 6 October 2020).
118. RIEGL Laser Measurement Systems GmbH. *RIEGL VZ-1000*; RIEGL Laser Measurement Systems GmbH: Vienna, Austria, 2017.
119. FARO Technologies Inc. *FARO Focus 3D Features, Benefits & Technical Specifications*; FARO Technologies, Inc.: Lake Mary, FL, USA, 2013. Available online: http://www.faro.com/en-us/products/3d-surveying/faro-focus3d/overview (accessed on 6 October 2020).
120. FARO Technologies Inc. *FARO Laser Scanner Focus 3D X 330 Features, Benefits & Technical Specifications*; FARO Technologies, Inc.: Lake Mary, FL, USA, 2013.
121. Heinzel, J.; Huber, M.O. Detecting tree stems from volumetric tls data in forest environments with rich understory. *Remote Sens.* **2016**, *9*, 9. [CrossRef]
122. Soma, M.; Pimont, F.; Allard, D.; Fournier, R.; Dupuy, J.-L. Mitigating occlusion effects in Leaf Area Density estimates from Terrestrial LiDAR through a specific kriging method. *Remote Sens. Environ.* **2020**, *245*, 111836. [CrossRef]
123. Stovall, A.E.L.; Shugart, H.H. Improved biomass calibration and validation with terrestrial lidar: Implications for future LiDAR and SAR missions. *IEEE J. Sel. Top. Appl. Earth Obs. Remote Sens.* **2018**, *11*, 3527–3537. [CrossRef]
124. Delagrange, S.; Rochon, P. Reconstruction and analysis of a deciduous sapling using digital photographs or terrestrial-LiDAR technology. *Ann. Bot.* **2011**, *108*, 991–1000. [CrossRef]
125. Schulze-Brüninghoff, D.; Hensgen, F.; Wachendorf, M.; Astor, T. Methods for LiDAR-based estimation of extensive grassland biomass. *Comput. Electron. Agric.* **2019**, *156*, 693–699. [CrossRef]
126. Cooper, S.D.; Roy, D.; Schaaf, C.; Paynter, I. Examination of the potential of terrestrial laser scanning and structure-from-motion photogrammetry for rapid nondestructive field measurement of grass biomass. *Remote Sens.* **2017**, *9*, 531. [CrossRef]
127. Bremer, M.; Wichmann, V.; Rutzinger, M. Multi-temporal fine-scale modelling of Larix decidua forest plots using terrestrial LiDAR and hemispherical photographs. *Remote Sens. Environ.* **2018**, *206*, 189–204. [CrossRef]
128. Pfeifer, N.; Winterhalder, D.; Gorte, B.G.H. Three-dimensional reconstruction of stems for assessment of taper, sweep and lean based on laser scanning of standing trees. *Scand. J. For. Res.* **2004**, *19*, 571–581. [CrossRef]
129. Tan, K.; Cheng, X. Correction of incidence angle and distance effects on TLS intensity data based on reference targets. *Remote Sens.* **2016**, *8*, 251. [CrossRef]
130. Zheng, G.; Moskal, L.M. Retrieving Leaf Area Index (LAI) Using Remote Sensing: Theories, Methods and Sensors. *Sensors* **2009**, *9*, 2719–2745. [CrossRef]
131. Beyer, R.; Bayer, D.; Letort, V.; Pretzsch, H.; Cournède, P.-H. Validation of a functional-structural tree model using terrestrial Lidar data. *Ecol. Model.* **2017**, *357*, 55–57. [CrossRef]
132. LaRue, E.A.; Wagner, F.W.; Fei, S.; Atkins, J.W.; Fahey, R.T.; Gough, C.; Hardiman, B.S. Compatibility of Aerial and Terrestrial LiDAR for Quantifying Forest Structural Diversity. *Remote Sens.* **2020**, *12*, 1407. [CrossRef]
133. Oveland, I.; Hauglin, M.; Gobakken, T.; Naesset, E.; Maalen-Johansen, I. Automatic estimation of tree position and stem diameter using a moving terrestrial laser scanner. *Remote Sens.* **2017**, *9*, 350. [CrossRef]
134. Grotti, M.; Calders, K.; Origo, N.; Puletti, N.; Alivernini, A.; Ferrara, C.; Chianucci, F. An intensity, image-based method to estimate gap fraction, canopy openness and effective leaf area index from phase-shift terrestrial laser scanning. *Agric. For. Meteorol.* **2020**, *280*, 107766. [CrossRef]
135. Kaasalainen, S.; Jaakkola, A.; Kaasalainen, M.; Krooks, A.; Kukko, A. Analysis of incidence angle and distance effects on terrestrial laser scanner intensity: Search for correction methods. *Remote Sens.* **2011**, *3*, 2207–2221. [CrossRef]
136. Xiangyu, W.; Donghui, X.; Guangjian, Y.; Wuming, Z.; Yan, W.; Yiming, C. 3D reconstruction of a single tree from terrestrial LiDAR data. In Proceedings of the International Geoscience and Remote Sensing Symposium (IGARSS), Quebec City, QC, Canada, 13–18 July 2014; IEEE: New York, NY, USA, 2014; pp. 796–799. [CrossRef]
137. Kong, J.; Ding, X.; Liu, J.; Yan, L.; Wang, J. New Hybrid Algorithms for estimating tree stem diameters at breast height using a two dimensional terrestrial laser scanner. *Sensors* **2015**, *15*, 15661–15683. [CrossRef] [PubMed]
138. Kelbe, D.; Van Aardt, J.; Romanczyk, P.; Van Leeuwen, M.; Cawse-Nicholson, K. Single-scan stem reconstruction using low-resolution terrestrial laser scanner data. *IEEE J. Sel. Top. Appl. Earth Obs. Remote Sens.* **2015**, *8*, 3414–3427. [CrossRef]
139. Yang, X.; Strahler, A.H.; Schaaf, C.B.; Jupp, D.L.; Yao, T.; Zhao, F.; Wang, Z.; Culvenor, D.S.; Newnham, G.J.; Lovell, J.L.; et al. Three-dimensional forest reconstruction and structural parameter retrievals using a terrestrial full-waveform lidar instrument (Echidna®). *Remote Sens. Environ.* **2013**, *135*, 36–51. [CrossRef]
140. Bazezew, M.N.; Hussin, Y.A.; Kloosterman, E. Integrating Airborne LiDAR and Terrestrial Laser Scanner forest parameters for accurate above-ground biomass/carbon estimation in Ayer Hitam tropical forest, Malaysia. *Int. J. Appl. Earth Obs. Geoinf.* **2018**, *73*, 638–652. [CrossRef]
141. Riegl. Training Material for RIEGL VZ-400 8. *Project Planning*. 2014. Available online: www.riegl.com: (accessed on 18 December 2020).
142. Greaves, H.E.; Vierling, L.A.; Eitel, J.U.H.; Boelman, N.T.; Magney, T.S.; Prager, C.M.; Griffin, K.L. Estimating aboveground biomass and leaf area of low-stature Arctic shrubs with terrestrial LiDAR. *Remote Sens. Environ.* **2015**, *164*, 26–35. [CrossRef]
143. Hu, C.; Pan, Z.; Li, P. A 3D point cloud filtering method for leaves based on manifold distance and normal estimation. *Remote Sens.* **2019**, *11*, 198. [CrossRef]

144. Han, X.-F.; Jin, J.S.; Wang, M.-J.; Jiang, W.; Gao, L.; Xiao, L. A review of algorithms for filtering the 3D point cloud. *Signal Process. Image Commun.* **2017**, *57*, 103–112. [CrossRef]
145. Kirton, A.; Scholes, B.; Verstraete, M.M.; Archibald, S.; Mennell, K.; Asner, G.; Nickless, A.; Scholes, R.; Asner, G.P. Detailed structural characterisation of the savanna flux site at Skukuza, South Africa. In Proceedings of the 2009 IEEE International Geoscience and Remote Sensing Symposium, Cape Town, South Africa, 12–17 July 2009; Volume 2, p. II-186. [CrossRef]
146. Calders, K.; Armston, J.; Newnham, G.; Herold, M.; Goodwin, N.R. Implications of sensor configuration and topography on vertical plant profiles derived from terrestrial LiDAR. *Agric. For. Meteorol.* **2014**, *194*, 104–117. [CrossRef]
147. Burt, A.; Disney, M.I.; Calders, K. Extracting individual trees from lidar point clouds using treeseg. *Methods Ecol. Evol.* **2018**, *10*, 438–445. [CrossRef]
148. Yurtseven, H.; Çoban, S.; Akgul, M.; Akay, A.O. Individual tree measurements in a planted woodland with terrestrial laser scanner. *Turk. J. Agric. For.* **2019**, *43*, 192–208. [CrossRef]
149. Côté, J.-F.; Fournier, R.A.; Luther, J.E.; Van Lier, O.R. Fine-scale three-dimensional modeling of boreal forest plots to improve forest characterization with remote sensing. *Remote Sens. Environ.* **2018**, *219*, 99–114. [CrossRef]
150. Yang, X.; Schaaf, C.; Strahler, A.; Li, Z.; Wang, Z.; Yao, T.; Zhao, F.; Saenz, E.; Paynter, I.; Douglas, E.S.; et al. Studying canopy structure through 3-D reconstruction of point clouds from full-waveform terrestrial lidar. In Proceedings of the 2013 IEEE International Geoscience and Remote Sensing Symposium-IGARSS, Melbourne, VIC, Australia, 21–26 July 2013; IEEE: New York, NY, USA, 2013; pp. 3375–3378.
151. Lau, A.; Bentley, L.P.; Martius, C.; Shenkin, A.; Bartholomeus, H.; Raumonen, P.; Malhi, Y.; Jackson, T.D.; Herold, M. Quantifying branch architecture of tropical trees using terrestrial LiDAR and 3D modelling. *Trees Struct. Funct.* **2018**, *32*, 1219–1231. [CrossRef]
152. Lau, A.; Calders, K.; Bartholomeus, H.M.; Martius, C.; Raumonen, P.; Herold, M.; Vicari, M.B.; Sukhdeo, H.; Singh, J.; Goodman, R.C. Tree Biomass equations from terrestrial LiDAR: A Case study in Guyana. *Forests* **2019**, *10*, 527. [CrossRef]
153. Isenburg, M. LAStools-Efficient LiDAR Processing Software. 2014. Available online: https://rapidlasso.com/lastools/ (accessed on 28 October 2020).
154. MathWorks. MATLAB-MathWorks-MATLAB & Simulink. 2020. Available online: https://www.mathworks.com/products/matlab.html (accessed on 28 October 2020).
155. The R Foundation. R: The R Project for Statistical Computing. 2018. Available online: https://www.r-project.org/ (accessed on 28 October 2020).
156. Python Software Foundation. Welcome to Python.org. 2020. Available online: https://www.python.org/ (accessed on 28 October 2020).
157. Leica Geosystems. *Leica Cyclone 3D Point Cloud Processing Software.* Available online: http://leica-geosystems.com/products/laser-scanners/software/leica-cyclone (accessed on 28 January 2021).
158. FARO Technologies Inc. SCENE—The Most Intuitive Data Scan Software I FARO Technologies. 2020. Available online: https://www.faro.com/products/construction-bim/faro-scene/ (accessed on 28 October 2020).
159. RIEGL Laser Measurement Systems GmbH. RIEGL—RiSCAN PRO. 2020. Available online: http://www.riegl.com/products/software-packages/riscan-pro/ (accessed on 28 October 2020).
160. Computree Group. The Computree Platform I Computree—Official Site. 2018. Available online: http://computree.onf.fr/?page_id=42 (accessed on 28 October 2020).
161. Girardeau-Montaut, D. *CloudCompare*; Électricité de France S.A. (EDF) R&D: Paris, France, 2003.
162. Popovas, D.; Mikalauskas, V.; Šlikas, D.; Valotka, S.; Šorys, T. Individual tree parameters estimation from terrestrial laser scanner data. In Proceedings of the 10th International Conference Environmental Engineering, ICEE, Vilnius, Lithuania, 27–28 April 2017; pp. 27–28.
163. Moorthy, I.; Miller, J.R.; Hu, B.; Chen, J.; Li, Q. Retrieving crown leaf area index from an individual tree using ground-based lidar data. *Can. J. Remote Sens.* **2008**, *34*, 320–332. [CrossRef]
164. Putman, E.B.; Popescu, S.; Eriksson, M.; Zhou, T.; Klockow, P.A.; Vogel, J.G.; Moore, G.W. Detecting and quantifying standing dead tree structural loss with reconstructed tree models using voxelized terrestrial lidar data. *Remote Sens. Environ.* **2018**, *209*, 52–65. [CrossRef]
165. Kato, A.; Kajiwara, K.; Honda, Y.; Watanabe, M.; Enoki, T.; Yamaguchi, Y.; Kobayashi, T. Efficient field data collection of tropical forest using terrestrial laser scanner. In Proceedings of the 2014 IEEE Geoscience and Remote Sensing Symposium, Quebec City, QC, Canada, 13–18 July 2014; pp. 816–819. [CrossRef]
166. Xi, Z.; Hopkinson, C.; Chasmer, L. Automating plot-level stem analysis from terrestrial laser scanning. *Forests* **2016**, *7*, 252. [CrossRef]
167. Chen, S.; Feng, Z.; Chen, P.; Khan, T.U.; Lian, Y. Nondestructive estimation of the above-ground biomass of multiple tree species in boreal forests of china using terrestrial laser scanning. *Forests* **2019**, *10*, 936. [CrossRef]
168. Zhou, J.; Zhou, G.; Wei, H.; Zhang, X. Estimation of the Plot-Level Forest Parameters from Terrestrial Laser Scanning Data. In Proceedings of the IGARSS 2018-2018 IEEE International Geoscience and Remote Sensing Symposium, Valencia, Spain, 22–27 July 2018; IEEE: New York, NY, USA, 2018; pp. 9014–9017.
169. Tao, S.; Wu, F.; Guo, Q.; Wang, Y.; Li, W.; Xue, B.; Hu, X.; Li, P.; Tian, D.; Li, C.; et al. Segmenting tree crowns from terrestrial and mobile LiDAR data by exploring ecological theories. *ISPRS J. Photogramm. Remote Sens.* **2015**, *110*, 66–76. [CrossRef]
170. Vierling, L.A.; Xu, Y.; Eitel, J.U.H.; Oldow, J.S. Shrub characterization using terrestrial laser scanning and implications for airborne LiDAR assessment. *Can. J. Remote Sens.* **2013**, *38*, 709–722. [CrossRef]

171. Zhang, W.; Wan, P.; Wang, T.; Cai, S.; Chen, Y.; Jin, X.; Yan, G. A Novel Approach for the Detection of Standing Tree Stems from Plot-Level Terrestrial Laser Scanning Data. *Remote Sens.* **2019**, *11*, 211. [CrossRef]
172. Gollob, C.; Ritter, T.; Wassermann, C.; Nothdurft, A. Influence of scanner position and plot size on the accuracy of tree detection and diameter estimation using terrestrial laser scanning on forest inventory plots. *Remote Sens.* **2019**, *11*, 1602. [CrossRef]
173. Reddy, R.S.; Jha, C.S.; Rajan, K.S. Automatic Tree Identification and Diameter Estimation Using Single Scan Terrestrial Laser Scanner Data in Central Indian Forests. *J. Indian Soc. Remote Sens.* **2018**, *46*, 937–943. [CrossRef]
174. FAO. Knowledge Reference for National Forest Assessments—Modeling for Estimation and Monitoring. 2020. Available online: http://www.fao.org/forestry/17109/en/ (accessed on 3 December 2020).
175. Kaasalainen, S.; Krooks, A.; Liski, J.; Raumonen, P.; Kaartinen, H.; Kaasalainen, M.; Puttonen, E.; Anttila, K.; Mäkipää, R. Change detection of tree biomass with terrestrial laser scanning and quantitative structure modelling. *Remote Sens.* **2014**, *6*, 3906–3922. [CrossRef]
176. Disney, M.I.; Vicari, M.B.; Burt, A.; Calders, K.; Lewis, S.L.; Raumonen, P.; Wilkes, P. Weighing trees with lasers: Advances, challenges and opportunities. *Interface Focus* **2018**, *8*, 20170048. [CrossRef] [PubMed]
177. Raumonen, P.; Kaasalainen, M.; Åkerblom, M.; Kaasalainen, S.; Kaartinen, H.; Vastaranta, M.; Holopainen, M.; Disney, M.I.; Lewis, P. Fast Automatic Precision Tree Models from Terrestrial Laser Scanner Data. *Remote Sens.* **2013**, *5*, 491–520. [CrossRef]
178. Madhibha, T.P. ASSESSMENT OF ABOVE GROUND BIOMASS WITH TERRESTRIAL LiDAR USING 3D QUANTITATIVE STRUCTURE MODELLING IN TROPICAL RAIN FOREST OF AYER HITAM FOREST RESERVE, MALAYSIA. University of Twente Faculty of Geo-Information and Earth Observation (ITC). Available online: http://www.itc.nl/library/papers_2016/msc/nrm/madhibha.pdf (accessed on 28 February 2016).
179. Olofsson, K.; Holmgren, J. Single tree stem profile detection using terrestrial laser scanner data, flatness saliency features and curvature properties. *Forests* **2016**, *7*, 207. [CrossRef]
180. Drake, J.B.; Knox, R.G.; Dubayah, R.O.; Clark, D.B.; Condit, R.; Blair, J.B.; Hofton, M. Above-ground biomass estimation in closed canopy Neotropical forests using lidar remote sensing: Factors. *Glob. Ecol. Biogeogr.* **2003**, *12*, 147–159. [CrossRef]
181. Li, A.; Glenn, N.F.; Olsoy, P.J.; Mitchell, J.J.; Shrestha, R. Aboveground biomass estimates of sagebrush using terrestrial and airborne LiDAR data in a dryland ecosystem. *Agric. For. Meteorol.* **2015**, *213*, 138–147. [CrossRef]
182. Olsoy, P.J.; Mitchell, J.J.; Levia, D.F.; Clark, P.E.; Glenn, N.F. Estimation of big sagebrush leaf area index with terrestrial laser scanning. *Ecol. Indic.* **2016**, *61*, 815–821. [CrossRef]
183. McElhinny, C.; Gibbons, P.; Brack, C.; Bauhus, J. Forest and woodland stand structural complexity: Its definition and measurement. *For. Ecol. Manag.* **2005**, *218*, 1–24. [CrossRef]
184. Côté, J.-F.; Fournier, R.A.; Frazer, G.W.; Niemann, K.O. A fine-scale architectural model of trees to enhance LiDAR-derived measurements of forest canopy structure. *Agric. For. Meteorol.* **2012**, *166–167*, 72–85. [CrossRef]
185. Maltamo, M. Estimation of timber volume and stem density based on scanning laser altimetry and expected tree size distribution functions. *Remote Sens. Environ.* **2004**, *90*, 319–330. [CrossRef]
186. Decuyper, M.; Mulatu, K.A.; Brede, B.; Calders, K.; Armston, J.; Rozendaal, D.M.; Mora, B.; Clevers, J.G.; Kooistra, L.; Herold, M.; et al. Assessing the structural differences between tropical forest types using Terrestrial Laser Scanning. *For. Ecol. Manag.* **2018**, *429*, 327–335. [CrossRef]
187. Kelbe, D.; Romanczyk, P.; Van Aardt, J.; Cawse-Nicholson, K. Reconstruction of 3D tree stem models from low-cost terrestrial laser scanner data. *Laser Radar Technol. Appl. XVIII* **2013**, *8731*, 873106. [CrossRef]
188. Paynter, I.; Saenz, E.; Genest, D.; Peri, F.; Erb, A.; Li, Z.; Wiggin, K.; Muir, J.; Raumonen, P.; Schaaf, E.S.; et al. Observing ecosystems with lightweight, rapid-scanning terrestrial lidar scanners. *Remote Sens. Ecol. Conserv.* **2016**, *2*, 174–189. [CrossRef]
189. Kelbe, D.; Romanczyk, P. Automatic extraction of tree stem models from single terrestrial lidar scans in structurally heterogeneous forest environments. *Proc. Silvilaser* **2012**, *2012*, 1–8.
190. Mukuralinda, A.; Kuyah, S.; Ruzibiza, M.; Ndoli, A.; Nabahungu, N.L.; Muthuri, C. Allometric equations, wood density and partitioning of aboveground biomass in the arboretum of Ruhande, Rwanda. *Trees For. People* **2021**, *3*, 100050. [CrossRef]
191. Wang, D.; Hollaus, M.; Puttonen, E.; Pfeifer, N. Fast and robust stem reconstruction in complex environments using terrestrial laser scanning. *Int. Arch. Photogramm. Remote Sens. Spat. Inf. Sci.* **2016**, *41*, 411–417. [CrossRef]
192. Tian, J.; Dai, T.; Li, H.; Liao, C.; Teng, W.; Hu, Q.; Ma, W.; Xu, Y. A Novel Tree Height Extraction Approach for Individual Trees by Combining TLS and UAV Image-Based Point Cloud Integration. *Forests* **2019**, *10*, 537. [CrossRef]
193. Martens, S.N.; Ustin, S.L.; Rousseau, R.A. Estimation of tree canopy leaf area index by gap fraction analysis. *For. Ecol. Manag.* **1993**, *61*, 91–108. [CrossRef]
194. Lau, A.; Bartholomeus, H.; Herold, M.; Martius, C.; Malhi, Y.; Patrick Bentley, L.; Shenkin, A.; Raumonen, P. Application of terrestrial LiDAR and modelling of tree branching structure for plantscaling models in tropical forest trees. *Proc. SilviLaser* **2015**, *2015*, 96–98.
195. Raumonen, P.; Kaasalainen, S.; Kaartinen, H. Approximation of Volume and Branch Size Distribution of Trees from Laser Scanner Data. *ISPRS—Int. Arch. Photogramm. Remote Sens. Spat. Inf. Sci.* **2012**, *5*, 79–84. [CrossRef]
196. Levick, S.R.; Asner, G.P. The rate and spatial pattern of treefall in a savanna landscape. *Biol. Conserv.* **2013**, *157*, 121–127. [CrossRef]

 *remote sensing*

Article

Exploring the Variability of Tropical Savanna Tree Structural Allometry with Terrestrial Laser Scanning

Linda Luck [1,2,*], Lindsay B. Hutley [1], Kim Calders [3] and Shaun R. Levick [2]

[1] Research Institute for the Environment & Livelihoods, College of Engineering, IT and Environment, Charles Darwin University, Casuarina, NT 0909, Australia; lindsay.hutley@cdu.edu.au

[2] CSIRO Land and Water, PMB 44, Winnellie, NT 0822, Australia; Shaun.Levick@csiro.au

[3] CAVElab—Computational and Applied Vegetation Ecology, Faculty of Bioscience Engineering, Ghent University, 9000 Gent, Belgium; Kim.Calders@UGent.be

* Correspondence: linda.luck@cdu.edu.au

Received: 31 August 2020; Accepted: 19 November 2020; Published: 27 November 2020

Abstract: Individual tree carbon stock estimates typically rely on allometric scaling relationships established between field-measured stem diameter (DBH) and destructively harvested biomass. The use of DBH-based allometric equations to estimate the carbon stored over larger areas therefore, assumes that tree architecture, including branching and crown structures, are consistent for a given DBH, and that minor variations cancel out at the plot scale. We aimed to explore the degree of structural variation present at the individual tree level across a range of size-classes. We used terrestrial laser scanning (TLS) to measure the 3D structure of each tree in a 1 ha savanna plot, with coincident field-inventory. We found that stem reconstructions from TLS captured both the spatial distribution pattern and the DBH of individual trees with high confidence when compared with manual measurements (R^2 = 0.98, RMSE = 0.0102 m). Our exploration of the relationship between DBH, crown size and tree height revealed significant variability in savanna tree crown structure (measured as crown area). These findings question the reliability of DBH-based allometric equations for adequately representing diversity in tree architecture, and therefore carbon storage, in tropical savannas. However, adoption of TLS outside environmental research has been slow due to considerable capital cost and monitoring programs often continue to rely on sub-plot monitoring and traditional allometric equations. A central aspect of our study explores the utility of a lower-cost TLS system not generally used for vegetation surveys. We discuss the potential benefits of alternative TLS-based approaches, such as explicit modelling of tree structure or voxel-based analyses, to capture the diverse 3D structures of savanna trees. Our research highlights structural heterogeneity as a source of uncertainty in savanna tree carbon estimates and demonstrates the potential for greater inclusion of cost-effective TLS technology in national monitoring programs.

Keywords: allometry; biomass; carbon; cost-effective; LiDAR; TLS

---

1. Introduction

Savanna vegetation structure and biomass are shaped by the scale-dependent interaction of resource and disturbance drivers. Whilst limited by abiotic resource factors such as climatic conditions, moisture availability [1] and topoedaphic controls [2,3], savanna vegetation is also modulated by disturbance factors such as fire regime (frequency, severity) and storm damage, herbivory and human utilization [4–7]. The interaction and relative importance of these drivers differ at contrasting scales and their impact on above-ground biomass (AGB) can be localised and patchy, varying even at individual tree level [5,8]. A commonly referenced estimate approximates savanna vegetation at 30% of global production; however, the range of 1–12 t C ha^{-1} yr^{-1} net primary productivity [9] provided with this estimate indicates large unresolved uncertainty and considerable opportunity to improve those estimates [10]. Global carbon models have long been challenged by poor representation of the structural heterogeneity in tropical savannas [11] and our understanding of savanna carbon dynamics remains limited [12–14].

At the individual tree level, resource constraints and disturbance drivers may result in strong variability of individual tree architecture and therefore AGB. Fire and herbivory are the two main drivers dramatically altering the vertical structure of savanna trees in African savannas [15,16]. In the tropical savannas of northern Australia however, termites in particular are a significant source of biomass consumption [17] and the activity of wood-eating termites results in stem and branch hollowing, increasing the susceptibility of woody vegetation to storm and fire damage. The resulting asymmetric tree architecture is particularly evident in the crown structure of large trees [18,19]. Tree crowns in north Australian savanna vegetation are estimated to account for up to nearly 45% of total tree AGB, where destructive sampling of a small number of trees (n = 48) has shown considerable variation in the proportion of AGB stored in the crown (Eucalypts: branch biomass 17.1% $\pm$ 7.6% and leaf biomass 3.6% $\pm$ 1.2%; non-Eucalypts: branch biomass 26.6% $\pm$ 11.7% and leaf biomass 5.5% $\pm$ 1.5%) [20]. However, allometric models commonly rely only on stem diameter (DBH), or DBH and height, to estimate AGB as these are relatively easy to measure in the field. Recent studies have shown that adding crown diameter as a predictive variable to estimate AGB provides more robust allometric models [21]. However, estimating crown dimensions by means of measuring vertical crown projection is generally omitted from field inventories due to time constraints and the inherent inaccuracy of such ground-based measures. Not accounting for variations in individual tree architecture could be particularly problematic in tropical savannas, where tree crown structure has the potential to vary significantly for a given DBH [22], however, the extent of heterogeneity in savanna crown architecture is yet to be established.

LiDAR scanning has become a well-recognised and rapidly developing approach to quantify forest structure in great detail, as it enables us to produce detailed 3D reconstructions of individual trees and their structure [23,24]. Of the range of LiDAR platforms, terrestrial laser scanning (TLS) has been increasingly used for investigating heterogeneity in tree and stand structure in space and time, due to its high point densities and scan collections from beneath the canopy [25,26]. TLS data allows us to extract traditional field measured variables such as DBH, stand basal area and tree canopy height with a high degree of precision [27]. In addition, TLS also enables new metrics, such as detailed vertical profiles [28] of tree and canopy structure [29–31] and individual crown-morphology to be quantified [32], thus providing a powerful tool to better quantify heterogeneity in tree crown structure in tropical savannas.

When applied in temperate Eucalypt open forest, TLS data was able to show a weakening allometric relationship between DBH and AGB with increasing DBH [33]. We hypothesise that in tropical savanna, given the cumulative effect of disturbance on tree architecture, variance in crown size is a major driver of this weakening relationship and a similar trend can therefore be observed between DBH and crown area. TLS now allows us to measure large numbers of individual trees to investigate the relationship between crown size and DBH as a potential source of uncertainty impacting the stability of savanna tree allometric relationships.

A major limiting factor to the adoption of TLS for vegetation surveys is the budgetary implications of purchasing high-quality scanners (>USD 100,000) generally used for geomorphological and ecological surveys [34]. Such scanners generally have small beam divergence, high signal-to-noise ratio and in-built GPS systems. These scanners also have a minimum scanning range of several hundred meters. However, for the purpose of vegetation surveys, scan points are generally located less than 50 m apart to avoid data loss through occlusion. Smaller, less precise and lower-powered TLS units generally used in architecture can technically obtain point clouds within that range, however, their efficacy in surveying savanna vegetation structure has not yet been investigated. While increasingly adopted in environmental research, the acquisition cost is a major barrier to the adoption of TLS scanning in vegetation monitoring programs, and the advantages of TLS need to be replicable using more affordable options to enable wider adoption [35].

As such, the aim of this study is to explore the potential of lower-cost (~USD 20,000) TLS in capturing diverse tree architecture in tropical savanna ecosystems. Using this technology, our primary research objective is to determine the degree of variability present in the tree crown area for a given DBH across many hundreds of individual trees. We approached this by coupling a lower-cost, light-weight TLS scanning system with traditional field-inventory surveys at a research site in the tropical savanna of northern Australia.

2. Materials and Methods

2.1. Study Site

The study was carried out in August 2018 in open woodland savanna at the Terrestrial Ecosystem Research Network (TERN) Savanna Supersite at Litchfield National Park in the Northern Territory, Australia [36] (Figure 1). Litchfield National Park (13.17°S, 130.79°E), located 100 km south of Darwin, is subject to a wet–dry tropical climate with the majority of annual rainfall (1750 mm average annual rainfall at Buley rockhole, 7.4 km [37]) occurring between November and April. The survey area was a 1 ha plot and is dominated by *Eucalyptus miniata* with moderate stem density of 492 trees ha^{-1} (>0.05 m DBH) and DBH values (up to 0.49 m). Most crowns are not overlapping, and leaves of local eucalypt species are largely near-vertically aligned, allowing good TLS coverage of tree crowns.

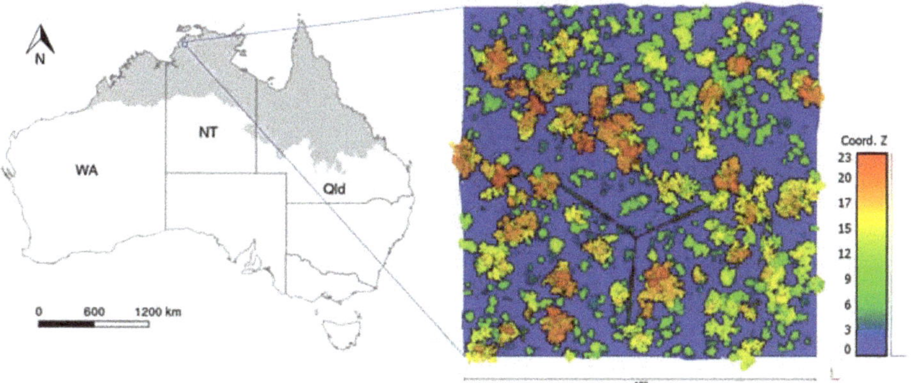

Figure 1. Study area. Left: The north Australian tropical zone spans across Western Australia (WA), the Northern Territory (NT), and Queensland (Qld). Right: canopy height model of the 1 ha study area from TLS (scale bar in m, color scale ranging from blue for lowest points through to red for highest points) showing heterogeneous distribution of tree size classes.

2.2. Data Acquisition

A manual field inventory survey was conducted at the 1 ha site where DBH, tree species and a tree health rating were recorded for every tree with DBH > 0.05 m. The trunks of multi-stemmed trees were measured individually if DBH exceeded 0.05 m. DBH was measured over-bark at 1.3 m above ground level. Tree health was assessed by a single assessor on a scale of 1 to 5 with 1 representing undamaged and 5 representing dead [38]. To correlate individual measurements to those obtained by the TLS survey, a stem map was produced using a real-time kinematic (RTK) positioning system (Leica GS16 with SmartLink). TLS point cloud data was collected during the same field campaign using a Leica BLK360 (wavelength 830 nm, maximum range 60 m at 78% albedo, beam divergence 0.4 mrad, range accuracy 4 mm at 10 m and 7 mm at 20 m [39]) in high point density collection mode (resolution 5 mm at 10 m, scan size approx. 65 mio points) taking less than four minutes for each scan (Figure 2). The plot was scanned in a regular grid fashion from 25 points with 25 m spacing between scan locations to minimise occlusion. A reflector was placed to the north of each scan to aid manual visual alignment of individual scans. Scanning took place in the morning and later afternoon to avoid strong midday breezes and was completed in one day.

(a)   (b)

Figure 2. (**a**) Leica BLK360 set up within the study area; (**b**) *Eucalyptus tetrodonta* sub-sampled to 2 points cm^{-2}, point size 1. The crown is irregularly shaped with evidence of storm and/or fire damage.

2.3. Point Cloud Processing

The raw data was sub-sampled into a uniform point-spacing using an octree filter of 0.01 m. The individual point clouds were visually aligned using CloudCompare v 2.10.2 (Zephyrus), and co-registered using the Multi-station Adjustment module in RIEGL RiSCAN PRO v 2.9. The co-registered point cloud was georeferenced to the WGS84/UTM52S coordinate system using CloudCompare where RTK points of the flux tower and anchor points for its three guy ropes were used as a reference. Point cloud segmentation and extraction of tree attributes (DBH, tree height, crown area) was performed using LiDAR360 v 4.0. Spatial analysis was performed using QGIS v 3.8.0 and statistical analysis using RStudio. An overview of the workflow is shown in Figure 3 and described in detail below. Crown area was chosen as a representative measurement for crown size as it is readily replicated and commonly used in airborne LiDAR surveys (e.g., [40,41]) and is therefore scalable.

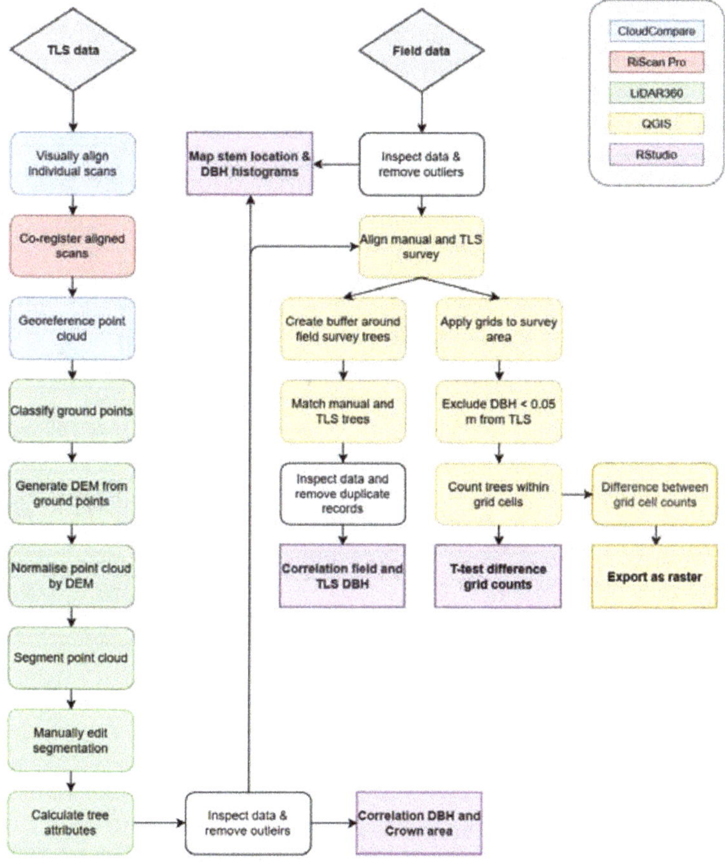

Figure 3. Workflow diagram for point cloud processing.

2.3.1. Point Cloud Segmentation

The co-registered point-cloud was filtered for outliers and ground points were classified based on morphometric properties. A digital elevation model (DEM) was generated from the ground points and was then used to normalise the remaining point cloud to height above ground level. A mean shift algorithm, implemented within LiDAR360 TLS Forest Point Cloud Segmentation tool, was used for automatic segmentation of the normalised point cloud into individual tree clouds. The Individual Tree Editor tool was then used to manually edit falsely classified trees (Figure 4). The resulting tree attribute dataset (n = 976) was inspected for errors and a DBH to height ratio was used to flag any potentially false DBH measurements. Flagged entries were inspected in the segmented point cloud and trees where DBH could not be reliably established due to foliage present at 1.3 m above ground were excluded from the dataset. This process led to the exclusion of 169 unusable segments. Individual tree structural attributes were recalculated after the completion of the manual editing and quality control steps. The automated segmentation and manual editing tools allow for a streamlined and user-friendly segmentation process.

Figure 4. The study site (1 ha plot) segmented into 976 individual trees. Colours are assigned randomly; automated segmentation was followed by fine scale user editing.

2.3.2. Matching of Field and TLS Survey

The manual and TLS survey data were overlaid as shapefiles in QGIS and corresponding trees were matched spatially. This was achieved by applying a 1 m buffer around points in the manual field survey data set. Offsets in the stem manual positioning due to signal drift was accounted for by moving TLS points into the buffer of corresponding manual points where a match could be confidently observed. Of the 492 trees manually surveyed, 450 could be matched in the TLS dataset. Attributes were then joined by spatial location and checked for duplicate records (indicating joins of two or more trees from one layer to one tree of the other).

To compare the spatial distribution of trees captured in both surveys, the plot was overlaid with a 10 m and a 20 m grid. For the TLS data, DBH values below 0.05 m DBH (less 0.07 m mean absolute error (MAE)) were excluded using the query builder. The number of trees located within each grid cell were counted for both surveys. The resulting polygons were converted to a raster and the difference between TLS and field tree count calculated using the raster calculator.

2.4. Statistical Analysis

All statistical analyses were performed in the R environment [42] v 3.6.0. Correlation of field and TLS-derived DBH was performed using linear regression (Figure 7). Visual inspection of the relationship between DBH and crown area suggested an inflection point beyond which the relationship between crown area and DBH changes. The location of this point was identified with a change-point regression using the chngpt package [43]. As the data are heteroscedastic (the variability of the outcome is not constant across the range of the predictor, violating the assumption of linear regression), a generalised least squares (GLS) was used to interpret the correlation before and after the change point (Figure 8). A paired sample *t*-test was then used to assess how closely spatial distribution of detected trees matched between the surveys. For each sub-plot the number of detected trees derived from each method was compared. For this analysis, trees with DBH smaller than 0.05 m were excluded from the TLS survey (less the mean average error from the field and TLS correlation), to match the sampling scope of the manual survey (Figure 6). All R code used is available in a github repository (see Supplementary Materials).

3. Results

3.1. Tree Counts and Distribution Obtained from Field and TLS

The TLS survey allowed for the capture of tree location and DBH measurements on a plot scale compatible with data collected using traditional field surveys. Visual inspection of the field inventory and TLS surveys showed similar patterns in the spatial distributions of DBH as captured by each survey method (Figure 5a,b) with a similar detection curve evident for both surveys (Figure 5c,d). Additionally, the TLS was able to detect a large number of small trees (DBH 0.02–0.04 m) which are often excluded from manual field inventory, as this sampling scale takes considerable time to undertake.

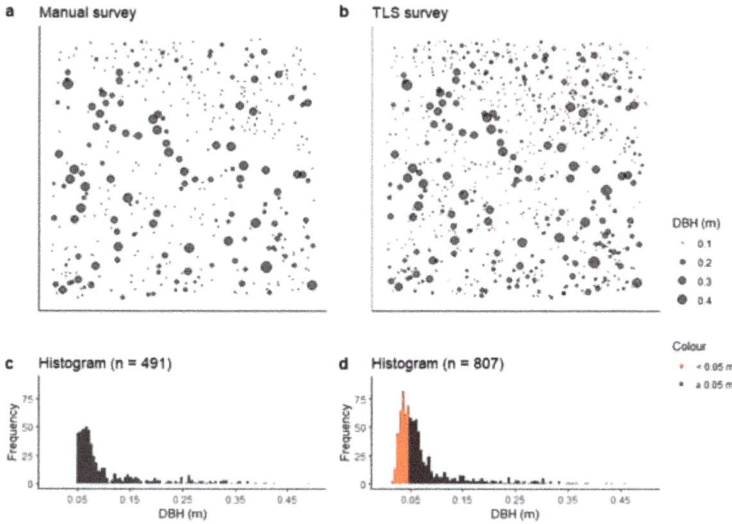

Figure 5. Spatial distribution and number of trees captured in both surveys conducted in August 2018. (**a**) distribution map of field survey; (**b**) distribution map of terrestrial laser scanning (TLS) survey, red denotes trees with stem diameter (DBH) < 0.05 m not captured in the manual survey (*n* grey = 449, *n* red = 358); (**c**) Histogram of field survey; (**d**) Histogram of TLS survey. TLS surveys are able to capture small trees that are impractical to include in manual field inventories.

Comparing the spatial distribution of tree detections in each survey highlights the potential impact of edge effect on small sub-plots (10 and 20 m^2). Across all sub-plots there was no significant difference between the TLS and field inventory for tree detection (Paired samples *t*-test 10 m grid: t(99) = −1.815, *p* = 0.07, 20 m grid: t(24) = −1.97, *p* = 0.06) (Figure 6). However, for a number of sub-plots there was clear over- or under- estimation observed and are likely due to edge effects and/or GPS drift. Using sub-plots as simulated by the 10 m and 20 m grids may be disadvantageous, as individual trees are forced into small plots and thus not necessarily representative of tree density in the area.

Figure 6. Difference between TLS tree count and field tree count. The average tree count per grid cell in the field survey is not significantly different from the average tree count in the TLS survey ($p > 0.05$). For this analysis only trees with DBH > 0.05 m (-0.007 m mean absolute error (MAE)) were used.

3.2. DBH Obtained from Manual Survey and TLS

The DBH values of individual trees derived from the TLS survey were strongly correlated with those obtained manually in the field. The RTK GPS tagging of trees during the field survey ensured precise matching of a large sample of trees captured in the field survey to corresponding TLS measurements (two outliers were excluded from the analysis). There was a strong correlation between DBH measured in the field and DBH derived from the TLS point cloud (Figure 7), with $R^2 = 0.98$ and MAE = 0.007 m (Spearman's $\rho = 0.93$, $p < 0.001$). The MAE shows the overall average of errors while the RMSE (root mean square error) highlights strong outliers as it increases with variance in the frequency distribution of error magnitudes. For this model, an RMSE value higher than the MAE indicates that the main source of error likely stems from a small number of strong outliers, rather than a systematic over or underestimation of field DBH by the TLS derived DBH.

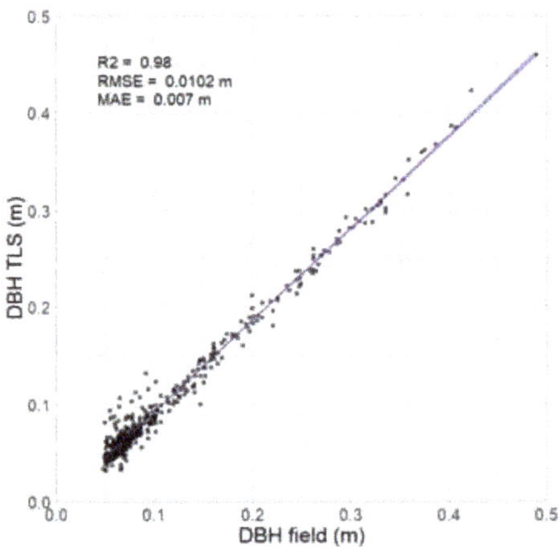

Figure 7. Correlation between field survey and concurrent TLS scan; ribbon showing 95% confidence interval. For every cm increase in TLS DBH there is an average 1.04 cm increase in field DBH (95% CI 1.03–1.05, $p < 0.001$) ($n = 448$).

3.3. Variability in Crown Architecture

Using TLS derived data, crown size (canopy area) was plotted as a function of DBH (Figure 8) which showed a strong positive relationship with heteroscedastic variance. Deviance from the mean increases along with maximum crown area, resulting in a non-linear relationship (Table 1). Our results align with previous observations made while calibrating the local allometric model [22].

A change-point regression shows a change point at DBH of 0.099 m (95%CI 0.056–0.14, $p < 0.001$). For trees below the change point, for every unit (1 m) increase in DBH there is a 88.55 unit increase in the crown area (95%CI 81.56–95.50); for trees above the change point, the increase in crown area per unit DBH increases to 240.87 and the 95%CI widens to 191.75–290. Trees with poor health ratings were not excluded, as this would obscure the true variability in crown size.

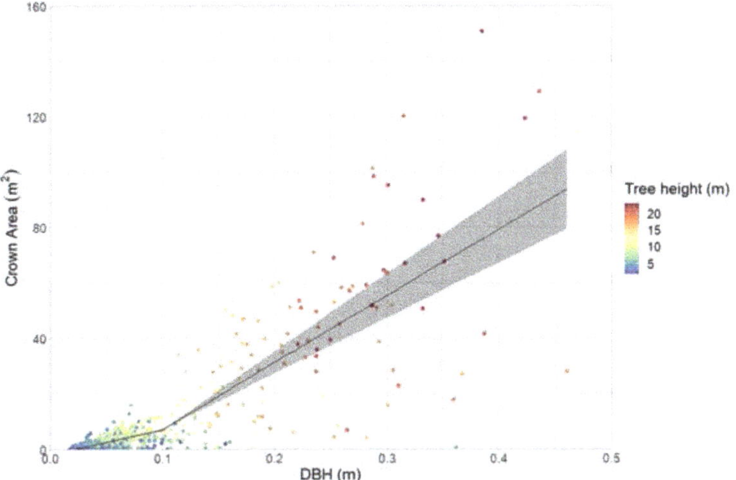

Figure 8. TLS derived crown area for a given DBH; ribbon showing 95% confidence interval. The spread of crown sizes increases with increasing DBH ($n = 807$).

Table 1. Variability in tree crown structural parameters as a function of DBH class, showing values of the mean and standard error of the mean.

DBH Range (m)	Canopy Height (m)		Crown Area (m²)	
	Mean	SE	Mean	SE
0.00–0.05	4.37	0.10	1.19	0.06
0.05–0.10	7.89	0.14	4.22	0.16
0.10–0.15	11.10	0.44	11.45	1.25
0.15–0.20	14.02	0.64	25.30	2.93
0.20–0.25	17.96	0.49	39.79	3.19
0.25–0.30	18.30	0.50	50.62	6.96
0.30–0.35	19.58	0.66	62.09	9.80
>0.35	17.89	1.56	64.64	18.33

4. Discussion

4.1. Heterogeneity at Landscape/Stand Level

Tropical savannas are disturbance-impacted ecosystems which result in stand-scale structural heterogeneity, introducing unquantified uncertainty to commonly used methods for estimating AGB that rely on small sub-samples of DBH-based allometry. When used in tropical savanna woodlands, TLS can provide accurate and detailed information that allows us to explore this source of uncertainty in allometric models. Furthermore, the large volume of samples obtainable using TLS allows us to negate further sources of uncertainty in traditional field surveys arising from the sampling limitations in the trees surveyed to represent a population when both calibrating and applying allometric models. Deriving stand size class distributions based on DBH measurements is commonplace in production forestry through to ecological surveys of stand biomass. Manual tree measurement and obtaining the geolocation data required to study size class distribution requires considerable resources and few studies include geolocation of individual trees within a study site [44]. Instead, a common field method for estimating the size class distribution is to extrapolate from sub-plots that may be 30 m × 30 m or less in size [45]. However, in tropical savannas, the assumption of spatial homogeneity underlying this approach is unlikely to be met, given inherent spatial variability at distances of 20 to 30 m (Figure 6). TLS data allows us to extract geolocation along with DBH for individual trees and is increasingly used to investigate ecological processes such as random vs clumped tree distributions, structural change over time following disturbance [25,30,46] and over annual to decadal scales, or woody encroachment and thickening [47]. We show that TLS data obtained using a lower-cost scanner enables fast and accurate acquisition of both, full 3D tree measurements and geographic positioning required to investigate landscape scale heterogeneity in tree size class distribution (Figures 1 and 5).

4.2. Accuracy of Lower-Cost TLS Scanner for Estimating DBH

Direct comparison of DBH obtained from a manual field survey and concurrent TLS survey showed a very close fit with a mean absolute error (MAE) of less than 0.01 m, suggesting that lower-cost TLS data can provide traditional forestry and ecological metrics such as stand basal area and size class distributions with high precision. Our analysis of DBH obtained in the field as compared to TLS (Section 3.2) indicates random variation as the main source of error rather than systemic over or under-prediction of the model (Figure 7). Random errors are common in manual data collection [48,49], while errors in TLS measurements are commonly caused by wind, random errors in point acquisition, registration error or occlusion. Errors in data processing occur when measuring DBH, as the model assumes a perfect circle, which is often not the case. Our study design aimed to minimise TLS acquisition error and occlusion. Our data shows an RMSE of 1.05 cm, representing a 50% reduction compared to a similar study in temperate Eucalypt open forest where a minimum distance of 40 m between scan points was used [33]. However, the two studies also differ in stand structure, equipment and software used and the impact of either of those factors is yet to be quantified. The accuracy achieved in this study also reflects results from other vegetation types, such as mixed forests in Austria [50] and Germany [51], deciduous forest in Iran [52] and India [53], Chinese fir plantations [54], or tropical forest of Malaysia [55]. However, none of these are structurally representative of tropical savannas and the suitability of TLS for obtaining DBH was yet to be established. We show that lower-cost TLS can replicate field measured DBH and tree height, establishing the suitability of TLS to reliably capture traditional measures of vegetation structure in tropical savanna vegetation.

4.3. Heterogeneity at the Individual Tree Level

Variability in tree crown structure introduces poorly quantified and potentially significant uncertainty in allometric models predicting AGB based on DBH. Previous research shows that in north Australian savannas the proportion of biomass stored in tree crowns can vary between 12% and 45% of total tree AGB (mean ± SD of the branch and foliage components) [20]. However, this estimate is based on a very small sample size of 48 destructively harvested individuals and is unlikely to capture the true extend of variability. Our TLS analysis of over 800 individual trees confirmed and quantified a significant spread of crown area for a given DBH, increasing with DBH (Figure 8). This is in contrast to previous research published exploring the relationship between DBH and canopy diameter [56] where a linear relationship (R^2 = 0.63) was established across a much larger area. However, investigation of the raw data shows a weaker relationship for trees surveyed within 150 km of Litchfield NP, indicating that local-scale variability arising from disturbance might exceed that of regional-scale trends. Furthermore, structural variability is not only found in overall crown size, but also architecture (Figure 9). While the addition of crown diameter as a predictive variable improves allometric models [21], the use of canopy area as a function of canopy diameter assumes a circular crown shape. Using lower-cost TLS we were able to highlight the impact of disturbance on tree architecture (Figure 9) and the potential effect on the stability of allometric relationships introduced by the assumption of homogeneity.

Figure 9. Two examples of typical mature *Eucalyptus miniata* individuals, both with a DBH of 0.38 m highlighting the dramatic difference in canopy structure, volume and biomass. Estimates of carbon storage of the left individual would be significantly overestimated using a manually measured DBH with allometry applied.

4.4. Further Applications—Capturing Change and Irregularities

The effect of fire as disturbance driver on the understory, including tree seedlings and saplings, and woody shrubs is a critical ecological process in savanna [57,58], however, detailed inventories of woody understory vegetation required to quantify this effect are not readily quantified using standard field inventory. TLS provides the potential for a more accurate inventory of understory density if high-resolution point clouds are generated. The ability to resolve the diameters of smaller stems (less than 0.05 m) provides crucial information on the recruitment potential of a stand and/or evidence of recent disturbance. This sampling detail is undertaken in some traditional inventory methods but can be extremely time consuming and costly. Furthermore, allometric equations have been calibrated for small to medium-sized trees with a small fraction of large individuals and are unlikely to be representative of seedlings and saplings.TLS allows us to capture understory vegetation such as seedlings, saplings and small shrubs (Figure 5) and use alternative volume-based analyses to estimate AGB. From the 3D point clouds, algorithms such as voxel or convex hull based approaches [59,60] can be used to estimate AGB of woody vegetation where metrics like DBH cannot easily be measured, or suitable allometric models are not available. When applied as successive scans, TLS provides detailed observations of structural changes at a stand level, such as mortality and recruitment, or at individual tree level, such as damage to branching structure and crown development [61,62]. The ability to quantify recruitment and mortality opens up the opportunity to explore disturbance dynamics such as fire impacts or storm damage impacts within the regeneration niche at unprecedented detail.

Multi-stemmed trees and shrubs can be challenging to survey manually, especially when a forking height is located below or at the designated DBH measuring height. To avoid this issue, some studies opt to measure diameter closer to the ground, e.g., 0.1 m or 0.15 m above ground level [63,64]. However, such a strong deviation from the norm will prevent comparison or integration of data from multiple studies and requires further calibration of specialised allometric models. TLS is not restricted to single-point measurements and can be used to extract a range of data to estimate AGB. For instance, Quantitative Structure Modelling (QSM) has gained popularity as an alternative to DBH-based allometric models. Individual tree AGB is estimated based on wood density and point cloud generated tree volume, avoiding the need for allometric relationships and standardised measurements [33]. The TLS approach provides the freedom to adapt sampling strategies to include stand elements that will not conform to traditional survey methods and tree allometric models.

Accounting for loss from termite consumption in susceptible tree species remains a challenge for biomass estimates based on crown and stem volumes [65]. However, if hollowing functions accounting for this loss can be developed, the use of TLS offers considerable advantages over manually sampled inventories given the sampling speed, precision, the range of metrics that can be estimated and the ability to accurately quantify structural change over time.

4.5. Negating the Cost Barrier

TLS provides clear benefits over traditional manual field surveys and is increasingly utilised in the research community. However, capital cost has been a major barrier to the wider adoption of this technology, constricting the application of research outcomes by the wider community [34,35]. We show that rapid developments in commercial scanners now enable us to cost-effectively acquire point clouds of sufficient quality for detailed vegetation surveys. We demonstrate the benefits of TLS to vegetation monitoring programs by combining traditional and new measurements obtained through TLS to explore a major source of uncertainty in traditional allometric models when applied in frequently disturbed woodlands such as tropical savannas. We also highlight and discuss further advantages of TLS to capture stand elements traditionally difficult to capture using field inventories. Limitations apply to the use of lower-cost scanners, particularly in regards to their range and therefore ability to survey trees above a height of 50 m [35], however, we have demonstrated that such commercial options provide data of sufficient quality to make state of the art vegetation survey methods more accessible.

5. Conclusions

Tropical savannas have been challenging to quantify in global carbon models. We show how heterogeneity in savanna tree structure introduces uncertainty into established protocols for estimating AGB and therefore carbon content across spatial scales. At plot scale this heterogeneity leads to sample plots rarely being representative of a larger area, while at tree scale the strength of DBH as an indicator of tree AGB is diminished by strong variability in tree architecture. Furthermore, the traditional exclusion of small understory trees (e.g. <0.05 m DBH) from field inventories omits a potentially important indicator for vegetation and carbon dynamics, including recruitment and regeneration. TLS provides a more holistic way to retrieve quantitative structural measurements that directly relate to biomass properties in tropical savanna. Using TLS we were able to reliably obtain traditional field attributes such as DBH and tree height, but also access new meaningful tree attributes such as crown size and extent, and spatial distribution of understory vegetation. However, adoption of this technology in monitoring programs has been limited due to acquisition costs of high-quality TLS scanners, and skepticism remains as to the utility of TLS in vegetation surveys. We show that in savanna vegetation a low-cost TLS scanner will produce high-quality LiDAR data and highlight the advantages and power in deriving meaningful tree attributes that will allow us to better quantify AGB at tree and plot scale. Establishing sampling protocols for highly detailed AGB surveys with reduced budgetary requirements may enable greater inclusion of TLS in national and international monitoring programs.

Supplementary Materials: The following are available online at https://github.com/LindaLuck/Luck2020Exploring, Analysis scripts for data analysis carried out in the R environment.

Author Contributions: Conceptualization: S.R.L. and L.L.; methodology: L.L., S.L; analysis: L.L.; resources: S.R.L., L.B.H.; data curation: L.L.; writing—original draft preparation: L.L.; writing—review and editing: S.R.L., L.B.H., K.C., L.L.; visualization: L.L.; supervision: S.R.L., L.B.H., K.C. All authors have read and agreed to the published version of the manuscript.

Funding: This research was funded by a Charles Darwin University (CDU) postgraduate RTP scholarship and a Commonwealth Scientific and Industrial Research Organisation (CSIRO) top-up scholarship.

Acknowledgments: We acknowledge the Ecosystem Processes facility in Australia's Terrestrial Ecosystem Research Network (TERN) and NT Parks and Wildlife. Jodie Hayward and Jon Schatz are thanked for their field support. We also thank Mirjam Kaestli for statistical support. This manuscript was greatly improved by valuable comments from Keryn Paul and Garry Cook.

Conflicts of Interest: The authors declare no conflict of interest.

Abbreviations

The following abbreviations are used in this manuscript:

AGB	above-ground biomass
DBH	diameter at breast height
DEM	Digital Elevation Model
GPS	Global Positioning System
LiDAR	Light detecting and ranging
MAE	mean absolute error
TERN	Terrestrial Ecosystem Research Network
TLS	Terrestrial laser scanning
QGIS	geographic information system (GIS) software
QSM	Quantitative Structure Model
RMSE	root mean square error
RTK	Real-time kinematic positioning
UTM	Universal Transverse Mercator
WGS	World Geodetic System

References

1. Scholes, R.; Dowty, P.; Caylor, K.; Parsons, D.; Frost, P.; Shugart, H. Trends in savanna structure and composition along an aridity gradient in the Kalahari. *J. Veg. Sci.* **2002**, *13*, 419–428. [CrossRef]
2. Veenendaal, E.M.; Torello-Raventos, M.; Feldpausch, T.R.; Domingues, T.; Gerard, F.; Schrodt, F.; Saiz, G.; Quesada, C.; Djagbletey, G.; Ford, A.; et al. Structural, physiognomic and above-ground biomass variation in savanna-forest transition zones on three continents-how different are co-occurring savanna and forest formations? *Biogeosciences* **2015**, *12*, 2927–2951. [CrossRef]
3. Levick, S.R.; Rogers, K.H. Context-dependent vegetation dynamics in an African savanna. *Landsc. Ecol.* **2011**, *26*, 515–528. [CrossRef]
4. Williams, R.; Duff, G.; Bowman, D.; Cook, G. Variation in the composition and structure of tropical savannas as a function of rainfall and soil texture along a large-scale climatic gradient in the Northern Territory, Australia. *J. Biogeogr.* **1996**, *23*, 747–756. [CrossRef]
5. Lehmann, C.E.R.; Anderson, T.M.; Sankaran, M.; Higgins, S.I.; Archibald, S.; Hoffmann, W.A.; Hanan, N.P.; Williams, R.J.; Fensham, R.J.; Felfili, J.; et al. Savanna Vegetation-Fire-Climate Relationships Differ Among Continents. *Science* **2014**, *343*, 548–552. [CrossRef]

6. Hutley, L.B.; Setterfield, S.A. Savannas. In *Encyclopaedia of Ecology*, 2nd ed.; Faith, B.D., Ed.; Elsevier B.V.: Oxford, UK, 2019; Volume 2, pp. 623–633. [CrossRef]

7. Hutley, L.B.; Evans, B.J.; Beringer, J.; Cook, G.D.; Maier, S.W.; Razon, E. Impacts of an extreme cyclone event on landscape-scale savanna fire, productivity and greenhouse gas emissions. *Environ. Res. Lett.* **2013**, *8*, 045023. [CrossRef]

8. Oliveras, I.; Malhi, Y. Many shades of green: The dynamic tropical forest–savannah transition zones. *Philos. Trans. R. Soc. Biol. Sci.* **2016**, *371*. [CrossRef]

9. Grace, J.; José, J.S.; Meir, P.; Miranda, H.S.; Montes, R.A. Productivity and carbon fluxes of tropical savannas. *J. Biogeogr.* **2006**, *33*, 387–400. [CrossRef]

10. Kanniah, K.D.; Beringer, J.; Hutley, L.B. Environmental controls on the spatial variability of savanna productivity in the Northern Territory, Australia. *Agric. For. Meteorol.* **2011**, *151*, 1429–1439. [CrossRef]

11. Schimel, D.S. Terrestrial ecosystems and the carbon cycle. *Glob. Chang. Biol.* **1995**, *1*, 77–91. [CrossRef]

12. Stevens, N.; Lehmann, C.E.; Murphy, B.P.; Durigan, G. Savanna woody encroachment is widespread across three continents. *Glob. Chang. Biol.* **2017**, *23*, 235–244. [CrossRef] [PubMed]

13. Wigley, B.J.; Bond, W.J.; Hoffman, M.T. Thicket expansion in a South African savanna under divergent land use: local vs. global drivers? *Glob. Chang. Biol.* **2010**, *16*, 964–976. [CrossRef]

14. Williams, R.J.; Hutley, L.B.; Cook, G.D.; Russell-Smith, J.; Edwards, A.; Chen, X. Assessing the carbon sequestration potential of mesic savannas in the Northern Territory, Australia: Approaches, uncertainties and potential impacts of fire. *Funct. Plant Biol.* **2004**, *31*, 415–422. [CrossRef] [PubMed]

15. Moncrieff, G.R.; Chamaillé-Jammes, S.; Higgins, S.I.; O'Hara, R.B.; Bond, W.J. Tree allometries reflect a lifetime of herbivory in an African savanna. *Ecology* **2011**, *92*, 2310–2315. [CrossRef]

16. Levick, S.R.; Asner, G.P.; Kennedy-Bowdoin, T.; Knapp, D.E. The relative influence of fire and herbivory on savanna three-dimensional vegetation structure. *Biol. Conserv.* **2009**, *142*, 1693–1700. [CrossRef]

17. Jamali, H.; Livesley, S.; Hutley, L.B.; Fest, B.; Arndt, S. The relationships between termite mound CH4/CO2 emissions and internal concentration ratios are species specific. *Biogeosciences* **2013**, *10*, 2229–2240. [CrossRef]

18. Davies, A.B.; Parr, C.L.; Janse Van Rensburg, B. Termites and Fire: Current Understanding and Future Research Directions for Improved Savanna Conservation. *Austral Ecol.* **2010**, *35*, 482–486. [CrossRef]

19. Midgley, J.J.; Lawes, M.J.; Chamaillé-Jammes, S. Savanna woody plant dynamics: The role of fire and herbivory, separately and synergistically. *Aust. J. Bot.* **2010**, *58*, 1–11. [CrossRef]

20. Chen, X. Carbon Balance of a Eucalypt Open Forest Savanna of Northern Australia. Ph.D. Thesis, Northern Territory University, Casuarina, Australia, 2002.

21. Jucker, T.; Caspersen, J.; Chave, J.; Antin, C.; Barbier, N.; Bongers, F.; Dalponte, M.; van Ewijk, K.Y.; Forrester, D.I.; Haeni, M.; et al. Allometric equations for integrating remote sensing imagery into forest monitoring programmes. *Glob. Chang. Biol.* **2017**, *23*, 177–190. [CrossRef]

22. Williams, R.J.; Zerihun, A.; Montagu, K.D.; Hoffman, M.; Hutley, L.B.; Chen, X. Allometry for estimating aboveground tree biomass in tropical and subtropical eucalypt woodlands: Towards general predictive equations. *Aust. J. Bot.* **2005**, *53*, 607–619, [CrossRef]

23. Dassot, M.; Colin, A.; Santenoise, P.; Fournier, M.; Constant, T. Terrestrial laser scanning for measuring the solid wood volume, including branches, of adult standing trees in the forest environment. *Comput. Electron. Agric.* **2012**, *89*, 86–93. [CrossRef]

24. Newnham, G.J.; Armston, J.D.; Calders, K.; Disney, M.I.; Lovell, J.L.; Schaaf, C.B.; Strahler, A.H.; Danson, F.M. Terrestrial laser scanning for plot-scale forest measurement. *Curr. For. Rep.* **2015**, *1*, 239–251. [CrossRef]

25. Zimbres, B.; Shimbo, J.; Bustamante, M.; Levick, S.; Miranda, S.; Roitman, I.; Silvério, D.; Gomes, L.; Fagg, C.; Alencar, A. Savanna vegetation structure in the Brazilian Cerrado allows for the accurate estimation of aboveground biomass using terrestrial laser scanning. *For. Ecol. Manag.* **2020**, *458*, 117798, [CrossRef]

26. Malhi, Y.; Jackson, T.; Patrick Bentley, L.; Lau, A.; Shenkin, A.; Herold, M.; Calders, K.; Bartholomeus, H.; Disney, M.I. New perspectives on the ecology of tree structure and tree communities through terrestrial laser scanning. *Interface Focus* **2018**, *8*, 20170052. [CrossRef]

27. Strahler, A.H.; Jupp, D.L.; Woodcock, C.E.; Schaaf, C.B.; Yao, T.; Zhao, F.; Yang, X.; Lovell, J.; Culvenor, D.; Newnham, G.; et al. Retrieval of forest structural parameters using a ground-based lidar instrument (Echidna). *Can. J. Remote Sens.* **2008**, *34*, S426–S440. [CrossRef]

28. Calders, K.; Armston, J.; Newnham, G.; Herold, M.; Goodwin, N. Implications of sensor configuration and topography on vertical plant profiles derived from terrestrial LiDAR. *Agric. For. Meteorol.* **2014**, *194*, 104–117. [CrossRef]

29. Hardiman, B.S.; LaRue, E.A.; Atkins, J.W.; Fahey, R.T.; Wagner, F.W.; Gough, C.M. Spatial variation in canopy structure across forest landscapes. *Forests* **2018**, *9*, 474. [CrossRef]

30. Singh, J.; Levick, S.R.; Guderle, M.; Schmullius, C.; Trumbore, S.E. Variability in fire-induced change to vegetation physiognomy and biomass in semi-arid savanna. *Ecosphere* **2018**, *9*, e02514. [CrossRef]

31. Singh, J.; Levick, S.R.; Guderle, M.; Schmullius, C. Moving from plot-based to hillslope-scale assessments of savanna vegetation structure with long-range terrestrial laser scanning (LR-TLS). *Int. J. Appl. Earth Obs. Geoinf.* **2020**, *90*, 102070. [CrossRef]

32. Kunz, F.; Gamauf, A.; Zachos, F.E.; Haring, E. Mitochondrial phylogenetics of the goshawk Accipiter gentilis superspecies. *J. Zool. Syst. Evol. Res.* **2019**, *57*, 942–958. [CrossRef]

33. Calders, K.; Newnham, G.; Burt, A.; Murphy, S.; Raumonen, P.; Herold, M.; Culvenor, D.; Avitabile, V.; Disney, M.; Armston, J.; et al. Nondestructive estimates of above-ground biomass using terrestrial laser scanning. *Methods Ecol. Evol.* **2015**, *6*, 198–208. [CrossRef]

34. Calders, K.; Adams, J.; Armston, J.; Bartholomeus, H.; Bauwens, S.; Patrick Bentley, L.; Chave, J.; Danson, F.M.; Demol, M.; Disney, M.; et al. Terrestrial Laser Scanning in Forest Ecology: Expanding the Horizon. *Remote Sens. Environ.* **2020**, in press. [CrossRef]

35. Disney, M.; Burt, A.; Calders, K.; Schaaf, C.; Stovall, A. Innovations in ground and airborne technologies as reference and for training and validation: Terrestrial laser scanning (TLS). *Surv. Geophys.* **2019**, *40*, 937–958. [CrossRef]

36. Karan, M.; Liddell, M.; Prober, S.M.; Arndt, S.; Beringer, J.; Boer, M.; Cleverly, J.; Eamus, D.; Grace, P.; Van Gorsel, E.; et al. The Australian SuperSite Network: A continental, long-term terrestrial ecosystem observatory. *Sci. Total. Environ.* **2016**, *568*, 1263–1274. [CrossRef] [PubMed]

37. Australian Bureau of Meteorology. *Monthly Rainfall, Station No. 014021*; Australian Bureau of Meteorology: Melbourne, Australia. Available online: http://www.bom.gov.au/climate/data/ (accessed on 27 November 2020)

38. Cook, G.D. Fire management and minesite rehabilitation in a frequently burnt tropical savanna. *Austral Ecol.* **2012**, *37*, 686–692. [CrossRef]

39. Leica Geosystems AG. *Leica BLK360 User Manual*; Leica Geosystems AG: St. Gallen, Switzerland, 2017.

40. Aubry-Kientz, M.; Dutrieux, R.; Ferraz, A.; Saatchi, S.; Hamraz, H.; Williams, J.; Coomes, D.; Piboule, A.; Vincent, G. A comparative assessment of the performance of individual tree crowns delineation algorithms from ALS data in tropical forests. *Remote Sens.* **2019**, *11*, 1086. [CrossRef]

41. Barnes, C.; Balzter, H.; Barrett, K.; Eddy, J.; Milner, S.; Suárez, J.C. Individual tree crown delineation from airborne laser scanning for diseased larch forest stands. *Remote Sens.* **2017**, *9*, 231. [CrossRef]

42. R Core Team. *R: A Language and Environment for Statistical Computing*; R Core Team: Vienna, Austria, 2013.

43. Fong, Y.; Huang, Y.; Gilbert, P.B.; Permar, S.R. chngpt: Threshold regression model estimation and inference. *BMC Bioinform.* **2017**, *18*, 454. [CrossRef]

44. Sasaki, T.; Konno, M.; Hasegawa, Y.; Imaji, A.; Terabaru, M.; Nakamura, R.; Ohira, N.; Matsukura, K.; Seiwa, K. Role of mycorrhizal associations in tree spatial distribution patterns based on size class in an old-growth forest. *Oecologia* **2019**, *189*, 971–980, [CrossRef]

45. Moore, J.R.; Zhu, K.; Huntingford, C.; Cox, P.M. Equilibrium forest demography explains the distribution of tree sizes across North America. *Environ. Res. Lett.* **2018**, *13*, 84019, [CrossRef]

46. Levick, S.R.; Richards, A.E.; Cook, G.D.; Schatz, J.; Guderle, M.; Williams, R.J.; Subedi, P.; Trumbore, S.E.; Andersen, A.N. Rapid response of habitat structure and above-ground carbon storage to altered fire regimes in tropical savanna. *Biogeosciences* **2019**, *16*, 1493–1503. [CrossRef]

47. Cuni-Sanchez, A.; White, L.J.T.; Calders, K.; Jeffery, K.J.; Abernethy, K.; Burt, A.; Disney, M.; Gilpin, M.; Gomez-Dans, J.L.; Lewis, S.L. African Savanna-Forest Boundary Dynamics: A 20-Year Study. *PLoS ONE* **2016**, *11*, e0156934. [CrossRef] [PubMed]

48. Paul, K.I.; Larmour, J.S.; Roxburgh, S.H.; England, J.R.; Davies, M.J.; Luck, H.D. Measurements of stem diameter: implications for individual-and stand-level errors. *Environ. Monit. Assess.* **2017**, *189*, 416. [CrossRef] [PubMed]

49. Bruce, D. Evaluating Accuracy of Tree Measurements made with Optical Instruments. *For. Sci.* **1975**, *21*, 421–426.
50. Wang, D.; Hollaus, M.; Puttonen, E.; Pfeifer, N. Automatic and Self-Adaptive Stem Reconstruction in Landslide-Affected Forests. *Remote Sens.* **2016**, *8*, 974, [CrossRef]
51. Bienert, A.; Georgi, L.; Kunz, M.; Maas, H.G.; Oheimb, G. Comparison and Combination of Mobile and Terrestrial Laser Scanning for Natural Forest Inventories. *Forests* **2018**, *9*, 395, [CrossRef]
52. Pazhouhan, I.; Najafi, A.; Kamkar-Rouhani, A.; Vahidi, J.; Najafi, A. Extraction of Individual Tree Parameters by Using Terrestrial Laser Scanner Data in Hyricanian. *Forest* **2018**, [CrossRef]
53. Reddy, R.S.; Rakesh, C.S.; Rajan, K.S. Automatic estimation of tree stem attributes using terrestrial laser scanning in central Indian dry deciduous forests. *Curr. Sci.* **2018**, *114*, 201. [CrossRef]
54. Sun, H.; Wang, G.; Lin, H.; Li, J.; Zhang, H.; Ju, H. Retrieval and Accuracy Assessment of Tree and Stand Parameters for Chinese Fir Plantation Using Terrestrial Laser Scanning. *IEEE Geosci. Remote Sens. Lett.* **2015**, *12*, 1993–1997, [CrossRef]
55. Beyene, S.M.; Hussin, Y.A.; Kloosterman, H.E.; Ismail, M.H. Forest Inventory and Aboveground Biomass Estimation with Terrestrial LiDAR in the Tropical Forest of Malaysia. *Can. J. Remote. Sens.* **2020**, *46*, 1–16. [CrossRef]
56. Cook, G.D.; Liedloff, A.C.; Cuff, N.J.; Brocklehurst, P.S.; Williams, R.J. Stocks and dynamics of carbon in trees across a rainfall gradient in a tropical savanna. *Austral Ecol.* **2015**, *40*, 845–856. [CrossRef]
57. Setterfield, S.A. Seedling establishment in an Australian tropical savanna: effects of seed supply, soil disturbance and fire. *J. Appl. Ecol.* **2002**, *39*, 949–959. [CrossRef]
58. Bond, W.J.; Cook, G.D.; Williams, R.J. Which trees dominate in savannas? The escape hypothesis and eucalypts in northern Australia. *Austral Ecol.* **2012**, *37*, 678–685. [CrossRef]
59. Olsoy, P.J.; Glenn, N.F.; Clark, P.E.; Derryberry, D.R. Aboveground total and green biomass of dryland shrub derived from terrestrial laser scanning. *ISPRS J. Photogramm. Remote Sens.* **2014**, *88*, 166–173. [CrossRef]
60. Wallace, L.; Hillman, S.; Reinke, K.; Hally, B. Non-destructive estimation of above-ground surface and near-surface biomass using 3D terrestrial remote sensing techniques. *Methods Ecol. Evol.* **2017**, *8*, 1607–1616. [CrossRef]
61. Asner, G.P.; Levick, S.R.; Kennedy-Bowdoin, T.; Knapp, D.E.; Emerson, R.; Jacobson, J.; Colgan, M.S.; Martin, R.E. Large-scale impacts of herbivores on the structural diversity of African savannas. *Proc. Natl. Acad. Sci. USA* **2009**, *106*, 4947–4952. [CrossRef]
62. Asner, G.P.; Levick, S.R. Landscape-scale effects of herbivores on treefall in African savannas. *Ecol. Lett.* **2012**, *15*, 1211–1217. [CrossRef]
63. Dohn, J.; Augustine, D.J.; Hanan, N.P.; Ratnam, J.; Sankaran, M. Spatial vegetation patterns and neighborhood competition among woody plants in an East African savanna. *Ecology* **2017**, *98*, 478–488, [CrossRef]
64. Paul, K.I.; Roxburgh, S.H.; Chave, J.; England, J.R.; Zerihun, A.; Specht, A.; Lewis, T.; Bennett, L.T.; Baker, T.G.; Adams, M.A.; et al. Testing the generality of above-ground biomass allometry across plant functional types at the continent scale. *Glob. Chang. Biol.* **2016**, *22*, 2106–2124, [CrossRef]
65. Cook, G.D.; Liedloff, A.C.; Eager, R.; Chen, X.; Williams, R.; O'Grady, A.P.; Hutley, L.B. The estimation of carbon budgets of frequently burnt tree stands in savannas of northern Australia, using allometric analysis and isotopic discrimination. *Aust. J. Bot.* **2005**, *53*, 621–630. [CrossRef]

Publisher's Note: MDPI stays neutral with regard to jurisdictional claims in published maps and institutional affiliations.

remote sensing

MDPI

Article

Leveraging TLS as a Calibration and Validation Tool for MLS and ULS Mapping of Savanna Structure and Biomass at Landscape-Scales

Shaun R. Levick [1,*], Tim Whiteside [2], David A. Loewensteiner [2], Mitchel Rudge [3] and Renee Bartolo [2]

1   CSIRO Land and Water, PMB 44, Winnellie, NT 0822, Australia
2   Department of Agriculture, Water and the Environment, Supervising Scientist Branch, Eaton, NT 0820, Australia; Tim.Whiteside@awe.gov.au (T.W.); David.Loewensteiner@awe.gov.au (D.A.L.); Renee.Bartolo@awe.gov.au (R.B.)
3   Centre for Mined Land Rehabilitation, Sustainable Minerals Institute, University of Queensland, Brisbane, QLD 4072, Australia; mitchel.rudge@uq.edu.au
*   Correspondence: shaun.levick@csiro.au

Abstract: Savanna ecosystems are challenging to map and monitor as their vegetation is highly dynamic in space and time. Understanding the structural diversity and biomass distribution of savanna vegetation requires high-resolution measurements over large areas and at regular time intervals. These requirements cannot currently be met through field-based inventories nor spaceborne satellite remote sensing alone. UAV-based remote sensing offers potential as an intermediate scaling tool, providing acquisition flexibility and cost-effectiveness. Yet despite the increased availability of lightweight LiDAR payloads, the suitability of UAV-based LiDAR for mapping and monitoring savanna 3D vegetation structure is not well established. We mapped a 1 ha savanna plot with terrestrial-, mobile- and UAV-based laser scanning (TLS, MLS, and ULS), in conjunction with a traditional field-based inventory (n = 572 stems > 0.03 m). We treated the TLS dataset as the gold standard against which we evaluated the degree of complementarity and divergence of structural metrics from MLS and ULS. Sensitivity analysis showed that MLS and ULS canopy height models (CHMs) did not differ significantly from TLS-derived models at spatial resolutions greater than 2 m and 4 m respectively. Statistical comparison of the resulting point clouds showed minor over- and under-estimation of woody canopy cover by MLS and ULS, respectively. Individual stem locations and DBH measurements from the field inventory were well replicated by the TLS survey (R^2 = 0.89, RMSE = 0.024 m), which estimated above-ground woody biomass to be 7% greater than field-inventory estimates (44.21 Mg ha^{-1} vs. 41.08 Mg ha^{-1}). Stem DBH could not be reliably estimated directly from the MLS or ULS, nor indirectly through allometric scaling with crown attributes (R^2 = 0.36, RMSE = 0.075 m). MLS and ULS show strong potential for providing rapid and larger area capture of savanna vegetation structure at resolutions suitable for many ecological investigations; however, our results underscore the necessity of nesting TLS sampling within these surveys to quantify uncertainty. Complementing large area MLS and ULS surveys with TLS sampling will expand our options for the calibration and validation of multiple spaceborne LiDAR, SAR, and optical missions.

Keywords: biomass; carbon; LiDAR; TLS

Citation: Levick, S.R.; Whiteside, T.; Loewensteiner, D.A.; Rudge, M.; Bartolo, R. Leveraging TLS as a Calibration and Validation Tool for MLS and ULS Mapping of Savanna Structure and Biomass at Landscape-Scales. *Remote Sens.* **2021**, *13*, 257. https://doi.org/10.3390/rs13020257

Received: 16 December 2020
Accepted: 11 January 2021
Published: 13 January 2021

Publisher's Note: MDPI stays neutral with regard to jurisdictional claims in published maps and institutional affiliations.

1. Introduction

Savanna vegetation structure is shaped by the interaction of climate, soils and a variety of disturbance agents acting at multiple spatio-temporal scales [1,2]. The actions of fire, herbivores, termites, and cyclones have marked effects on the structure of savanna tree crowns, leading to high degree of structural diversity [3–5]. Mapping and monitoring the structure of savanna ecosystems is therefore challenging not only because of their heterogeneous spatial patterning, but also because of disturbance driven variability in individual tree crown structure [6]. Structural attributes such as height, canopy diameter, and projected foliage cover are commonly assessed through field inventories—but these measures are often subjective and prone to sampling errors [7]. Robust assessment of structural change in tropical savannas requires methods that can account for spatial and temporal heterogeneity, and a high level of accuracy and precision in structural attribute measurement.

Remote sensing is very attractive from a monitoring and management perspective, offering large area assessment at ever increasing spatial, temporal and spectral resolutions [8,9]. Despite many advances in spaceborne imaging technology, savanna ecosystems remain challenging from an earth observation perspective due to their inherent mix of discontinuous woody cover and herbaceous vegetation [10]. Light-detection-and-ranging (LiDAR), especially airborne laser scanning (ALS), has emerged as a prominent technology for mapping and monitoring vegetation in a variety of ecosystems across the globe [11]. ALS is particularly well suited to the measurement of savanna ecosystem structure where it can delineate subtle topographic and vegetation characteristics [12]. The inclusion of ALS in ecological studies has advanced the field of savanna ecology, revealing new insights into the controls and drivers of carbon storage [13,14], and providing pathways for understanding how vegetation structure influences the ecology and diversity of various fauna [15,16].

Nonetheless, despite the many advantages of ALS, certain ecological processes operate at even finer spatial scales than can be assessed from traditional airborne platforms. Detecting stress, sub-canopy structural changes, and recruitment dynamics are all examples of processes that sit at the limits of airborne surveying. Terrestrial laser scanning (TLS) has emerged as the gold standard for fine-scale 3D reconstruction of above-ground elements, and is increasingly being used in forestry and ecological surveying [17,18]. One of the main constraints of TLS is the time-intensive nature of data acquisition and post-processing, and the relatively small spatial extent that can feasibly be covered [19]. Newer sensors are overcoming some of these barriers by increasing the speed and ranging distance, and enabling larger area coverage in suitable environmental conditions [20]. Reduced size and weight of modern LiDAR sensors has broadened the field of UAV-based laser scanning (ULS), offering a high degree of acquisition flexibility and increased spatial resolution due to lower and slower flying speeds [21–23]. In conjunction with these developments, handheld and mobile laser scanning (MLS) systems have also been gaining traction, enabling greater freedom of movement and increased speed of acquisition in comparison to TLS. MLS enables navigation in confined spaces, and the continuous mapping approach can reduce occlusion and provide thorough coverage in complex environments [24–26].

Given these recent advances in LiDAR sensor and platform technologies, choosing the right solution for a particular mapping and monitoring problem has become more challenging. The relative strengths and weaknesses of different systems needs to be explored across a range of ecosystems to best match system performance with the specific research or management objectives. The objective of this study was to assess the degree of divergence and complementarity of three different laser scanning systems for capturing the 3D structure of tropical savanna woodland structure. We explore the relative strengths and weaknesses of terrestrial laser scanning (TLS), mobile laser scanning (MLS) and UAV laser scanning (ULS) in a vegetation monitoring context, by addressing two key questions:

1. Are there significant differences in the representation of woody canopy height and cover captured by TLS, MLS and ULS?
2. How reliably can tree stem DBH and above-ground woody biomass be mapped from TLS, MLS and ULS?

2. Materials and Methods

2.1. Study Site

This study was undertaken in northern Kakadu National Park in the Northern Territory of Australia (Figure 1). The climate is tropical, with maximum and minimum temperature averaging 34.1 °C and 22.6 °C respectively (Köppen climate classification = Aw). Rainfall is highly seasonal, with the vast majority of the annual 1557 mm yr^{-1} falling predominantly in the December—March wet season [27]. The vegetation is representative of much of the Australian tropical savanna zone, with an over-storey dominated by *Eucalyptus tetrodonta*, *Eucalyptus miniata*, and *Erythrophleum chlorostachys*. The under-storey is characterized by annual grasses, particularly Sorghum species. Our data collection campaign took place on 23/24 July 2019, well into the dry season and one month after a fire passed through—the occurrence of which is typical every 1–2 years in this system [28]. The herbaceous grass layer was reduced by the fire, but small shrubs were still present.

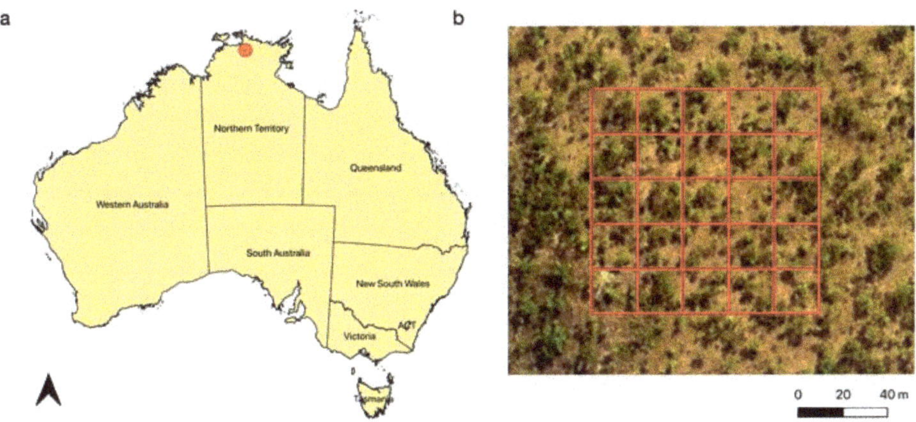

Figure 1. Location of the study site in northern Australia (**a**), and an aerial view of vegetation cover within the 20 × 20 m subplots of the 1 ha study area (**b**).

2.2. Field Inventory

A series of long-term monitoring plots were established in the savanna woodlands of the Jabiru region in 2017 to provide reference sites for understanding natural system dynamics and informing the closure criteria of the adjacent Ranger uranium mine [29]. Each plot measures 100 m × 100 m, and the four corners and center are marked with permanent posts. We selected one of these sites for our study and each woody plant with a stem diameter > 0.03 m and that was > 1.5 m in height was inventoried in the field. Survey tapes were laid out in a 20 m × 20 m grid to guide the field crew, and each individual tree was: (i) tagged with a RTK dGPS (Leica GS16 with Precise Point Positioning); (ii) identified to species level; and (iii) measured with a DBH tape at 1.3 m above ground level. A total of 521 trees consisting of 572 stems were inventoried in the field.

2.3. Laser Scanning

The selected site was surveyed on 24 July 2019 with three different LiDAR systems, operating from terrestrial, mobile and UAV platforms (Table 1). Terrestrial laser scanning (TLS) was conducted from a surveying tripod at 1.8 m agl, using a Riegl VZ-2000i system with integrated RTK-GNSS (Figure 2). Sixteen scan locations were established, using a regular grid spacing of 30 m. The scanner was operated at 600 kHz with an angular sampling resolution of 30 mdeg. The scanner communicated with a RTK base station (Emlid RS2) established at the site and operating over a LoRa network. Real-time positioning error of the scanner at each scan location was < 0.05 m.

Figure 2. Site conditions at the time of the LiDAR surveys. Occlusion from the grass layer was minimal due to the effects of a fire one month prior.

Mobile laser scanning (MLS) was conducted with a GreenValley LiBackpack D50. The unit consists of two Velodyne VLP-16 sensors, one operating vertically and one horizontally. The system uses Simultaneous Location and Mapping (SLAM) technology for co-registration. The MLS system was carried as a backpack by a field technician who walked slowly through the site ensuring consistent coverage and loop closure. The trajectory of the mobile unit was displayed in real time to the operator via a tablet to ensure full coverage of the site.

UAV laser scanning (ULS) point clouds were collected using the Nextcore® system (www.nextcore.co) which uses a Quanergy M8 discrete return LiDAR sensor integrated with a Spatial Dual (www.advancednavigation.com) INS including IMU and dual antenna RTK GPS, mounted on a DJI Matrice 600 Pro. The UAV was flown in a grid pattern at 3.5 ms^{-1} at approximately 40 m agl with a line spacing of 18 m. The sensor pulse rate was 1200 kHz, which resulted in average point densities of 2000 points per m^2 within the one hectare plot.

Table 1. Instrument, surveying, and processing characteristics of the three LiDAR systems. Reported point densities are after the application of a 0.01 m sub-sampling spatial filter. Note: TLS = terrestrial laser scanning, MLS = mobile laser scanning, ULS = UAV laser scanning.

	TLS	**MLS**	**ULS**
Instrument	Riegl VZ-2000i	LiBackPack D50	Nextcore Version 1
Manufacturer	Riegl	Green Valley International	Nextcore
Sensor	Riegl	Velodyne	Quanergy M8
Platform	Tripod	Backpack	Drone (DJI M600)
Acquisition time	120 min	20 min	15 min
Co-registration processing time	60 min	20 min	20 min
Filtered point density (m^2)	7340	2401	2015

2.4. Point-Cloud Processing

Raw processing of the TLS dataset was conducted in Riegl's RiSCAN Pro software suite (v2.9). Scan data were projected into the WGS84 coordinate system (UTM Zone 53S) and filtered for noise based on reflectance and deviation characteristics. The Multi-Station Adjustment (MSA) plugin was used to finely co-register the individual scans, which were already well positioned given the integrated GNSS with real-time correction.

Raw MLS data were converted to a point cloud incorporating the IMU data from the backpack using proprietary SLAM algorithms in the LiBackpack software from GreenValley International. The point cloud was geo-referenced in LiDAR360 by detecting six pyramid structures that were placed throughout the site as ground controls. The 3D RTK-GNSS location of the peaks of each pyramid were recorded during the field campaign.

ULS point clouds were PPK geo-registered using the proprietary Nextcore® software. Each of the flight-lines underwent a statistical outlier removal (SOR) filter in CloudCompare. The number of neighbors used to compute mean distance was set to 10, and the standard deviation multiplier was set to 2. There was still some systematic misalignment of flight-lines following the PPK correction. To correct for this, each flight-line was sequentially registered to its neighbor using the Iterative Closest Point (ICP) algorithm implemented in CloudCompare. The random sample limit was set to 250,000 and RMS error difference threshold set to 1×10^{-08}. After registration, flight-line point clouds were merged into a single point cloud.

Co-registered point clouds from each platform were exported to the ASPRS LAS format version 1.4 for further analysis. A 0.01 m spacing filter was applied to all three clouds to remove duplicate points and ensure an even distribution of points across the landscape.

2.5. Structural Analysis

2.5.1. Canopy Characterization

The MLS, and ULS point clouds were finely co-registered to the TLS point cloud using the Iterative Closest Point (ICP) algorithm in CloudCompare [30]. The random sample limit was set to 500,000 and the RMS error difference threshold set to 1×10^{-05}, with 100% overlap specified. Ground point classification was conducted with the *lasground_new* tool in the LAStools suite [31], using -wilderness and -extra_fine settings, and the point clouds were subsequently normalized to height above ground using *lasheight*. Canopy height models (CHMs) were generated with *lasgrid* at multiple spatial resolutions using the 95th percentile height. CHM resolution increased in 0.25 m increments from 0.25 m to 5 m resolution to enable a sensitivity analysis of the resolution at which the combinations of the three scanning systems were significantly different. Differences in canopy height distribution derived from the three laser scanning systems was tested across the range of CHM resolutions with the Kolmogorov–Smirnov test [32] in the *stats* package in R (version 4.0.3).

The 0.25 m resolution CHMs derived from the three scanning systems were used to assess the percentage of woody canopy cover at multiple height class thresholds that represented key strata in the savanna woodland (0.25 m, 3 m, 7 m, 12 m). For each height threshold, median woody canopy cover for 20 m × 20 m subplots embedded in the 1 ha site (n = 25) was compared between TLS-MLS and TLS-ULS with linear regression.

2.5.2. Individual Tree Delineation and Measurement

Individual trees segmentation was applied to the point clouds derived from the three scanning systems using the approach described by [33] and implemented in LiDAR360 (v4.1, GreenValley International) (Figure 3). Results were manually inspected post-segmentation and edited to correct any instances of over- or under-segmentation. Following this quality control checking, the point clouds representing individual trees were processed further in 3DForest (v0.5) to derive individual tree attributes—tree height, crown diameter, crown area, crown volume and stem DBH [34]. An overview of the full processing pipeline is provided in Figure 4.

Figure 3. Cross-section through the TLS point cloud, displaying an example of the individual tree segmentation. Color scheme is a randomised palette.

Above-ground woody biomass (AGB) was estimated at the individual tree level using a set of locally calibrated allometric equations that included the dominant species at the site [35]. AGB was derived from the field inventory as a function of stem DBH (Equation (1), and from the TLS data as a function of stem DBH and tree height (Equation (2)).

$$ln(AGB) = \beta0 + \beta1 * ln(D) \tag{1}$$

$$ln(AGB) = \beta0 + \beta1 * ln(D) + \beta2 * ln(H)^2 \tag{2}$$

Multiple linear regression and Random Forest modeling (randomForest [36]), conducted in R, were used to assess the relationship between individual tree biomass (as determined from DBH allometry) and additional measures of tree 3D structure (canopy height, crown area, crown width, crown volume).

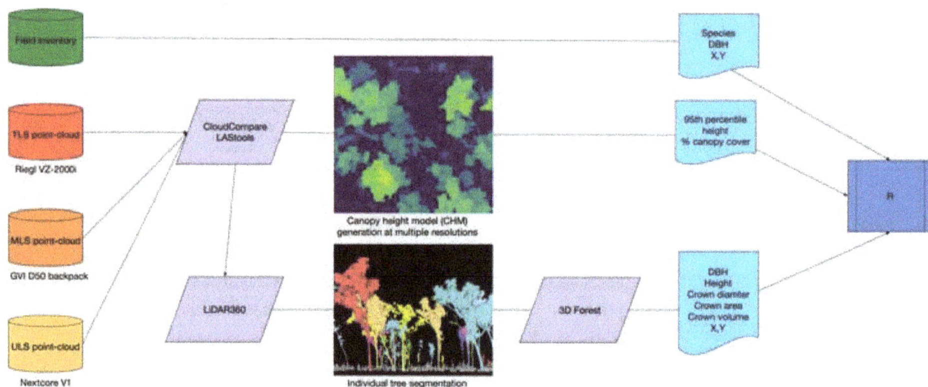

Figure 4. Overview of the processing workflow and software packages used at different stages of analysis.

3. Results

3.1. Woody Canopy Characterization

All three LiDAR sensing systems provided high quality point-cloud data which captured the three-dimensional structure of the savanna vegetation in rich detail (Figure 5). Viewing the 1 ha plot from an oblique angle, the three point clouds showed very similar structure and it is evident that all three systems captured the canopy patterns well (Figure 5a–c). A cross-sectional slice through the point clouds (5 × 30 m) revealed the narrower beam footprint and higher precision of the TLS instrument, with crisp trunk and branch detail, as well as good definition of material in the herbaceous layer (Figure 5d). The MLS and ULS point clouds were less resolute in comparison, exhibiting some minor noise around tree trunks and evidence of occlusion in the lower stems and herbaceous layer (Figure 5e–f).

The generation of canopy height model (CHM) raster layers confirmed that differences between the three scanning systems was minimal from a visual perspective and displayed very similar spatial patterns of canopy height (Figure 6a–c). Subtle differences were evident, however, and the canopy height distribution ((Figure 6d–f) differed significantly between all three systems at 0.25 m (KS-test, $p < 0.001$). Sensitivity analysis showed that these differences remained significant until a CHM resolution of 2 m was reached for the TLS-MLS comparisons, and 4 m for TLS-ULS and MLS-ULS comparisons (KS-test, $p > 0.05$).

Estimation of maximum canopy height was very consistent across the three systems, even at 0.25 m raster resolution. However, the MLS and ULS showed consistent bias in woody canopy cover estimation, over- and under-estimating the woody fraction respectively at multiple height intervals (Figure 7 and Table 2).

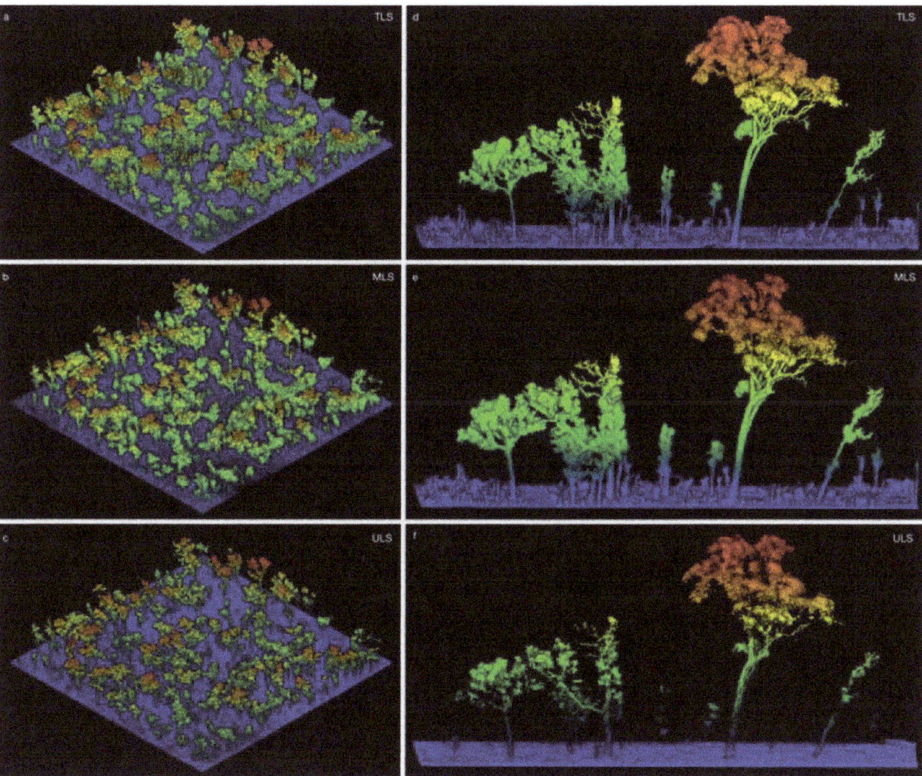

Figure 5. Oblique view of the point clouds derived from the three LiDAR platforms: (**a**) TLS, (**b**) MLS and (**c**) ULS. Cross-sectional view of the point clouds derived from: (**d**) TLS, (**e**) MLS, and (**f**) ULS.

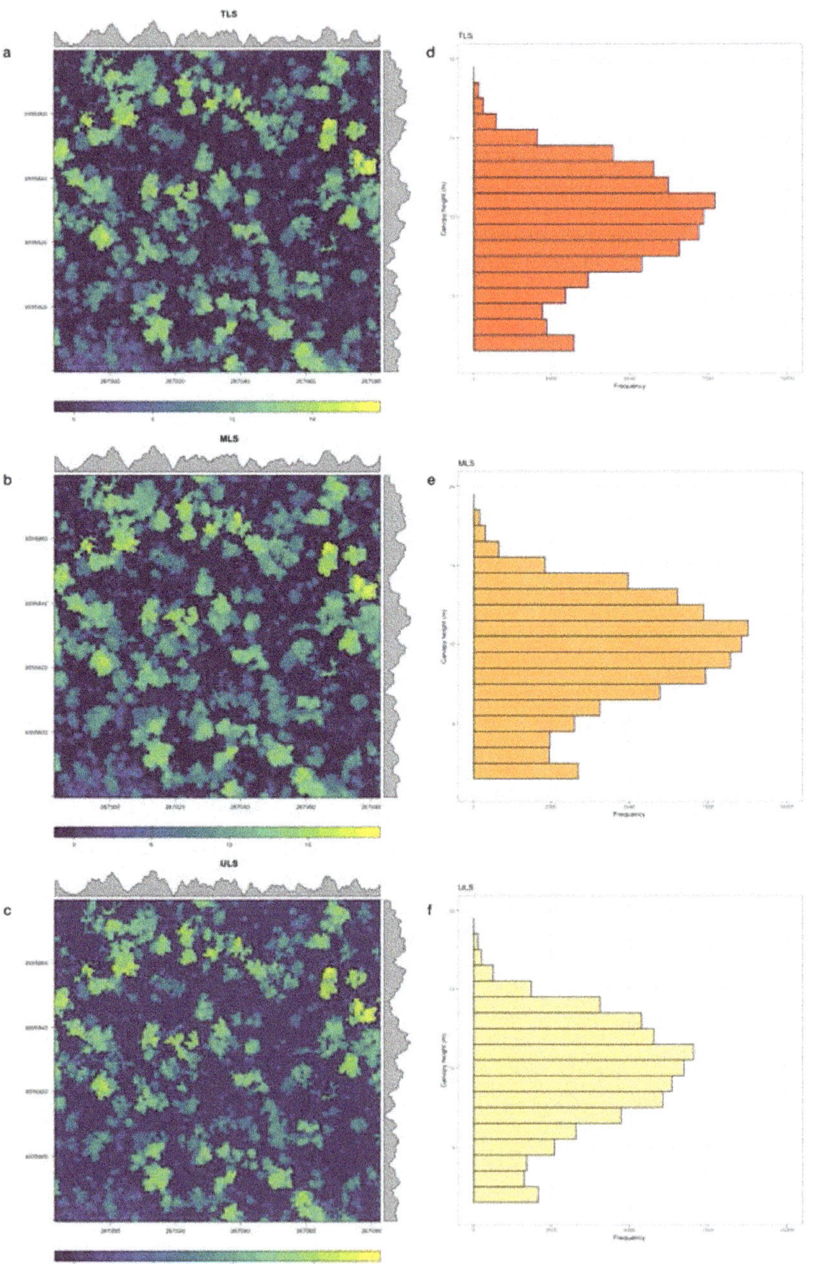

Figure 6. Canopy height models derived from the three co-incident LiDAR datasets, plotted at 0.25 m resolution (**a**–**c**) and their height class distributions (**d**–**f**).

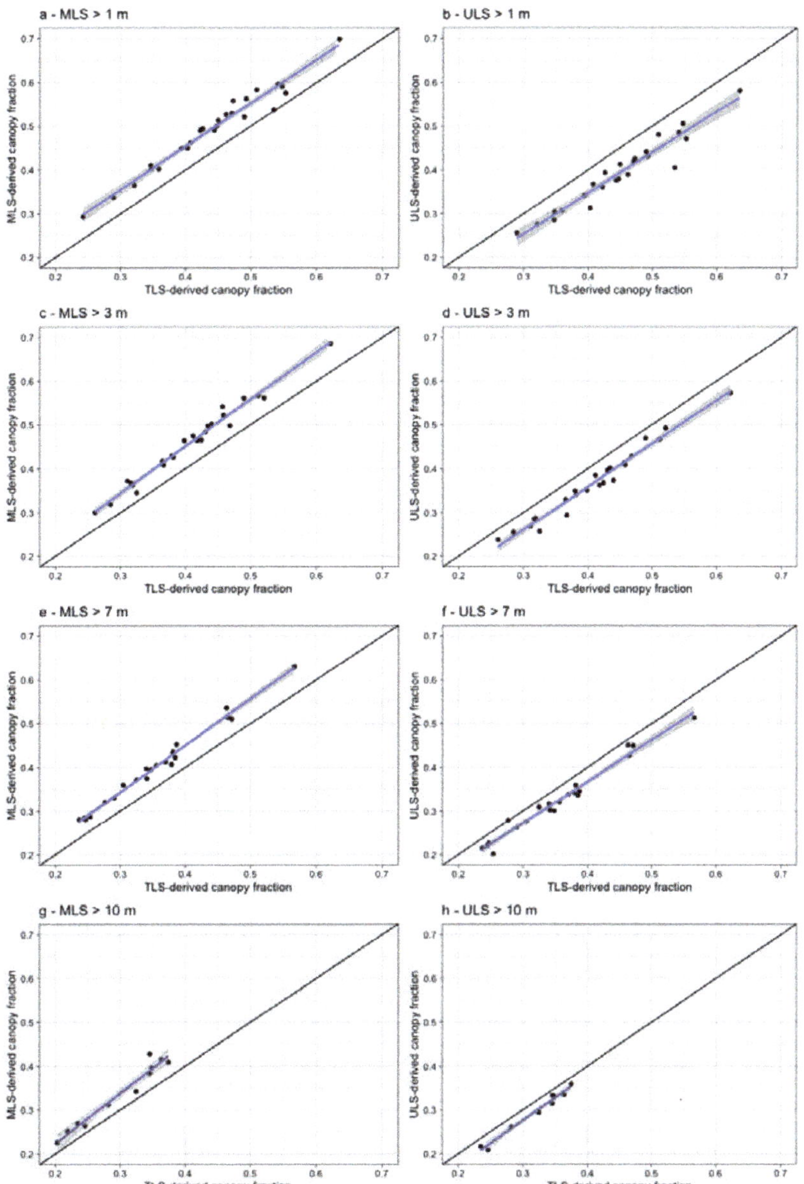

Figure 7. Canopy cover comparisons between MLS and ULS at different height class intervals, against a TLS reference. Solid black is the 1:1 fit.

Table 2. Summary statistics of woody canopy attributes collected by the three LiDAR systems at the 1 ha plot scale.

Attribute	TLS	MLS	ULS
CHM max height (m)	18.02	18.65	18.05
Woody canopy cover (%)	51.5	55.3	39.8
Woody canopy cover > 3 m (%)	42.4	46.1	36.9
Woody canopy cover > 5 m (%)	39.9	44.7	34.8
Woody canopy cover > 10 m (%)	18.9	22.1	17.9

3.2. Individual Tree Characterization

Assessment of 3D structure at the individual tree level highlighted some further distinctions between TLS, MLS and ULS characterization of 3D structure. The TLS capture provided a very clean representation stem, branch and leaf structure, with points spaced close enough together to give the impression of a continuous surface (Figure 8a). Extracting a point-cloud slice from 1.25–1.35 m (to encompass the traditional DBH measurement height) revealed a clear ring of points that allowed for DBH measurement via cylinder fitting (Figure 8d).

The MLS scan captured the general size and shape of individual trees very well, including small under-storey plants, but loss of detail was evident compared to TLS (Figure 8b). The MLS point cloud exhibited more noise around tree stems, which is particularly clear in the thick ring of points extracted at DBH measurement height (Figure 8e). The ULS also captured the general shape of individual trees very well, and although many returns were received from tree stems the density of points in the 1.25–1.35 m range was much lower than from TLS and MLS (Figure 8c,f).

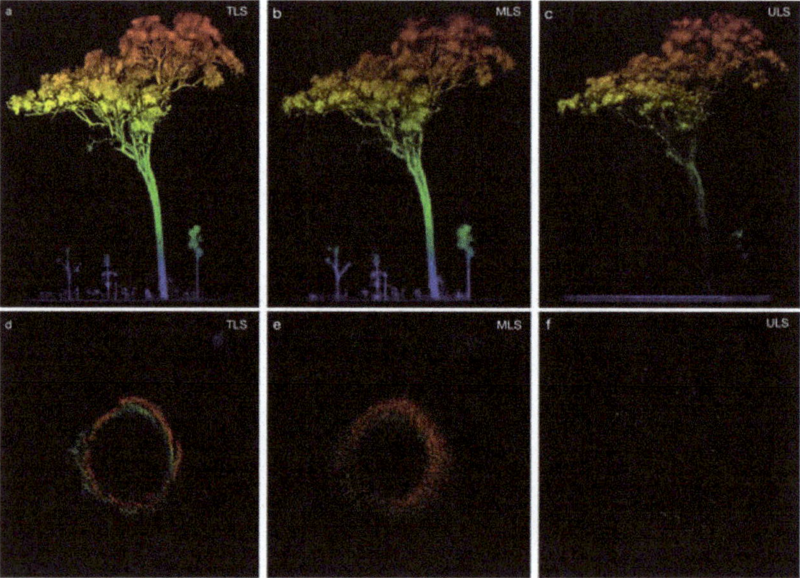

Figure 8. Example of a single large tree as captured by the three LiDAR systems (**a–c**). Horizontal slice cut through each stem at 1.25–1.35 m above ground level, showing the pattern and density of points available for DBH fitting from each system (**d–f**).

Considering this, we used a broader portion of the tree stems (1–2 m) for cylinder fitting and DBH estimation. Comparison of the DBH distributions against those obtained from the field inventory showed that the TLS estimates closely mirrored the field-measured values (Figure 9a,b), but MLS and ULS estimates (Figure 9c,d) were skewed to larger values and were more normally distributed than the inverse-J patterns evident in the field and TLS data.

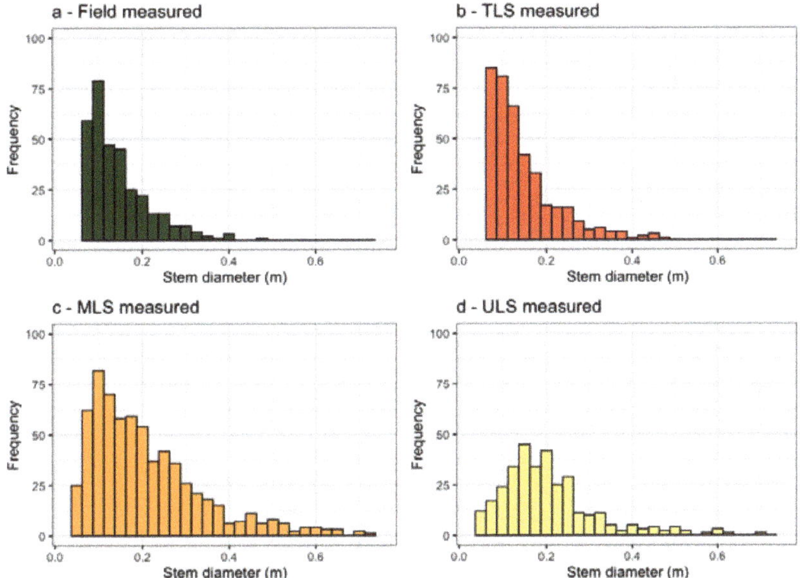

Figure 9. Distribution of DBH estimates from field inventory (**a**) and the three LiDAR approaches (**b–d**).

Linear regression between field-measured DBH and TLS-estimated DBH confirmed that the TLS could reliably extract individual tree DBH for a broad range of species (Figure 10). Taller tree species with larger sample sizes in the plot such as *Eucalyptus tetradonta*, *Corymbia porrecta* and *Xanthostemon paradoxus* showed the strongest fits (Figure 10), while smaller statured species with narrower DBH ranges exhibited greater error (e.g., Figure 10).

Based on the DBH representation findings discussed above, we restricted our analysis of above-ground woody biomass to the field and TLS datasets. Our field inventory returned a total of 572 stems, corresponding to 41.08 Mg ha^{-1} of above-ground woody biomass and 20.54 Mg C ha^{-1} from DBH-based allometry (Equation (1)). Segmentation of the TLS dataset identified almost 100 more stems (n = 669) than measured in the field survey, which were mostly in the smaller size classes (Figure 9b). TLS-estimated biomass, using DBH allometry according to Equation (2), was 7/% greater than the field-inventory estimate at 44.21 Mg ha^{-1} of above-ground woody biomass and 22.10 Mg C ha^{-1}.

Mapping out the distribution of individual tree biomass spatially showed that field records and TLS predictions were very closely co-aligned (Figure 11a–b). The TLS tree location mapping showed very few instances of omission from field-mapped locations, and a number of commissions which were predominantly located at the very boundary of the plot (Figure 11c). Omissions in the field-inventory mapping along the plot boundary (e.g., south-western corner) are indicative of positioning errors associated with the manual boundary line placement in the field.

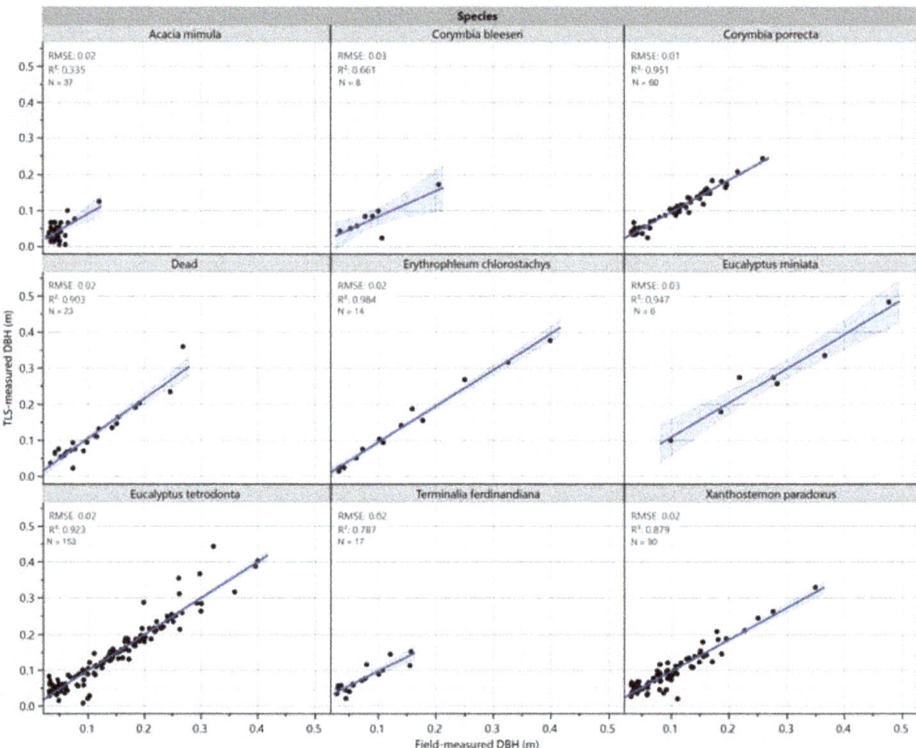

Figure 10. Comparison of TLS-derived DBH with field-measured values for the dominant woody species.

3.3. Allometric Scaling

Given that the MLS and ULS point clouds from our acquisitions were not suitable for direct DBH/AGB retrieval (Figure 9), we explored the possibility of predicting AGB in the MLS and ULS datasets through allometric scaling of tree crown parameters. MLS and ULS captured the general crown structure of individual trees very well (Table 3), and the relationship between tree height and crown area that was sensed by the three scanning systems was very consistent (Figure 12). However, both multiple linear regression and Random Forest modeling failed to establish a reliable relationship between canopy attributes and stem DBH/AGB (R^2 = 0.34, RMSE = 0.075 m).

Table 3. Summary statistics of individual tree attributes collected by the three LiDAR systems.

Attribute	TLS	MLS	ULS
Crown volume max (m^3)	428.37	475.29	406.37
Crown volume mean (m^3)	19.94	23.49	41.67
Crown volume CV (m^3)	2.33	2.29	1.6
Woody biomass (Mg ha^{-1})	44.21	-	-
Woody biomass (Mg C ha^{-1})	22.10	-	-

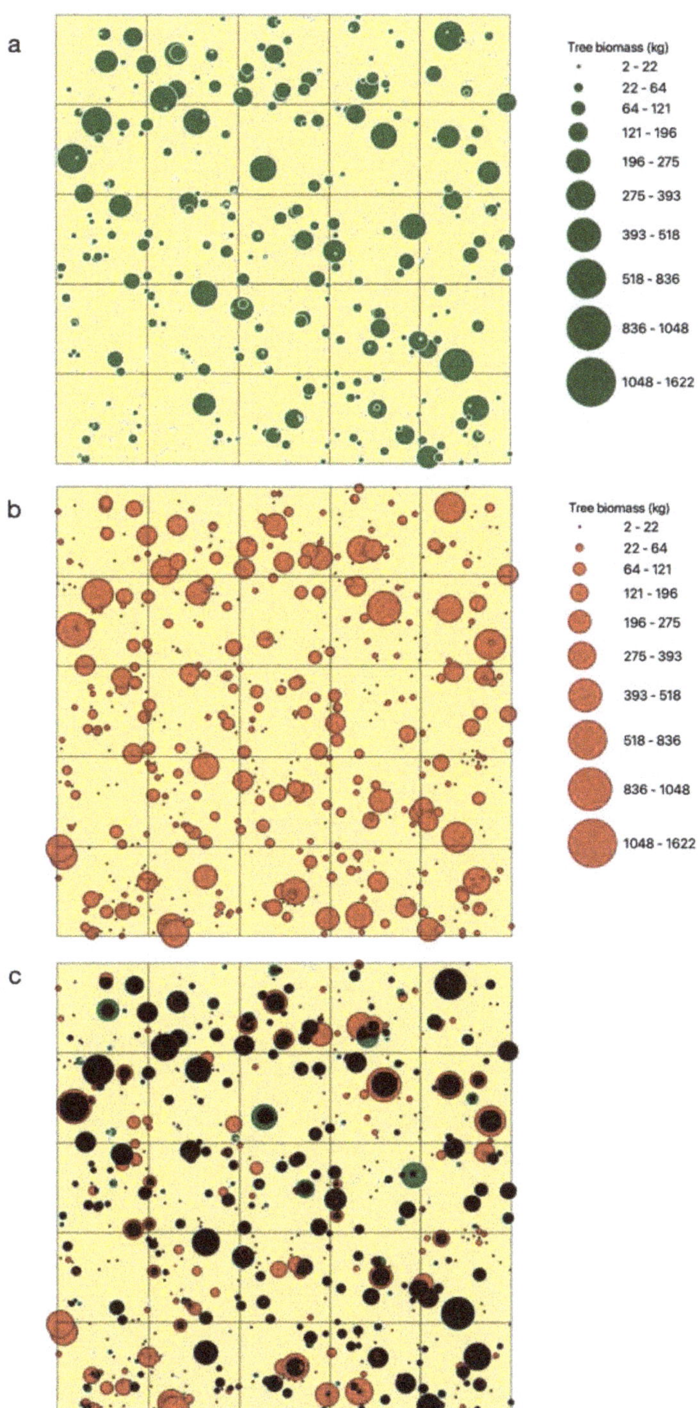

Figure 11. Spatial distribution of tree biomass as mapped through field inventory (**a**, n = 572) and from TLS (**b**, n = 669) using DBH-based allometric equations. Agreement, omissions, and commissions between field and TLS mappings are shown in (**c**) with overlay blending.

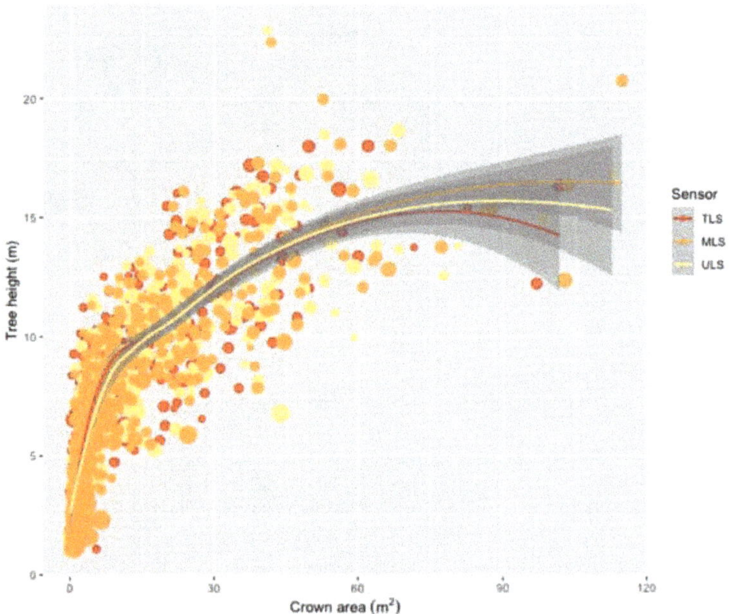

Figure 12. Relationship between individual tree height and crown area, as determined from the three different LiDAR systems.

4. Discussion

Laser-based technologies have transformed our ability to measure and monitor savanna vegetation structure. The level of detail and efficiency exhibited in this study from all three sensors is very encouraging from a long-term monitoring and management perspective—providing rigorous baselines from which to assess current state and future dynamics.

4.1. Efficient Monitoring of Habitat Structure

TLS has emerged as the gold standard for 3D characterization of vegetation structure in ecosystems around the globe [17,19]. The high quality results obtained here from TLS are unsurprising, but it is worth noting that the scan position density that we employed (30 m spacing, 16 scans ha^{-1}) was considerably lower than current recommendations in the literature which suggest aiming for 10 m spacing [37]. The TLS literature is heavily skewed towards forested ecosystems where high density scanning is critical for minimizing occlusion. In savanna woodlands and shrublands, which occupy 20% of the global terrestrial surface, the probability of occlusion is much lower and long-range scanners allow for wider scan position placement [20]. This is important in the context of habitat monitoring, as TLS is often considered too time intensive for broad extent and repeat coverage sampling.

Nonetheless, the time required for our TLS survey, and subsequent processing, was still substantially longer than that required for MLS and ULS captures (Table 1). The efficiency of MLS and ULS in terms of time are clear, but do they provide sufficient structural information for ecosystem monitoring? From a canopy height model perspective there is minimal difference among the three scanning approaches and systems. The TLS CHM could not be matched by MLS or ULS at the 0.25 m resolution, but was indistinguishable from MLS at 2 m resolution and ULS at 4 m resolution. For many ecological applications 2–4 m is more than sufficient for providing insight into how canopy growth is tracking over time, and the extended geographic coverage may often outweigh the benefits of finely detailed local structural representation.

The minor under-estimation of woody canopy cover by ULS was consistent across the subplots of our study site. Similar findings have recently emerged from dry sclerophyll forests in Tasmania, where ULS underestimated canopy cover by a few percent in comparison to TLS at the plot scale [38]. Here we show that this under-estimation is a consistent trend across multiple height classes, indicating that the source of this bias is not only under-storey, but also small canopy features such as the edges of branches for example. MLS on the other-hand consistently over-estimated canopy cover in comparison to TLS. This is contrary to the few published studies exploring MLS characterization of woody canopy—which have mostly been conducted in forested ecosystems where tall dense canopy restricts access of the laser beam through to the canopy surface [21,39,40]. In the open savanna woodland of our study, ground-based TLS and MLS can penetrate canopies well with good fields of view. In this instance we attribute the over-estimation of canopy cover by MLS to the nosier point cloud that it produced in comparison to TLS, and these results could likely be improved through further post-processing of the MLS point cloud.

4.2. Individual Trees and Biomass Scaling

Given that MLS and ULS were able to characterize individual tree height and crown attributes very well, albeit with minor bias, we had anticipated that estimation of above-ground biomass would have been possible from these point clouds through allometric relationships with crown attributes. However, our TLS data showed that DBH-AGB did not scale in a predictable manner with tree height, crown diameter, area, and volume ($R^2 = 0.36$, RMSE = 0.075 m). This was initially surprising in light of how well AGB scales as a function of tree height and crown diameter in many ecoregions around the globe [41], but it is an important reminder of how local-scale variability is averaged out in regional and global syntheses. Savanna eucalypts have notoriously complex crown structures, exacerbated by the actions of fire, termites and cyclone damage [42]. As shown recently in a similar savanna setting in Litchfield National Park, savanna trees exhibit a wide range of crown size variability for a given DBH [6]. In addition to confounding the estimation of DBH and AGB from crown attributes, this variability also brings into question the reliability of using DBH allometry in the first instance for biomass estimation in these systems. The growing use of TLS for building Quantitative Structure Models (QSMs) of trees and accounting for biomass volumetrically has started revealing errors associated with DBH-based allometry in a variety of ecosystem types [43,44] and needs greater exploration in different savanna settings.

4.3. Limitations and Future Directions

TLS is a mature field, with well established protocols for data acquisition and processing [17]. TLS can be deployed in many different ecosystems with high confidence placed in data capture and structural representation, with efficient open-source tools for tree segmentation and structural modeling [45]. Our results obtained from MLS and ULS are encouraging and capture key elements of the savanna woody structure thoroughly. As expected both MLS and ULS did not detect some of the finer-scale ecosystem elements as well as TLS, but are more than sufficient for addressing many ecological questions.

A challenge that arises when comparing different platforms is that the end results are influenced heavily by the acquisition and pre- and post-processing parameters. As such, meaningful generalisations are difficult to make. TLS has the advantage of being a mature technology, with a large user-base and established collection and pre- and post-processing protocols. MLS and ULS are at earlier stages of development and uptake. Furthermore, it is a rapidly expanding industry and a wide variety of sensors and platform exist, from very high-end survey-grade systems (e.g., RIEGL RiCOPTER flying a RIEGL VUX-1) to lighter weight and lower cost options (e.g., DJI M600 with a Velodyne or M8 sensor). Some MLS systems rely solely on SLAM for positioning, while others have integrated RTK-GNSS, so results and noise levels will vary. Additionally, the actual flight/walk patterns and speeds have a marked impact on point-cloud characteristics—and optimal acquisition parameters may be ecosystem specific. Fortunately, since the point-cloud outputs from MLS and ULS share many features of TLS point clouds (at lower densities) analysis of these products can make use of and further advance breakthroughs in TLS processing pipelines.

A key challenge moving forward with MLS and ULS surveying will be to optimize the acquisition and processing parameters in a robust and repeatable manner. We suggest that TLS be leveraged at the start of baseline and long-term monitoring programs to provide a benchmark against which the MLS and/or ULS acquisitions can be optimized. This will enable MLS and ULS to be applied over larger areas and at more frequent time intervals, with intermittent re-calibration against nested TLS collections in key spatial locations and temporal stages throughout the monitoring program. Furthermore, establishing the scaling uncertainty of TLS measurements through to large area ULS surveying will greatly benefit the earth observation community by assisting in the calibration and validation of spaceborne optical, LiDAR and SAR missions (e.g., GEDI, ICESAT-2, NISAR, BIOMASS). When considering the faster acquisition time and the ability to collect data over larger areas, MLS and ULS systems could prove particularly useful for expanding the geographic range and representation of current calibration and validation libraries, with known uncertainties from the nested TLS models.

Lastly, while we have primarily focused here on comparing the individual point clouds and derived products produced through TLS, MLS and ULS surveying, we recognize that much could be gained through their integration. For example, a fused point cloud derived from both TLS and ULS could provide rich vegetation detail from the oblique TLS perspective and comprehensive terrain characterization from the aerial ULS viewpoint.

5. Conclusions

From a canopy modeling perspective, the differences between the TLS, MLS, and ULS were negligible for most applications, but there was a large difference in the acquisition time, with MLS and ULS offering distinct advantages. The trade-off comes through at the individual tree and stem modeling level, where TLS can reproduce DBH estimates from the field with high accuracy and precision, and capture smaller stems than are typically measured in field surveys. Comparisons among TLS, MLS and ULS are challenging as there are a large range of parameters that can be modified across a wide range of applications. TLS surveying is relatively mature, whereas optimal sampling and processing strategies for MLS and ULS are less mature, but rapidly developing. Future research should focus on using TLS as a benchmarking tool to optimize the acquisition parameters of MLS and ULS for a given research and/or management objective. Importantly, this needs to be done across a range of different ecosystem types as the strengths and weaknesses of each sensing system, and platform, will vary with site-specific conditions.

TLS provides a holistic representation of 3D structure that cannot be obtained through traditional field inventory and provides an avenue for optimizing MLS and ULS acquisition parameters, generating correction coefficients for MLS and ULS canopy products, and exploring new allometric models linking point-cloud metrics directly to above-ground woody biomass.

Author Contributions: Conceptualization, S.R.L. and R.B.; Data curation, T.W. and D.A.L.; Formal analysis, S.R.L. and T.W.; Methodology, S.R.L.,T.W., D.A.L. and M.R.; Project administration, R.B.; Visualization, S.R.L.; Writing—original draft, S.R.L.; Writing—review & editing, S.R.L., T.W., D.A.L., M.R. and R.B. All authors have read and agreed to the published version of the manuscript.

Funding: This research was not externally funded.

Institutional Review Board Statement: Not applicable.

Informed Consent Statement: Not applicable.

Data Availability Statement: Data available on request due to geographic restrictions.

Acknowledgments: We thank Jon Schatz and Jaylen Nicholson for conducting the field inventory, with assistance from Pierre Grandclément. Linda Luck and Stephanie Johnson provided assistance with data processing. This manuscript benefited greatly from the comments of two anonymous reviewers.

Conflicts of Interest: The authors declare no conflict of interest.

Abbreviations

The following abbreviations are used in this manuscript:

AGB	Above-ground biomass
DBH	Diameter at breast height
CHM	Canopy height model
GEDI	Global Ecosystem Dynamics Investigator
GPS	Global Positioning System
LiDAR	Light-detection-and-ranging
MLS	Mobile Laser Scanning
RTK	Real-time Kinematic
SAR	Synthetic Aperture Radar
SLAM	Simultaneous Location and Mapping
TLS	Terrestrial Laser Scanning
ULS	UAV Laser Scanning
UAV	Unmanned Aerial Vehicle

References

1. Gillson, L. Evidence of Hierarchical Patch Dynamics in an east African savanna? *Landsc. Ecol.* **2005**, *19*, 883–894. [CrossRef]
2. Levick, S.R.; Rogers, K.H. Context-dependent vegetation dynamics in an African savanna. *Landsc. Ecol.* **2011**, *26*, 515–528. [CrossRef]
3. Levick, S.R.S.; Asner, G.G.P.; Kennedy-Bowdoin, T.T.; Knapp, D.D.E. The relative influence of fire and herbivory on savanna three-dimensional vegetation structure. *Biol. Conserv.* **2009**, *142*, 1693–1700. [CrossRef]
4. Moncrieff, G.R.; Chamaillé-Jammes, S.; Higgins, S.I.; O'Hara, R.B.; Bond, W.J. Tree allometries reflect a lifetime of herbivory in an African savanna. *Ecology* **2011**, *92*, 2310–2315. [CrossRef]
5. Woolley, L.A.; Murphy, B.P.; Radford, I.J.; Westaway, J.; Woinarski, J.C. Cyclones, fire, and termites: The drivers of tree hollow abundance in northern Australia's mesic tropical savanna. *For. Ecol. Manag.* **2018**, *419–420*, 146–159. [CrossRef]
6. Luck, L.; Hutley, L.B.; Calders, K.; Levick, S.R. Exploring the Variability of Tropical Savanna Tree Structural Allometry with Terrestrial Laser Scanning. *Remote Sens.* **2020**, *12*, 3893. [CrossRef]
7. Paul, K.I.; Larmour, J.S.; Roxburgh, S.H.; England, J.R.; Davies, M.J.; Luck, H.D. Measurements of stem diameter: Implications for individual- and stand-level errors. *Environ. Monit. Assess.* **2017**, *189*. [CrossRef]
8. Wulder, M.A.; Coops, N.C.; Roy, D.P.; White, J.C.; Hermosilla, T. Land Cover 2.0. *Int. J. Remote Sens.* **2018**, *39*, 4254–4284 [CrossRef]
9. Geller, G.N.; Halpin, P.N.; Helmuth, B.; Hestir, E.L.; Skidmore, A.; Abrams, M.J.; Aguirre, N.; Blair, M.; Botha, E.; Colloff, M.; et al. Remote Sensing for Biodiversity. In *The GEO Handbook on Biodiversity Observation Networks*; Walters, M., Scholes, R.J., Eds.; Springer International Publishing: Cham, Switzerland, 2017; pp. 187–210. [CrossRef]
10. Hill, M.J.; Hanan, N.P. *Ecosystem Function in Savannas: Measurement and Modeling at Landscape to Global Scales*; CRC Press, Taylor and Francis Group: New York, NY, USA, 2010; pp. 1–612. [CrossRef]
11. Lefsky, M.A.; Cohen, W.B.; Parker, G.; Harding, D.J. Lidar Remote Sensing for Ecosystem Studies. *BioScience* **2002**, *52*, 19–30. [CrossRef]
12. Levick, S.; Rogers, K. Structural biodiversity monitoring in savanna ecosystems: Integrating LiDAR and high resolution imagery through object-based image analysis. In *Object-Based Image Analysis*; Springer: Berlin/Heidelberg, Germany, 2008; pp. 477–491. [CrossRef]
13. Colgan, M.S.; Asner, G.P.; Levick, S.R.; Martin, R.E.; Chadwick, O.A. Topo-edaphic controls over woody plant biomass in South African savannas. *Biogeosciences* **2012**, *9*, 1809–1821. [CrossRef]
14. Levick, S.R.; Richards, A.E.; Cook, G.D.; Schatz, J.; Guderle, M.; Williams, R.J.; Subedi, P.; Trumbore, S.E.; Andersen, A.N. Rapid response of habitat structure and above-ground carbon storage to altered fire regimes in tropical savanna. *Biogeosciences* **2019**, *16*, 1493–1503. [CrossRef]
15. Davies, A.B.; Asner, G.P. Advances in animal ecology from 3D-LiDAR ecosystem mapping. *Trends Ecol. Evol.* **2014**, *29*, 681–691. [CrossRef]
16. Stobo-Wilson, A.M.; Murphy, B.P.; Cremona, T.; Carthew, S.M.; Levick, S.R. Illuminating den-tree selection by an arboreal mammal using terrestrial laser scanning in northern Australia. *Remote Sens. Ecol. Conserv.* **2020**. [CrossRef]

17. Calders, K.; Adams, J.; Armston, J.; Bartholomeus, H.; Bauwens, S.; Bentley, L.P.; Chave, J.; Danson, F.M.; Demol, M.; Disney, M.; et al. Terrestrial laser scanning in forest ecology: Expanding the horizon. *Remote Sens. Environ.* **2020**, *251*, 112102. [CrossRef]

18. Liang, X.; Hyyppä, J.; Kaartinen, H.; Lehtomäki, M.; Pyörälä, J.; Pfeifer, N.; Holopainen, M.; Brolly, G.; Francesco, P.; Hackenberg, J.; et al. International benchmarking of terrestrial laser scanning approaches for forest inventories. *ISPRS J. Photogramm. Remote Sens.* **2018**, *144*, 137–179. [CrossRef]

19. Newnham, G.J.; Armston, J.D.; Calders, K.; Disney, M.I.; Lovell, J.L.; Schaaf, C.B.; Strahler, A.H.; Mark Danson, F. Terrestrial laser scanning for plot-scale forest measurement. *Curr. For. Rep.* **2015**, *1*, 239–251. [CrossRef]

20. Singh, J.; Levick, S.R.; Guderle, M.; Schmullius, C. Moving from plot-based to hillslope-scale assessments of savanna vegetation structure with long-range terrestrial laser scanning (LR-TLS). *Int. J. Appl. Earth Obs. Geoinf.* **2020**, *90*, 102070. [CrossRef]

21. Bauwens, S.; Bartholomeus, H.; Calders, K.; Lejeune, P. Forest inventory with terrestrial LiDAR: A comparison of static and hand-held mobile laser scanning. *Forests* **2016**, *7*, 127. [CrossRef]

22. Corte, A.P.D.; Rex, F.E.; de Almeida, D.R.A.; Sanquetta, C.R.; Silva, C.A.; Moura, M.M.; Wilkinson, B.; Zambrano, A.M.A.; da Cunha Neto, E.M.; Veras, H.F.; et al. Measuring individual tree diameter and height using gatoreye high-density UAV-lidar in an integrated crop-livestock-forest system. *Remote Sens.* **2020**, *12*, 863. [CrossRef]

23. Kellner, J.R.; Armston, J.; Birrer, M.; Cushman, K.C.; Duncanson, L.; Eck, C.; Falleger, C.; Imbach, B.; Král, K.; Krůček, M.; et al. New Opportunities for Forest Remote Sensing Through Ultra-High-Density Drone Lidar. *Surv. Geophys.* **2019**, *40*, 959–977. [CrossRef]

24. Brack, C.; Schaefer, M.; Jovanovic, T.; Crawford, D. Comparing terrestrial laser scanners' ability to measure tree height and diameter in a managed forest environment. *Aust. For.* **2020**, *83*, 161–171. [CrossRef]

25. Velas, M.; Spanel, M.; Sleziak, T.; Habrovec, J.; Herout, A. Indoor and outdoor backpack mapping with calibrated pair of velodyne lidars. *Sensors* **2019**, *19*, 3944. [CrossRef]

26. Stal, C.; Verbeurgt, J.; De Sloover, L.; De Wulf, A. Assessment of handheld mobile terrestrial laser scanning for estimating tree parameters. *J. For. Res.* **2020**, 1–11. [CrossRef]

27. Prior, L.D.; Whiteside, T.G.; Williamson, G.J.; Bartolo, R.E.; Bowman, D.M. Multi-decadal stability of woody cover in a mesic eucalypt savanna in the Australian monsoon tropics. *Austral Ecol.* **2020**, *45*, 621–635. [CrossRef]

28. Beringer, J.; Hutley, L.B.; Abramson, D.; Arndt, S.K.; Briggs, P.; Bristow, M.; Canadell, J.G.; Cernusak, L.A.; Eamus, D.; Edwards, A.C.; et al. Fire in Australian savannas: From leaf to landscape. *Glob. Chang. Biol.* **2015**, *21*, 62–81. [CrossRef]

29. Hernandez-Santin, L.; Rudge, M.L.; Bartolo, R.E.; Whiteside, T.G.; Erskine, P.D. Reference site selection protocols for mine site ecosystem restoration. *Restor. Ecol.* **2021**, *29*. [CrossRef]

30. Girardeau-Montaut, D. CloudCompare—Open Source Project. Available online: http://www.cloudcompare.org/ (accessed on 10 January 2020).

31. rapidlasso GmbH. LAStools—Efficient LiDAR Processing Software. 2020. Available online: https://rapidlasso.com/lastools/ (accessed on 12 January 2020).

32. Massey, F.J. The Kolmogorov-Smirnov Test for Goodness of Fit. *J. Am. Stat. Assoc.* **1951**, *46*, 68–78. [CrossRef]

33. Tao, S.; Wu, F.; Guo, Q.; Wang, Y.; Li, W.; Xue, B.; Hu, X.; Li, P.; Tian, D.; Li, C.; et al. Segmenting tree crowns from terrestrial and mobile LiDAR data by exploring ecological theories. *ISPRS J. Photogramm. Remote Sens.* **2015**, *110*, 66–76. [CrossRef]

34. Trochta, J.; Kruček, M.; Vrška, T.; Kraâl, K. 3D Forest: An application for descriptions of three-dimensional forest structures using terrestrial LiDAR. *PLoS ONE* **2017**, *12*, e0176871. [CrossRef]

35. Williams, R.J.; Zerihun, A.; Montagu, K.D.; Hoffman, M.; Hutley, L.B.; Chen, X. Allometry for estimating aboveground tree biomass in tropical and subtropical eucalypt woodlands: Towards general predictive equations. *Aust. J. Bot.* **2005**, *53*, 607–619. [CrossRef]

36. Breiman, L. Random forests. *Mach. Learn.* **2001**, *45*, 5–32. [CrossRef]

37. Wilkes, P.; Lau, A.; Disney, M.; Calders, K.; Burt, A.; Gonzalez de Tanago, J.; Bartholomeus, H.; Brede, B.; Herold, M. Data acquisition considerations for Terrestrial Laser Scanning of forest plots. *Remote Sens. Environ.* **2017**, *196*, 140–153. [CrossRef]

38. Hillman, S.; Wallace, L.; Lucieer, A.; Reinke, K.; Turner, D.; Jones, S. A comparison of terrestrial and UAS sensors for measuring fuel hazard in a dry sclerophyll forest. *Int. J. Appl. Earth Obs. Geoinf.* **2021**, *95*, 102261.

39. Bienert, A.; Georgi, L.; Kunz, M.; Maas, H.G.; von Oheimb, G. Comparison and combination of mobile and terrestrial laser scanning for natural forest inventories. *Forests* **2018**, *9*, 395. [CrossRef]

40. Gollob, C.; Ritter, T.; Nothdurft, A. Comparison of 3D point clouds obtained by terrestrial laser scanning and personal laser scanning on forest inventory sample plots. *Data* **2020**, *5*, 103. [CrossRef]

41. Jucker, T.; Caspersen, J.; Chave, J.; Antin, C.; Barbier, N.; Bongers, F.; Dalponte, M.; van Ewijk, K.Y.; Forrester, D.I.; Haeni, M.; et al. Allometric equations for integrating remote sensing imagery into forest monitoring programmes. *Glob. Chang. Biol.* **2017**, *23*, 177–190. [CrossRef]

42. Cook, G.D.; Goyens, C.M. The impact of wind on trees in Australian tropical savannas: Lessons from Cyclone Monica. *Austral Ecol.* **2008**, *33*, 462–470. [CrossRef]

43. Disney, M.I.; Boni Vicari, M.; Burt, A.; Calders, K.; Lewis, S.L.; Raumonen, P.; Wilkes, P. Weighing trees with lasers: Advances, challenges and opportunities. *Interface Focus* **2018**, *8*, 20170048. [CrossRef]

44. Disney, M.; Burt, A.; Wilkes, P.; Armston, J.; Duncanson, L. New 3D measurements of large redwood trees for biomass and structure. *Sci. Rep.* **2020**, *10*, 1–11. [CrossRef]
45. Raumonen, P.; Kaasalainen, M.; Åkerblom, M.; Kaasalainen, S.; Kaartinen, H.; Vastaranta, M.; Holopainen, M.; Disney, M.; Lewis, P. Fast Automatic Precision Tree Models from Terrestrial Laser Scanner Data. *Remote Sens.* **2013**, *5*, 491–520. [CrossRef]

Article

Optical and SAR Remote Sensing Synergism for Mapping Vegetation Types in the Endangered Cerrado/Amazon Ecotone of Nova Mutum—Mato Grosso

Flávia de Souza Mendes [1,*], Daniel Baron [2], Gerhard Gerold [1], Veraldo Liesenberg [3] and Stefan Erasmi [4]

1   Physical Geography Department, University of Göttingen, Goldschmidtstr. 5, 37077 Göttingen, Germany; ggerold@gwdg.de
2   Landesamt für Vermessung und Geoinformation Schleswig-Holstein, Dezernat 22, Mercatorstraße 1, 24106 Kiel, Germany; Daniel.Baron@LVermGeo-landsh.de
3   Forest Engineering Department, Santa Catarina State University, Av. Luiz de Camões 2090, 88.520-000 Lages, Santa Catarina, Brazil; veraldo.liesenberg@udesc.br
4   Cartography, GIS & Remote Sensing Department, University of Göttingen, Goldschmidtstr. 5, 37077 Göttingen, Germany; serasmi@gwdg.de
*   Correspondence: fdesouz@uni-goettingen.de

Received: 18 April 2019; Accepted: 10 May 2019; Published: 15 May 2019

Abstract: Mapping vegetation types through remote sensing images has proved to be effective, especially in large biomes, such as the Brazilian Cerrado, which plays an important role in the context of management and conservation at the agricultural frontier of the Amazon. We tested several combinations of optical and radar images to identify the four dominant vegetation types that are prevalent in the Cerrado area (i.e., cerrado denso, cerradão, gallery forest, and secondary forest). We extracted features from both sources of data such as intensity, grey level co-occurrence matrix, coherence, and polarimetric decompositions using Sentinel 2A, Sentinel 1A, ALOS-PALSAR 2 dual/full polarimetric, and TanDEM-X images during the dry and rainy season of 2017. In order to normalize the analysis of these features, we used principal component analysis and subsequently applied the Random Forest algorithm to evaluate the classification of vegetation types. During the dry season, the overall accuracy ranged from 48 to 83%, and during the dry and rainy seasons it ranged from 41 up to 82%. The classification using Sentinel 2A images during the dry season resulted in the highest overall accuracy and kappa values, followed by the classification that used images from all sensors during the dry and rainy season. Optical images during the dry season were sufficient to map the different types of vegetation in our study area.

Keywords: Cerrado; Amazon; vegetation type; optical; sar; synergism; mapping

---

1. Introduction

The Cerrado biome is considered as being among the most extensive and diverse ecosystems in the Neotropics and is a hotspot in the context of biodiversity [1]. It is also one of the most threatened ecosystems in South America, with over 40% of the biome converted to agriculture and the remainder highly fragmented [2]. Despite the threat to the Brazilian Cerrado, studies on this ecosystem are few and recent.

The Cerrado biome is the second largest complex vegetation present in Brazil and occupies about 200 million hectares, of which the largest territory is in the state of Mato Grosso [3]. This large distribution of the Cerrado biome in Brazil covers three main vegetation types: grassland, savannas, and forest formations, which results in indeterminate boundary and a gradient of biomass, height, and tree cover. This large variance in different types of vegetation in the Cerrado is responsible for the high biodiversity in this biome. The three areas of biodiversity in Cerrado, the South–Southeast, Central Plateau, and Northeast areas are mainly separated by the altitude and latitude [4]. The heterogeneity of the vegetation types is also seen in the microclimate variability and in different types of soils, e.g., mostly latosol, red-yellow latosol, red latosol, quartz-neosols, and argisols [5–7]. In addition, the amount of biomass and carbon storage is differently distributed in the biome, depending on the vegetation type and soil [8–10]. This large biodiversity and floristic heterogeneity in Cerrado was and is decreasing due to deforestation since the 1980s, which can lead to a loss or decrease in ecosystem services [11].

Cerrado is the most deforested biome in Brazil due to the high agricultural impact, particularly caused by the world market-oriented production of soy, cotton, and sugarcane. Deforestation is facilitated by its flat topography, easy management of the soil for agricultural activities, and high mechanization [12]. For pasture and agricultural activities, Cerrado has become a more viable alternative to the Amazon despite its poor soil quality. Despite a lack of consistent deforestation records, a few studies have looked at rates of deforestation in the Cerrado biome. Machado et al. [13] analyzed the deforestation rates from two different sources. They found that from 1985 to 1993, Cerrado lost 1.5% of its total vegetation area annually. From 1993 to 2002, the rate of deforestation per year decreased to 0.67%. Starting from 2002, several Brazilians institutes, such as the Instituto Brasileiro do Meio Ambiente e dos Recursos Naturais Renováveis (IBAMA) and National Institute for Space Research (INPE), started to monitor the rate of deforestation in the Cerrado biome. From 2002 to 2017, Cerrado lost around 0.8% of its total vegetation area per year. The ease of deforesting in the Cerrado biome created a hotspot region for deforestation at the boundary with Amazon biomes.

Current regulation and restrictions in ecosystem preservation have driven deforestation and cover changes into the forest–savanna transition zone, as in West Africa [14] and South America [15]. Janssen at al. [14] projected an increase of tree cover losses from 20 to 85% in Ghana. In South America, the transitional zone is known as "arc of the deforestation" (AOD). The AOD is located in the frontier states of Mato Grosso, Pará, and Rondônia, and it accounts for 85% of the areas that were deforested between 1996 to 2005 [16,17]. In these transition areas, the laws that support the protection of forests are even weaker and unmanaged. One example is the environmental legislation that defines the amount of natural vegetation that has to be preserved (80% in the Amazon, 20% in the Cerrado biome) [18]. Additionally, this forest–savanna boundary comprises a mixture of floristic characteristics from both adjacent regions, which increases the complexity in mapping the ecotone between the Amazon and Cerrado. Marques et al. [19] showed that the official boundary between Cerrado and Amazon conducted by the Brazilian Institute of Geography and Statistics (IBGE) is not accurate, and in some areas, the length of the transition zone was miscalculated by 245.5%. This problem is likely to misinterpret the mapping of land use and consequently decrease the accuracy of vegetation classification. Moreover, the problem of low accuracy in the mapping of the boundary between Cerrado and Amazon affects the calculation of wood density, and therefore biomass estimation as well [20]. To overcome these problems, Brazil needs to improve the monitoring system of deforestation and land use change (LUC), especially for the Cerrado biome.

The field monitoring of Cerrado is a time-consuming challenge, given the large size of the biome. Hence, the use of remote sensing facilitates the monitoring of the status and changes in land cover and land use at large scales. Most studies with remote sensing to monitor the differentiation of the vegetation types in Cerrado use optical sensors, mainly in the savanna and grassland formations where there is low signal saturation. These studies mostly use the Normalized Difference Vegetation Index (NDVI) [21–23], Spectral Linear Mixture Model (SLMM) [24,25], and phenological profiles [26]. Additionally, Müller et al. [27] demonstrated the challenges in mapping land use in Cerrado, essentially a result of its high diversity. The study showed a considerable uncertainty in the classification of cropland and pastures areas. The same problem was reported by Sano et al. [28], whose study reports a spectral similarity between cropland, pasture, and natural savanna vegetation, which can increase the uncertainty when mapping. Moreover, Ministry of the Environment (MMA) [29] mapped the land use in the whole Cerrado biome and the study showed that one of the biggest challenges for this area was to map the different types of vegetation, due to the strong seasonality of natural vegetation. However, the optical sensors can extract the information from the canopy, but arboreal vegetation types have differences in vertical structure and tree cover, so that with optical sensors, uncertainties in the identification of forest savanna vegetation types increase. Additionally, optical images are affected by weather conditions (cloud cover).

Radar sensors have an important advantage compared to optical sensors: the ability of radiation to penetrate through cloud cover and considerable parts of the canopy of trees/forest stands due to the higher wavelengths compared to optical sensors. Thus, the resulting radar signals (amplitude/backscatter and, if available, coherence) provide information that can be used to describe the vertical structure of vegetation stands. This information can be used to better estimate forest structure variables such as canopy cover, tree density, tree height or others, as well as to stratify vegetation (e.g., different types of forest). The longer the wavelength, the deeper radar Bands penetrate dense vegetation, which increases its sensibility to perceive the differences that improve discrimination of vegetation types. Almeida-Filho and Shimabukuro [30] demonstrated that the L Band from the JERS-1 synthetic-aperture radar (SAR) can be used to detect cover changes in forested and non-forested areas in the Cerrado biome. Evans and Costa [31] also mapped six vegetation habitats in Brazil using L and C Bands using the backscattering information from the surface. In the same country, Saatchi et al. [32] mapped five land cover types using the JERS-1 mosaic, using texture measurement. Santos et al. [33], Sano et al. [34], and Mesquita et al. [35] had satisfactory results using radar images to discriminate the vegetation types in the Cerrado biome, especially with the L Band. The sensitivity of the radar sensor to perceive the differences in vegetation structures makes it useful for mapping different types of forests.

The savanna vegetation has one of the largest forest diversities. In this case, the combination of different satellite images (optical and radar) and spatial resolution (low, medium, and high) may help to improve the quality of satellite based monitoring concepts [32]. However, there is little information about how the synergy of different data can contribute to map forest vegetation types in Cerrado. Yet, the free availability and the development of new optical and radar sensors, such as Sentinel 2A, Sentinel 1A (both free) or ALOS2 and TanDEM-X, are increasing the use of both sensors (radar and optical) for vegetation mapping. Recent studies have shown that the synergy of radar and optical images improved vegetation type discrimination, especially in the Cerrado biome, where the greenness seasonality had a huge influence during the year [36–38].

The aforementioned studies concentrate on parts of Cerrado where the vegetation cover is mostly homogeneous and that are not located in transitional areas such as the the Arc of Deforestation. However, most of the deforestation and expansion of agricultural and pasture areas are concentrated in this region. Additionally, these regions have a mixture of vegetation type and species from Cerrado and Amazon, which makes the study in this area more complex. The few studies in this area are related to the land use and not the mapping of vegetation type, as in Zaiatz et al. [39], who evaluated the spatial and temporal dynamics of land use and cover of the Upper Teles Pires River Basin from 1986 to 2014. In order to overcome the lack of studies using both sensors to discriminate vegetation types, the aim of this study is to evaluate the use of optical and radar remote sensing for mapping the different types of vegetation in the transitional area between the Cerrado and Amazon biomes.

2. Materials and Methods

2.1. Study Sites

The study area was the result of an overlap between the satellite images selected for this study and it is located around the city of Nova Mutum, Mato Grosso, which includes the Cerrado and Amazon biomes. Nova Mutum is located in the north of Mato Grosso, Brazil, and it is part of the Alto Teles Pires River Basin (Figure 1). Its climate is classified as Aw (after the Köppen climate classification), with a clear seasonality of rainy season (October to April) and dry season (May to September). The annual average temperature is 24 °C and annual precipitation is approximately 2200 mm [40]. The topography is flat with maximum slopes of 3%. The soils in the the area are Oxisols (80%) and Entisols (20%) [41].

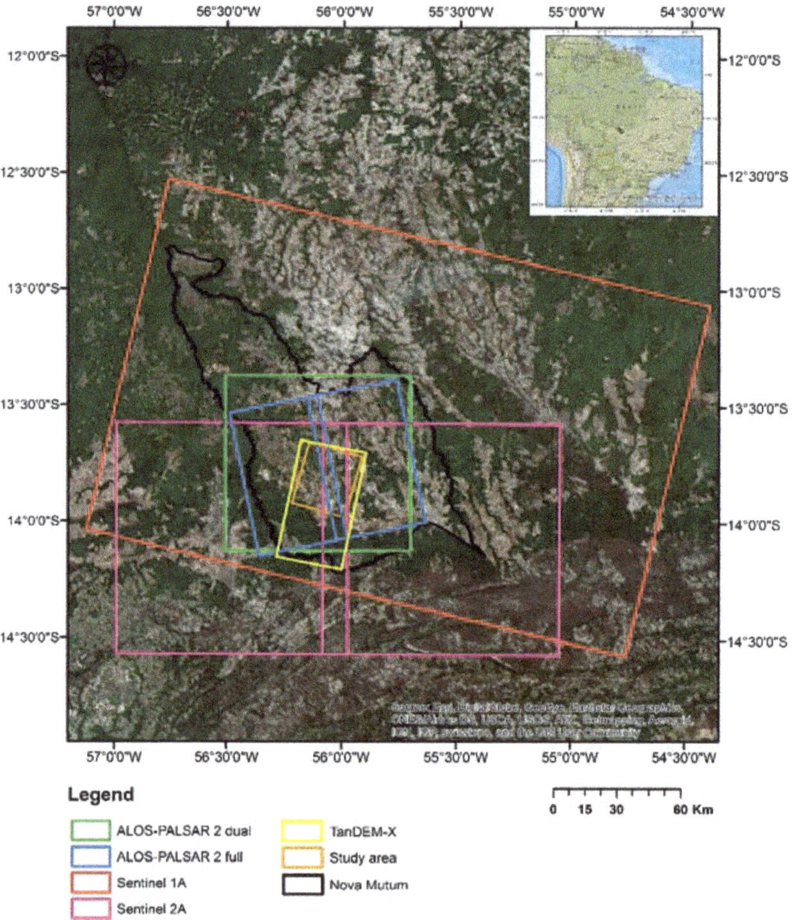

Figure 1. Location of the study area within the South America context. The scene footprint of different satellites are shown on top of a Google Earth image.

Nova Mutum is located in the AOD; this area covers 256 municipalities with the most intensive deforestation activities in an area of approximately 1,700,000 km^2 and it plays an important role in the context of deforestation in the frontier of Amazon and Cerrado. The AOD accounts for 75% of the deforestation in the Brazilian Amazon and the largest agricultural area [17]. Legislation, soil, relief, climate conditions, and the subsidies offered by the government have encouraged agricultural activity since 1970. Recently, the Brazilian government has established policies to decrease the rates of deforestation in these areas, such as the "Soy Moratorium" [42], which was an agreement with the major soybean traders not to purchase soybean that was planted in deforested areas after July 2006 in the Brazilian Amazon biome.

In general, the vegetation of Cerrado in Brazil covers three main different vegetation types: grassland, savannas, and forest formations. The forest formations consist of arboreal species in a continuous canopy and include the Gallery, Dry, and Open Forest. The savanna formation is characterized by a discontinuous herbaceous-shrub and tree canopy. The seven types of savanna formation are Dense Woodland, Woodland, Open Woodland, Park Woodland, Palm, Vereda, and Stone Woodland. The grassland formations include three vegetation types: Stone Grassland, Shrub Savanna, and Grassland. The first two types of grasslands are characterized by the large presence of shrubs with different types of soils. Figure 2 summarizes the distribution of the three different vegetation formations in the Cerrado biome. Each of these types has a high diversity, which is a consequence of the high variability of the soil and microclimates as well as the floristic evolution with plants from different Brazilian biomes [7].

Figure 2. Cerrado biome phytophysiognomies. The graphic depicts two vegetation formations and their subdivisions in the study area, except the Dry Forest. Source: Adapted from [43].

In our study, we mapped the four dominating vegetation types in Nova Mutum, cerradão (Open Forest), cerrado denso (Dense Woodland), gallery forest, and secondary forest. Cerradão and cerrado denso are located within the transition area of the Amazon and Cerrado biomes. As mentioned before, this area has a high deforestation rate, which can explain the presence of secondary forest. Cerradão has a crown cover between 50 and 90% and the height of the trees varies from 8 to 15 m. In general, the soils of Cerradão are well drained, deep, and have medium–low fertility. Cerrado denso has a crown cover between 5 to 70% and tree height varies from 5 to 8 m. The layers of shrubs and herbs are less dense compared to cerradão. In general, the soils of cerrado denso have medium to very clayey texture and are middle-well drained. The gallery forest has a crown cover between 70 to 95% and the height of the trees varies from 20 to 30 m. Secondary forests are formed after clear-cutting and have different structures, depending on the age of the succession. At the beginning of its succession time, these secondary forests are poor in biodiversity and have a simple structure, whereas in the next succession time, its structure depends on environmental factors such as soil, climate, and management [44].

2.2. Satellite Data

In order to analyze the use of optical and radar sensors to map the vegetation type in Cerrado, we used a set of images from four sensors (3 SARs and 1 optical). The PALSAR-2 aboard ALOS-2 from the Japan Aerospace Exploration Agency (JAXA); TanDEM-X (TerraSAR-X add-on for Digital Elevation Measurement) from the German Aerospace Center, DLR, and Astrium GmbH; Sentinel 1A and the optical Sentinel 2A from the European Union's Copernicus programme. Figure 3 shows the temporal coverage of each satellite image used in our analysis.

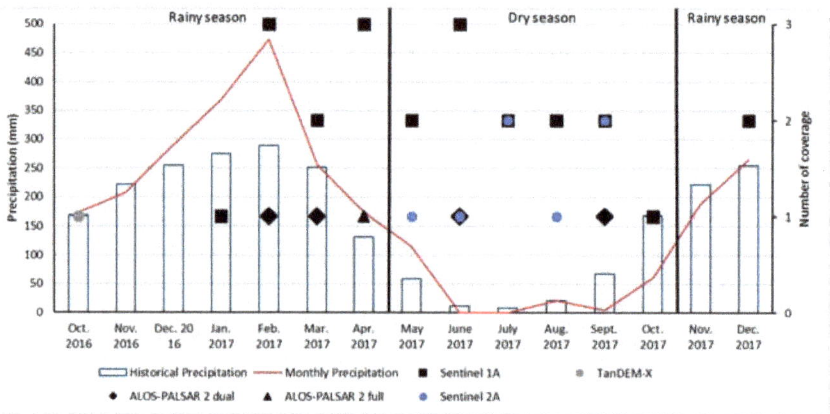

Figure 3. . Historical (1961–2017) and Monthly (2016–2017) precipitation values and acquisition dates of Sentinel 2A, ALOS PALSAR 2 full, ALOS PALSAR 2 dual, TanDEM-X and Sentinel 1A. Precipitation data were collected from the Diamatino fluviometric station located near to the study area.

We selected the satellite image data following some criteria. First, we selected images from 2017 since the field data were collected in 2017, except for the TanDEM-X image. Secondly, we selected the radar images on the dates of low precipitation, prior to the date of acquisition. Table 1 shows the date, polarization, orbit, and accumulated precipitation values three days before the acquisitions.

Table 1. Characteristics of the selected satellite images and the accumulated precipitation values three days before Radar acquisition.

Sensor	Date	Polarization	Orbit	Accumulated Precipitation Values 3 Days Before Radar Acquisition (mm)
	08.01.2017			3.3
	01.02.2017			11.5
	13.02.2017			58.3
	25.02.2017			30.2
	09.03.2017			27.2
	21.03.2017			24.0
	02.04.2017			16.8
	14.04.2017			0
	26.04.2017			33.3
	08.05.2017			0
	20.05.2017			0
Sentinel 1A	01.06.2017	Dual—VH and VV	Descending	0
	13.06.2017			0
	25.06.2017			0
	19.07.2017			0
	31.07.2017			0
	12.08.2017			0
	24.08.2017			0
	05.09.2017			0
	17.09.2017			0
	11.10.2017			0
	10.12.2017			88.7
	22.12.2017			0

Table 1. *Cont.*

Sensor	Date	Polarization	Orbit	Accumulated Precipitation Values 3 Days Before Radar Acquisition (mm)
TanDEM-X	20.10.2016	Single—HH		8.1
ALOS-PALSAR 2 dual	01.02.2017	Dual—HH and HV	Ascending	11.5
	26.03.2017		Descending	35.7
	04.06.2017			0
	13.09.2017		Ascending	0
ALOS-PALSAR 2 full	13.04.2017	Full	Ascending	0
	27.04.2017			0
Sentinel 2A	07.05.2017	-	-	0
	16.06.2017			0
	06.07.2017			0
	26.07.2017			0
	15.08.2017			19.2
	04.09.2017			0
	24.09.2017			0

2.3. Data Processing

2.3.1. Sentinel 2A

Seven coverages (using nine Bands altogether, from Band 2 to Band 8A, Band 11, and Band 12) of the Multispectral Instrument (MSI) on-board Sentinel-2A were processed using the ESA's Sentinel-2 toolbox in the ESA Sentinel Application Platform (SNAP). First, we applied atmospheric correction using Sen2cor, which is a L2A-processor for Sentinel-2 data that creates Bottom-Of-Atmosphere (BOA) reflectance images using Top-Of-Atmosphere (TOA) data [45]. Secondly, we resampled all the bands to a 10-m spatial resolution based on the geolocations obtained from Level-1C metadata. For the last step, we created a subset of our study area to speed up processing time, and lastly we mosaicked the images, as our study area was between two different orbits of the Sentinel 2A.

During the final step, we reduced the number of features by applying principal component analysis (PCA) on the spectral dataset due to the fact that some classification algorithms, such as Random Forest, cannot work well with high correlation data. Principal component analysis is a mathematical procedure that reduces a large amount of variables into principal components. The primary function of the PCA is to determine the extent of the correlation between multispectral bands and to remove it through an appropriate mathematical transformation [46]. We used the first principal component (PC1) of each one of the ten bands to aggregate only information that was essential to the classification process, as it explained most of the variance, e.g., PC1 of Band 2, PC1 of Band 3, and PC1 of Band 4. Overall, this resulted in a set of ten variables as input for classification for the dry season, and the dry and rainy season, respectively.

2.3.2. Sentinel 1A

Twenty-three coverages from Sentinel-1A IW Ground Range Detected (GRD) Level-1 product were processed using the ESA Sentinel Application Platform toolbox. First, each image was radiometrically calibrated to radar brightness values (β^0) [47]. Secondly, we applied the terrain flattening to correct any terrain variations in the images. Terrain flattering is an important step for the mapping of land use. Without the terrain flattening, an additional error into the coherency and covariance measurement could be created, due to the difference in the terrain and subsequently the brightness of the radar return [48]. During the third step, we coregistrated the 23 images based on the cross-correlation technique to guarantee that every pixel was correctly located in the same target of all images [49]. Once we had the images from the co-registration process, we separated them into two sub-processes.

During the first process, we applied the grey level co-occurrence matrix (GLCM) to extract second order statistical textures features. The GLCMs were extracted separately from every single date and polarization (VV and VH). These textures can be useful to improve land use classification in that it extracts intensity variations from the image involving the information of the neighbor pixels to identify specific clusters or objects [50]. Additionally, we applied the Refined Lee speckle filter (window size 5 × 5), after GLCMs extraction. This process is necessary to reduce the noise caused by random constructive and destructive interference from the radar signal [51]. For the second branch of processing, we only applied the Refined Lee speckle filter on the backscatter images. We applied the Range Doppler terrain correction in the last step. This process is necessary to geocode and correct the distortions in the image, which are caused by the topographical variations and the tilt of the sensor [52]. The entire process is illustrated in Figure 4.

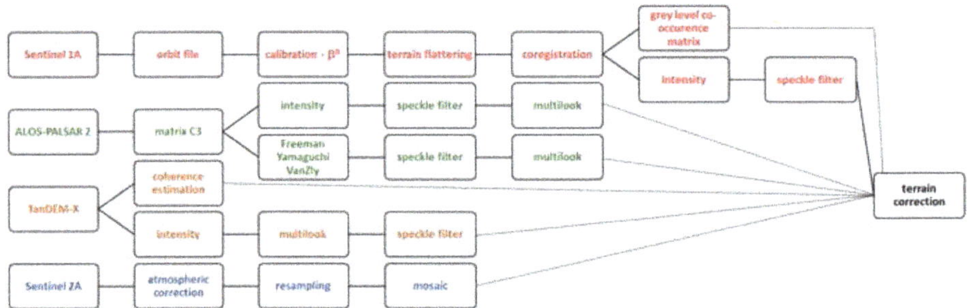

Figure 4. Flowchart of the proposal methodology.

We applied the PCA to reduce the number of features in the same way as explained above for Sentinel 2A. In the backscattering images, we applied the PCA for the two different seasons and two polarizations, which resulted in four PCs: (a) images of VV polarization in the rain and dry season; (b) images of VH polarization in the rain and dry season; (c) images of VV polarization during the dry season; and (d) images of VH polarization during the dry season. The same was applied for the texture images (ten textures: ASM, contrast, correlation, dissimilarity, energy, entropy, homogeneity, MAX, mean, variance. This resulted in a set of twenty two inputs in the dry season (one PC of VV, one PC of VH, ten PCs of VV textures, and ten PCs of VH textures) and a set of fourth four inputs in the rainy and dry season (one PC of VV in the dry season; one PC of VV in the rainy and dry season; one PC of VH in the dry season; one PC of VH in the rainy and dry season; ten PCs of VV textures in the dry season; ten PCs of VV in the rainy and dry season; ten PCs of VH textures in the dry season; ten PCs of VH in the rainy and dry season).

2.3.3. ALOS-PALSAR 2 (Dual and Full Polarimetric)

Four coverages of the dual polarization images were converted to covariance matrix C2 and one coverage of the full polarization images was converted to C3 matrix [53]. In this step, multilook with 4 looks in azimuth and 1 in range was applied to convert the image from single look complex to ground range detected. We applied the speckle filter on all images to reduce speckle noise. For that we used the Refined Lee adaptive filter (5 × 5 window), which is more efficient and whose results have less destructive averaging, having been largely used in the radar studies [51–54]. Here, we separated the images into two subprocesses. For the first, we kept the backscattering images. For the second, we calculated polarimetric decompositions. Polarimetric SAR decomposition is a useful method to map and discriminate the different targets on the surface, especially due to the signal of the target, which is a combination of speckle noise and random vector scattering effects [55]. In our study, we chose the Freeman–Durden, Yamaguchi, and VanZyl polarimetric decompositions. In general, these three decompositions are based on the covariance matrix that is divided into three scattering mechanism:

volume, double bounce and surface scatter [55]. Additionally, polarimetric compositions have been used before in mapping of vegetation showing the improvement in the vegetation classification in the Amazon and Cerrado [56]. We applied the Range Doppler terrain correction in all images.

Following the same process of the other images, we applied the PCA to reduce the number of features and consequently facilitated the further classification process. This resulted in a set of ten PCs, resulting in four of the dual polarimetric, PC1 of backscattering in each polarization (HH and VH) and each orbit (ascending and descending) and six of the full polarimetric: PC1 of backscattering in each polarization (HH, VV and VH) and PC1 of each scattering mechanism (volume, double bounce and surface scatter).

2.3.4. TanDEM-X

The TanDEM-X mission operates two X-Band satellites flying in close formation in order to acquire single-pass interferometric SAR data. The primary mission goal of the TanDEM-X mission was the generation of a global digital elevation model [57]. All TanDEM-X acquisitions are available to the science community on request in Coregistered Single look Slant range Complex (CoSSC) format. For the study area, we acquired one TAnDEM-X scene in standard (bistatic) mode with horizontal polarization (HH). The processed data was separated into two different parts. In the first part, we estimated the magnitude of coherence. Coherence describes the the degree of correlation between the two complex radar images [58]. It is a measure of quality of the phase measurement in interferometric SAR analysis and also used as a proxy for soil and vegetation structural parameters.

In the second part, we processed the intensity images. We performed multilook in both images (coherence and intensity) from the first and second part, with 4 looks in azimuth and 3 in range to reduce the noise. Additionally, we applied, as with the other images before, the speckle filter Refined Lee (window 5×5). For the last step, we applied the Range Doppler terrain correction. All images were processed to have a final product with spatial resolution of 10 m. The PCA was not applied for TanDEM-X due to the use of only one single date.

2.4. Classification

The process of image classification was separated into two steps. First, a forest mask was generated as a result of a forest/non-forest classification. At the second step, we classified the forest type within this forest mask. The area under investigation is covered by all images used in this study.

We used the RF algorithm implemented in R software for image classification. Random Forests is a supervised classification algorithm that uses multiples decision trees to get an accurate classification and prediction. The N numbers of trees are being built by the classifier, contributing each to the assignment of the most frequent class. This algorithm uses the bagging method to produce random samples of training sets for each random decision tree. Every tree uses a random subset from the original set. This original set was created from training samples, where two-thirds were used to train the classifier and one-third of them were used for validation. Two-thirds of the training sample were the out-of-bag (OOB) data and one-third of the training sample were the OOB error estimate [59]. Random Forest can be used for classification and regression and is an efficient tool due to measuring the relative importance of each feature. This variable importance measures the decrease of accuracy when a variable is removed from the classification. The higher a variable is ranked, the more it is contributing to the accuracy. Additionally, it has a lower probability to overfit compared to other models if there are enough trees. This method has many improvements: it does not require any input preparation, it is more stable using big data since it works well with variable non-linearity, it provides a pre-feature selection building the trees, and reduces the time required for the process. For remote sensing analysis, RF showed to be a stable and accurate algorithm, especially when it is applied to different types of sensors and large time series. The achievement of this method can be seen in recent studies such as References [59–61], which applied RF for vegetation mapping using different types of data. In this study, the RF models consisted of 1000 trees, and 70% of our samples were used for training the classifier and 30% for validation of the classification results.

2.4.1. Forest and Non-Forest

In order to classify the several vegetation types, we first needed to create a map of forest and non-forest areas. For accuracy assessment, we created 100 random points of 3.13 ha each and visually classified them using high-resolution imagery from Google Earth and the sensor Sentinel 2A (Figure 5). During the classification process, we used 70% of these points for training the classifier and 30% for validation of the classification results.

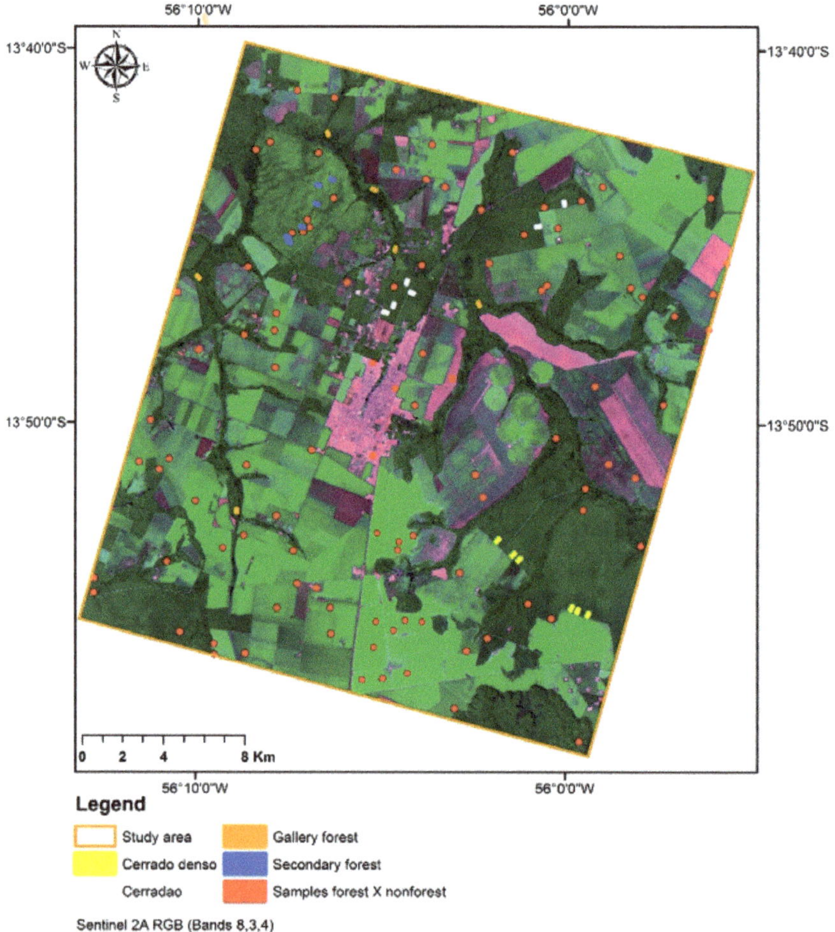

Figure 5. Distribution of Random Forest and non-forest samples and forest type samples (4 classes) in the study area, on top of a false color composition of Sentinel 2A Band 3, 4, and 8 (07 July 2017).

2.4.2. Forest Type

Forest-type mapping was only conducted in the areas masked as forests in the previous step. We created 24 reference areas equally distributed into four different vegetation classes (cerradão, cerrado denso, gallery, and secondary forest). Each one had an area of 14.265 ha (Figure 5). The polygons were classified based on field data collection (July 2017) and high-resolution imagery from

Google Earth and Sentinel 2A (26 July 2017). During the classification process using RF, we used 70% of the pixels in the 24 references areas for training and 30% for validation.

To analyze the synergy of optical and radar data for mapping Cerrado vegetation types, all possible combinations between optical and radar sensors were tested in two different scenarios, dry season, dry and rainy seasons (Table 2). In addition, we used the sensors separately and analyzed the SAR classifications. In total, 23 datasets were processed.

Table 2. Classification scheme (number in brackets shows the number of variables per input data set).

Dry Season (SAR and Optical)	Rainy + Dry Seasons (SAR and Optical)	Dry Season (SAR)	Rainy + Dry Seasons (SAR)
Sentinel 2A (10)	Sentinel 1A (44)	Sentinel 1A + TanDEM-X	Sentinel 1A + TanDEM-X
Sentinel 1A (22)	ALOS2 full (6)	Sentinel 1A + ALOS2 dual	Sentinel 1A + ALOS2 dual
TanDEM-X (3)	Sentinel 2A + Sentinel 1A	Sentinel 1A + ALOS2 full	Sentinel 1A + ALOS2 full
ALOS2 dual (4)	Sentinel 2A + ALOS2 full	TanDEM-X + ALOS2 full	TanDEM-X + ALOS2 full
Sentinel 2A + Sentinel 1A	All images	ALOS2 dual + ALOS2 full	ALOS2 dual + ALOS2 full
Sentinel 2A + TanDEM-X			
Sentinel 2A + ALOS2 dual			
All images			

For the classifications, which combined two or more sensors, e.g., Sentinel 2A and ALOS-PALSAR 2, we did not use all the features of each sensor. In this case, we selected the first three features based on variable importance, which was calculated during RF classification for the single sensor dataset, respectively. Variable importance shows the interaction between the variables/features and inserts them into an hierarchy within a level of contribution and importance for the classification.

For both classifications, forest/non-forest and vegetation type, we used the confusion matrix to analyze the performance of Random Forest classifications. The confusion matrix assesses the accuracy of the classification, showing the relation between classification result and sample site. Column values correspond to the sample site results, rows to the classification results, and diagonal to the correctly classified pixels. The general measurement showed in confusion matrices of q classes is the overall accuracy, which is a result of dividing the total number of pixels and the pixels that were correctly classified. Additionally, the kappa coefficient was largely used to measure the accuracy of the classification. The values of the kappa coefficient range from 0 to 1, where 0 means no relation between the classification results and the sample site results, and 1 means that both are identical [62].

Finally, for detailed analysis, we calculated both the user's and producer's accuracy. User's accuracy (U_i) is obtained considering the number of the correctly identified pixels of a given class (p_{ii}), divided by the total number of pixels of the class in the classified image ($p_{i.}$).

$$U_i = \frac{p_{ii}}{p_{i.}},\qquad(1)$$

On the other hand, producer's accuracy (P_j) is the number of correctly identified pixels (p_{jj}) divided by the total number of pixels in the reference image ($p_{.j}$). A detailed description of the classification assessment can be found in the literature [62,63].

$$P_j = \frac{p_{jj}}{p_{.j}},\qquad(2)$$

3. Results

3.1. Forest and Non-Forest

The two different combinations used for classifications, Sentinel 2A with ALOS-PALSAR 2 dual polarimetric and Sentinel 2A with ALOS-PALSAR 2 full polarimetric, showed similar high overall accuracy of 0.99 and 1, respectively. The variable importance showed similar results. In both cases, the

PC1 of Band 11 and Band 5 of the Sentinel 2A images had the highest contribution for the Random Forest classifier.

Based on this result, we created a mask of forest and non-forest areas, where 34% was forest and 66% was non-forest. (Figure 6). This mask was used in the next step for the forest type classification.

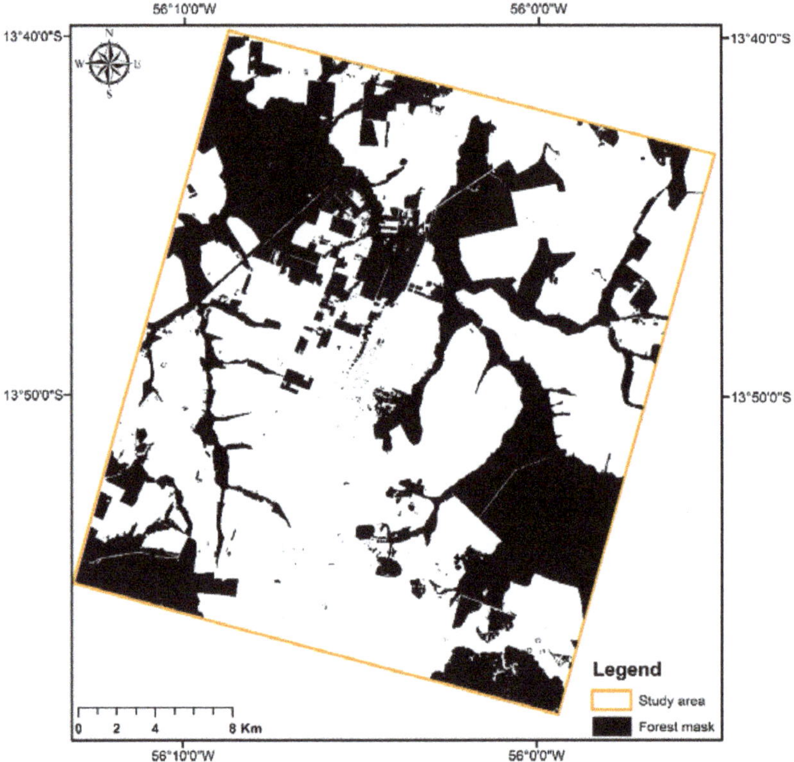

Figure 6. Forest mask extracted from the classification of Sentinel 2A with ALOS-PALSAR full polarimetric PC1 images.

3.2. Forest Type

3.2.1. Dry Season

The Table 3 shows the overall average accuracy (OAA), Kappa, confidence interval (CI) values, and variable importance of the classifications during the dry season.

Classifications using only a single radar sensor (Sentinel 1A, TanDEM-X and ALOS2 dual) had lower overall accuracy and kappa values compared to the classification that used two or more sensors. Sentinel 2A (S2) had with 82.60 % the highest overall accuracy and kappa values with 0.77. The variable importance shows the PC1 of Bands 11 and 12 were more important during the RF classification, followed by the PC1 of Bands 5, 4 and 2. Figure 7 shows the results of the S2 classification. A gradient is visible, with the north mostly comprising of areas of cerradão, which is closest to the Amazon biome, and whose south cerrado denso areas are prevailing. Additionally, it illustrates a large area of secondary forest in the northwest of the study area. Based on this map, Cerrado denso covers 34.50% of the Cerrado area, cerradão 28.70%, gallery forest 28.14% and secondary forest 8.66%.

Table 3. Overall accuracy, kappa, confidence interval 95%, overall average accuracy (OAA), and the three most important variables for the classifications based on the Random Forest variable importance for Sentinel 2A (S2), ALOS PALSAR 2 full (A2f), ALOS PALSAR 2 dual (A2d), TanDEM X (TX), and Sentinel 1A (S1). The parameters listed in the variable importance are the PC1 derived from the PCA, except for the images from TanDEM-X, as only one acquisition was available. Only data acquisitions during the dry season were considered.

Dry Season (SAR and Optical)	Overall Accuracy	Kappa	Conf. 95% OAA	Variable
Sentinel 2A	82.60%	0.77	2.29	Bands 11, 12, 5
Sentinel 1A	48.51%	0.31	2.76	Entropy VV, mean VV, variance VH
TanDEM-X	58.22%	0.44	2.78	Coherence, intensity master, intensity slave
ALOS2 dual	59.70%	0.46	2.73	HH descending, HV descending, HH ascending
Sentinel 2A + Sentinel 1A	79.90%	0.73	2.08	Bands 12, 11, 5 S2
Sentinel 2A + TanDEM-X	81.91%	0.76	2.14	Band 11 S2, Band 12 S2, coherence HH TX
Sentinel 2A + ALOS2 dual	74.91%	0.67	2.53	Band 11 S2, HV descending A2d, HH descending A2d
All images	80.33%	0.74	2.05	Band 11 S2, HV descending A2d, Band 12 S2

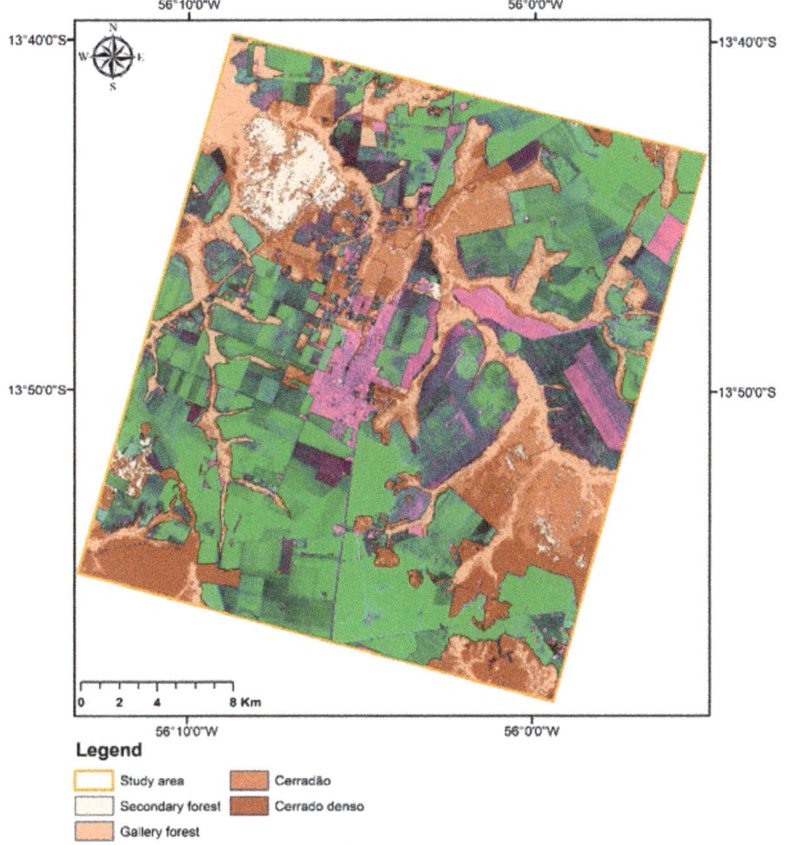

Figure 7. Classification results of the spatial distribution of the four Cerrado forest types using Sentinel 2A (dry season) on top of a false color composition of Sentinel 2A, Bands 3, 4, and 8 (07 July 2017).

The overall accuracy and kappa values of the Sentinel 1A (S1) classification had the lowest classification results using only single sensors with an overall accuracy of 48.51% and a kappa value of 0.31. Additionally, the PC1 of entropy and mean images of VV polarization and PC1 of variance image of VH polarization were more important to the RF classifier. The TanDEM-X classification also presented low accuracy and kappa values, 58.22% and 0.44, respectively. The coherence was more important than the intensity. The images from ALOS-PALSAR 2 dual and full polarimetric showed different results in the classification. In our study, the dual polarization images had a higher overall accuracy and kappa values, 59.70% and 0.46, respectively, compared to the full polarimetric images. However, we used four different dates of dual polarimetric images and one of full polarimetric image. This difference in the number of acquisitions from dual and full polarimetric images can cause a better accuracy for the dual polarization images.

The combinations of two or more sensors, in general, improved the extraction of the target's information, and consequently, the classification. The classification that used S2 and TanDEM-X showed the highest overall kappa values, 81.91% and 0.76. Variable importance shows the PC1 of Bands 11 and 12 of S2 and the coherence of TanDEM-X were more important to the RF classifier.

The S1 and S2 classifications had an overall accuracy and kappa value of 79.90% and 0.73. PC1 of Bands 11 and 12 of S2 and the PC1 of contrast of VH polarization of S1 had a high ranking in the variable importance. The classification that used all images from the dry season had a similar overall accuracy and kappa values compared to the S2 and TanDEM-X classification. The PC1 of Band 11, PC1 of ALOS-PALSAR 2 dual VH polarization descending orbit, and PC1 Band 12 images had the highest importance.

The highest accuracy for each of the four forest classes was obtained by different classification inputs, the highest producer's accuracy for cerrado denso class was achieved with S2 and S1 classification and the highest user's accuracy with the classification that used S2 images. The highest producer's accuracy for cerradão class was reached with the classification that used all images, and the user's accuracy was reached with the S2 and TanDEM-X classification. For the gallery forest, the highest producer's accuracy was obtained with the classification that used S1 images, and the user's accuracy was obtained using S2 images. The highest users' accuracy for secondary forest class was again reached with S2 images. The ALOS-PALSAR 2 dual polarimetric images resulted here in the best producer's accuracy (Figure 8).

3.2.2. Dry and Rainy Season

Table 4 summarizes the results for the classifications during the dry and rainy season. The classification of S1 images during the dry and rainy seasons had higher overall accuracy and kappa values compared to the S1 classification of the dry season, with 16% overall accuracy and a kappa of 33%. This result shows that the use of images combining the dry and rainy seasons improved the classification of S1 images. Here, the PC1 of entropy and of mean images of VV polarization as well as of the VH polarization contrast image were the three most important variables. The ALOS-PALSAR 2 full polarimetric classification showed a lower overall accuracy and kappa values compared to the ALOS-PALSAR 2 dual polarimetric classification during the dry season. Moreover, the volume polarimetric decomposition image was more important to the RF classifier.

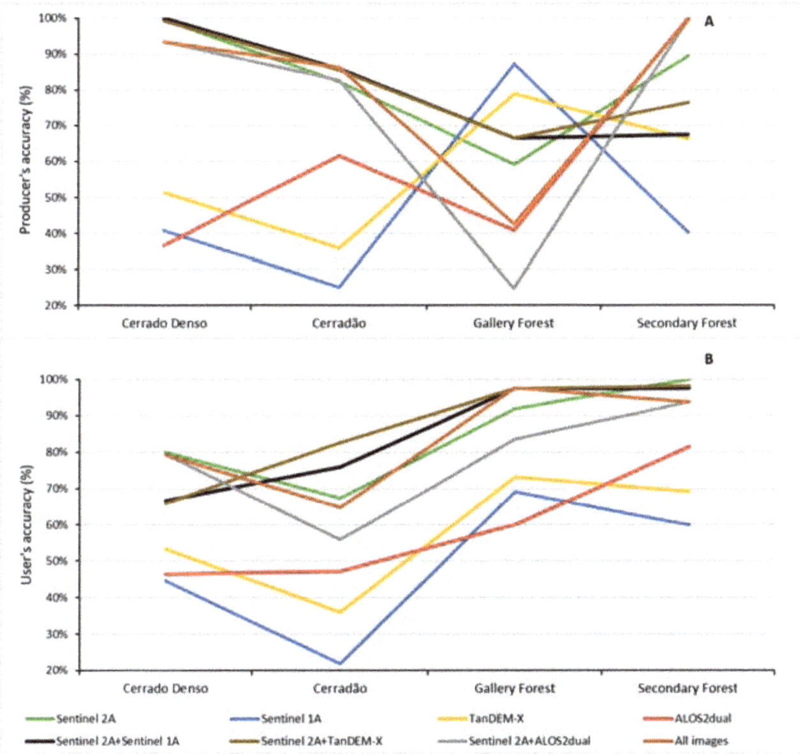

Figure 8. Producer's (**A**) and user's (**B**) accuracy of classifications based on single-sensor and optical/SAR-combinations for the four Cerrado types. Only data acquisitions during the dry season were considered.

Table 4. Overall accuracy, kappa values, confidence interval 95% OAA, the three most important variables for the classifications according to Random Forest variable importance for Sentinel 2A (S2), ALOS PALSAR 2 full (A2f), ALOS PALSAR 2 dual (A2d), TanDEM X (TX), and Sentinel 1A (S1). The parameters listed in the variable importance are the PC1 derived from the PCA, except for the images from TanDEM-X, as only one acquisition was available. All data acquisitions during the dry and rainy season were considered.

Rainy + Dry Season	Overall Accuracy	Kappa	Conf. 95% OAA	Variable Importance
Sentinel 1A	56.21%	0.42	2.87	Entropy VV, mean VV, contrast VH
ALOS2 full	41.26%	0.22	2.77	Volumetric, C22, C11
Sentinel 2A + Sentinel 1A	81.73%	0.76	2.05	Bands 12 S2, 11 S2, contract VH S1 rainy season
Sentinel 2A + ALOS2 full	79.98%	0.73	2.28	Bands 11, 12, 5 S2
All images	81.91%	0.76	1.97	Band 11 S2, HV descending A2d, Band 12 S2

For the dry and rainy season, the classifications that combined radar and optical sensors were more accurate. From each classification, which used more than on sensor, we selected the first three images with highest variable importance, totalling 15 images and used these images as input for all image classifications. This classification had the highest overall accuracy and kappa values (81.91% and 0.76) (Figure 9). The PC1 of Band 11 of S2, PC1 of ALOS-PALSAR 2 dual VH polarization at descending orbit and PC1 of Band 12 of S2 were the most important images that contributed to the classification of all images.

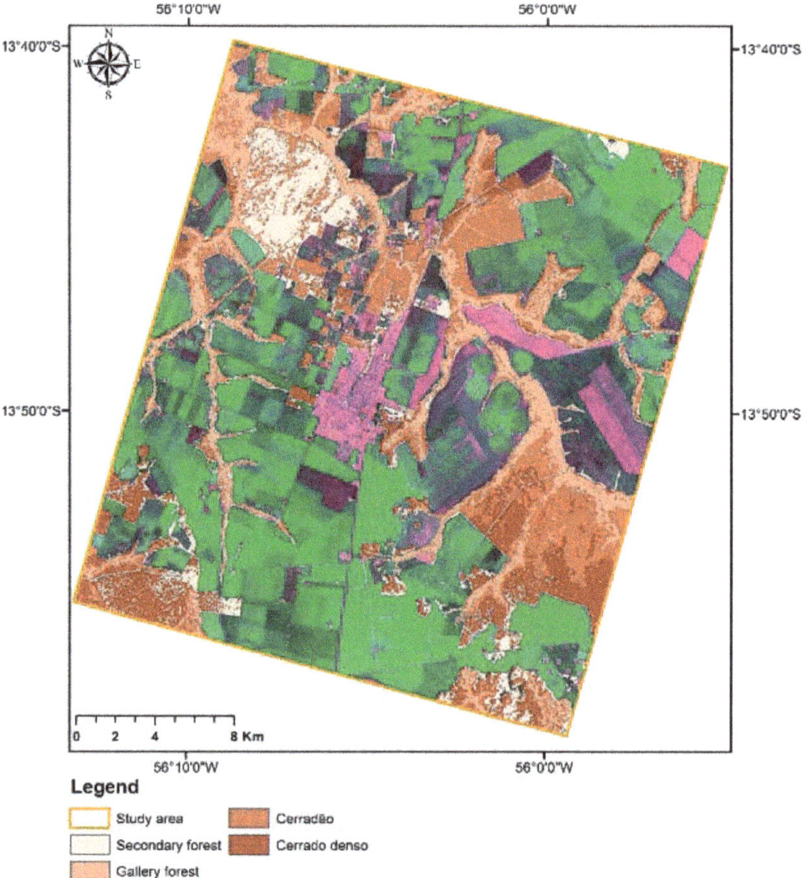

Figure 9. Classification of Cerrado forest type using all images from the dry and rainy season on top of a false color composition from Sentinel 2A, Bands 3, 4, and 8 (07 July 2017).

The S2 and S1 classifications showed a higher overall accuracy and kappa values, 81.73% 0.75, compared to the classification during the dry season. Variable importance showed that the PC1 of Bands 12 and 11 of S2 and the PC1 of contrast VH polarization were more important.

The highest producer's accuracy for the cerrado denso class was achieved with S2 and ALOS-PALSAR 2 full polarimetric classification, and the highest user's accuracy was achieved with the classification that used all images. The highest producer's and user's accuracy for cerradão class was reached with the classification that used S2 and S1. For the gallery and secondary forest, the highest user's accuracy was obtained using S2 and S1 images. The highest producer's accuracy for the gallery forest was achieved with S1 images and for the secondary forest class with all images (Figure 10).

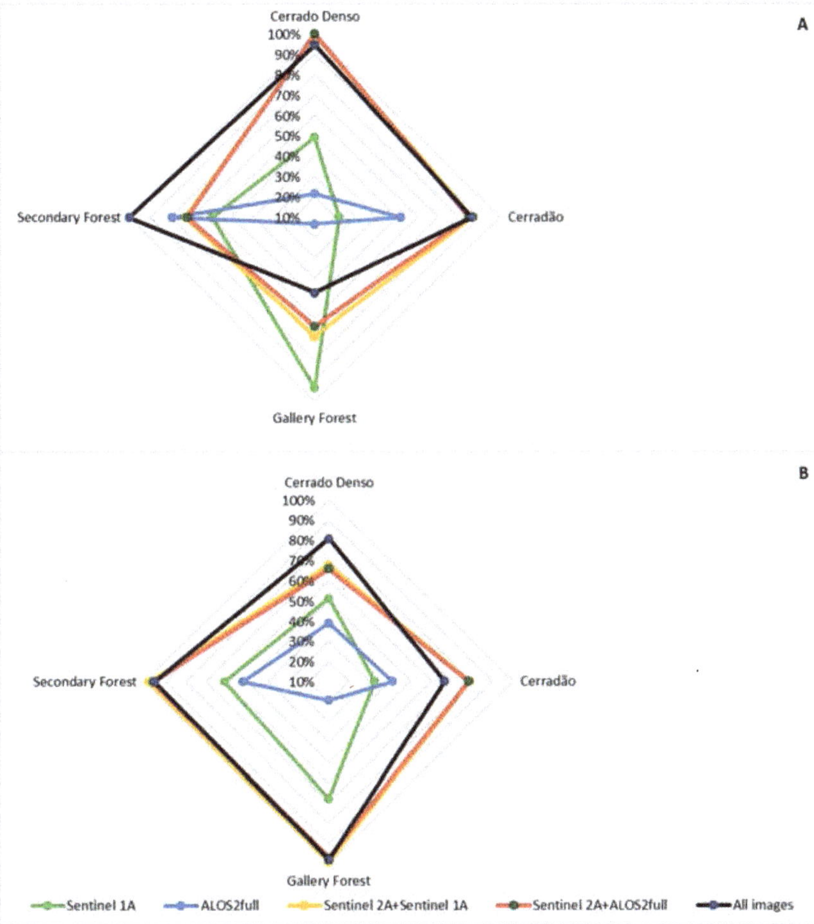

Figure 10. Producer's (**A**) and user's (**B**) accuracy of classifications based on single-sensor and optical/SAR-combinations for the four Cerrado types. All data acquisitions during the dry and rainy season were considered.

3.2.3. Radar Classification

We separately analyzed the radar classifications of Sentinel 1A, ALOS-PALSAR 2 dual/full polarimetric, and TanDEM-X (C Band, L Band, and X Band, respectively) for both seasons. Table 5 presents the results of these classifications during the dry season. The TanDEM-X in combination with ALOS-PALSAR 2 dual polarimetric classification achieved the highest overall accuracy and kappa values, 66.96% and 0.56. S1, and TanDEM-X had the lowest overall accuracy with 54.46% and 0.39. Furthermore, PC1 of ALOS-PALSAR 2 dual polarimetric VH descending orbit and HH descending orbit and coherence of TanDEM-X images were higher ranked in the variable importance list of Random Forests (Table 5).

Table 5. Overall accuracy, kappa values, confidence interval 95% OAA, and the three most important variables for the classifications based on the Random Forest variable importance for Sentinel 2A (S2), ALOS PALSAR 2 full (A2f), ALOS PALSAR 2 dual (A2d), TanDEM X (TX), and Sentinel 1A (S1). The parameters listed in the variable importance are the PC1 derived from the PCA except for the images from TanDEM-X, as only one acquisition was available.

Dry Season	Overall Accuracy	Kappa	Conf. 95% OAA	Variable Importance
Sentinel 1A + TanDEM-X	54.46%	0.39	2.66	Coherence HH TX, mean VV S1, entropy VV S1
Sentinel 1A + ALOS2 dual	62.33%	0.50	2.61	HV descending A2d, HH descending A2d, HV ascending A2d
TanDEM-X + ALOS2 dual	66.96%	0.56	2.61	HV descending A2d, HH descending A2d, coherence HH TX
Rainy + dry season				
Sentinel 1A + TanDEM-X	64.42%	0.53	0.67	Coherence HH TX, mean VV S1 dry season, entropy VV S1 dry season
Sentinel 1A + ALOS2 dual	66.61%	0.55	2.69	HV descending A2d, HH descending A2d, contrast VH S1 rainy season
Sentinel 1A + ALOS2 full	61.10%	0.48	2.78	Volumetric, entropy VV S1 dry seanson, mean VV S1 dry season
TanDEM-X + ALOS2 full	61.01%	0.48	2.81	Coherence HH TX, volumetric A2f, C22 A2f
ALOS2 dual + ALOS2 full	58.30%	0.44	2.81	HV descending A2d, HH descening A2d, HH ascending A2d

Combining the dry and rainy seasons, S1 and ALOS-PALSAR 2 dual polarimetric classification achieved the highest overall accuracy and kappa values, 66.61% and 0.55, respectively. Here, PC1 of ALOS-PALSAR 2 dual polarimetric VH descending orbit, HH descending orbit, and PC1 contrast of VH polarization images were more important. The ALOS-PALSAR 2 dual polarimetric and ALOS-PALSAR 2 full polarimetric classification showed the lowest overall accuracy and kappa values, 58.30% and 0.44, respectively.

Highest producer's and user's accuracy for cerrado denso and the cerradão class for the dry season were achieved with TanDEM-X and ALOS-PALSAR 2 dual polarimetric classification. This sensor combination also had the highest user's accuracy and producer's accuracy together with S1 and ALOS-PALSAR 2 dual polarimetric in the secondary forest. For the gallery forest, highest producer's and user's accuracies were achieved with S1 and TanDEM-X classification (Table 6). The radar sensors combinations presented a higher overall accuracy and kappa values compared to the single use of these sensors.

Table 6. Producer's and user's accuracy of classifications based only on combinations of SAR sensors for the four Cerrado types. Only data acquisitions during the dry season were considered.

Dry Season	Cerrado Denso	Cerradão	Gallery Forest	Secondary Forest
Producer's accuracy				
Sentinel 1A + TanDEM-X	40.21%	28.52%	94.46%	54.04%
Sentinel 1A + ALOS2 dual	38.11%	34.86%	76.12%	100.00%
TanDEM-X + ALOS2 dual	40.91%	59.86%	67.13%	100.00%
User's accuracy				
Sentinel 1A + TanDEM-X	50.88%	24.92%	73.19%	70.00%
Sentinel 1A + ALOS2 dual	53.96%	34.14%	70.97%	83.33%
TanDEM-X + ALOS2 dual	68.02%	47.22%	72.93%	82.37%

The dry and rainy season had similar results. Producer's and user's accuracy were for the gallery forest the highest using S1 and TanDEM-X, too. For cerrado denso, the best user's accuracy was achieved with S1 and TanDEM-X classification. Highest producer's accuracy was obtained by using S1 and ALOS-PALSAR 2 dual polarimetric images as input for the classification. This combination was also the best for the secondary forest, paired with ALOS-PALSAR 2 dual polarimetric and ALOS-PALSAR 2 full polarimetric. Here, TanDEM-X and ALOS-PALSAR 2 full polarimetric images reached the highest user's accuracies. The highest producer's and user's accuracy for cerradão class were achieved with ALOS-PALSAR 2 dual polarimetric and ALOS-PALSAR 2 full polarimetric classification (Table 7).

Table 7. Producer's and user's accuracy of classifications based only on combinations of SAR sensors for the four Cerrado types. All data acquisitions during the dry and rainy seasons were considered.

Rainy + Dry Seasons	Cerrado Denso	Cerradão	Gallery Forest	Secondary Forest
Producer's accuracy				
Sentinel 1A + TanDEM-X	53.85%	40.85%	94.81%	67.72%
Sentinel 1A + ALOS2 dual	55.24%	44.72%	66.44%	100.00%
Sentinel 1A + ALOS2 full	47.90%	36.97%	92.04%	67.02%
TanDEM-X + ALOS2 full	43.36%	42.61%	78.89%	78.95%
ALOS2 dual + ALOS2 full	41.96%	51.41%	40.14%	100.00%
User's accuracy				
Sentinel 1A + TanDEM-X	71.96%	37.54%	76.75%	73.11%
Sentinel 1A + ALOS2 dual	65.29%	44.88%	76.19%	77.66%
Sentinel 1A + ALOS2 full	59.57%	40.70%	68.56%	71.27%
TanDEM-X + ALOS2 full	56.36%	39.41%	67.66%	80.36%
ALOS2 dual + ALOS2 full	45.63%	46.79%	58.29%	77.03%

Furthermore, the polarization of radar sensors is shown to be an important factor for the Random Forest classification. The intensity of cross-polarized HV polarization PC1 images were one of the most important variables in 60% of the classification, which used radar sensors.

3.2.4. Summary of the Classification

The three highest overall accuracies and kappa values belonged to S2, S2 with TanDEM-X, and to the combinations of all images for the dry and rainy seasons. Nevertheless, the range of confidence interval shows different results compared to the overall accuracy and kappa values. The three narrowest ranges, which indicate good precision, belong to all images of the dry and rainy season, all images of the dry season and S2 with S1 from the dry and rainy season classifications (Table 5).

The variable importance for the classifications that combined optical and radar images showed that PC1 of Bands 11, 12, and 5 from S2, PC1 of ALOS-PALSAR 2 dual polarimetric VH descending orbit, PC1 of ALOS-PALSAR 2 dual polarimetric HH descending orbit, coherence of TanDEM-X, and the PC1 of contrast VH from the rainy season of Sentinel 1A images were the most important variables during the Random Forest classification.

4. Discussion

The results showed the importance of integrating satellite images from different sensors to classify the forest and non-forest area. The Program for the Estimation of Amazon Deforestation (PRODES) is the most important project that has been conducting satellite monitoring of deforestation in the Legal Amazon, producing annual deforestation rates in the region, using Landsat images (30 m spatial resolution). Comparing the data of forest areas from the PRODES project with the results of our work, it is possible to verify a high underestimation in the forest areas, mainly in the classes gallery forest and cerrado denso. The PRODES estimated an area of 12,702 ha of forest, and our work estimated an area of 27,326 ha. This difference can be associated to the different spatial resolution used in PRODES (30 m) and in our study (10 m).

Optical images are largely used to map vegetation types in the Cerrado biome. In our results, S2 classifications showed the highest overall accuracy and kappa values. The application of S2 images to map vegetation types in the Cerrado biome is new. In general, Landsat is the most common sensor used to discriminate vegetation types in the Cerrado. Nascimento and Sano [23] had 85% overall accuracy for mapping vegetation types in this biome. The authors used Landsat 7 ETM+ images to discriminate the Rupestrian Cerrado (Savanna formation) in the Chapada dos Veadeiros National Park in Goias

State, which can be difficult due to the spectral confusion with other types of Cerrado vegetation. The optical bands located in the red and NIR wavelengths showed high importance and contribution to the discrimination of vegetation type, as was visible in our results (Tables 3 and 4). Nascimento and Sano (2010) [23] agree on the importance of VIS and NIR regions for characterizing forest areas, as the vegetation has higher reflectance in this wavelength range and is thus more sensitive. Additionally, the number of optical images in ours and other studies helps the increase of discrimination power of different vegetation types, due to the unique spectral signatures of the plant during the year [64,65]. The optical data are certainly useful to map the vegetation type in Cerrado; however, these images are usually not available during the rainy season and the optical data cannot extract information from the structure of the forest [66]. Moreover, the availability of images in the rainy season would allow for a higher temporal resolution, which is crucial to better discriminate the vegetation types in the Cerrado biome due its high seasonality. Additionally, in dense areas of vegetation, the optical sensor is usually saturated due to the low optical depth penetration through these areas, affecting the mapping of the various vegetation types. There are important projects assessing the land use of the Cerrado biome, such as the TerraClass Cerrado project, which produced a map of the land use of the Cerrado biome. However, the project had great difficulties to discriminate the different types of vegetation, which is important for the preservation of biodiversity in this region. Nevertheless, TerraClass presents another step in the challenge of mapping the different types of vegetation in the Cerrado [29].

The use of radar images can be a solution to overcome the lack of image availability in the rainy season and the high saturation of optical images in areas of great biomass density. In our radar, classification results from the dry and rainy seasons, TanDEM-X (X Band) and ALOS-PALSAR 2 (L Band) dual polarimetric classification from the dry season showed the highest overall accuracy and kappa values. The influence of vegetation scattering mechanism dependencies is strongly dependent on the wavelength and polarization of the sensor. In the short/intermediate wavelengths, such as X and C Bands, backscattering represents the radiation interaction of canopy, leaves, branches, secondary branches, and part of volumetric scattering (inside crown). Longer wavelengths, such as the L and P Bands, have the capability for deeper penetration. Bigger vegetation components such as trunks, crown, ground, and branches interact with these lower wavelengths. According to the results for the dry season, L Band dual polarimetric images had the highest overall accuracy and kappa values were comparable to the classifications that used single sensor (X and C Bands). The study area is mostly forested. In these areas, radar signals are more likely to be saturated in the X and C Bands compared to the L Bands [67]. The polarization controls the types of components that interact with the radiation. In our study, the L Band cross-polarized HV polarization was the most important variable that contributed to the random classifier in the best classification. This agrees with the fact that cross-polarized images have direct relation with volumetric scattering, and are therefore sensitive to forest structure [68]. There are few studies in the Cerrado biome using only radar images. Sano et al. [34] used the L Band from JERS-1 SAR data to map the different types of vegetation by analyzing the backscattering coefficient values. The study could well separate the grassland, mixed grass/shrub/woodland, and woodland in the state of Distrito Federal.

The results of the CI 95% OAA showed the importance of the fusion between optical and radar data to map vegetation type in the Cerrado biome, since the confidence interval with the narrowest range belonged to the classification that used all images from the dry and rainy seasons, where the narrower the interval, the more accurate the classification. The Cerrado vegetation has one of the largest forest diversities, consequently the combination of different sensors (optical and radar) and spatial resolution (low, medium, and high) results in a great improvement in the accuracy [32]. Of the three classifications that obtained the highest values of accuracy and kappa, two used radar and optical images. This showed the importance of the integration of different sensors in improving the mapping of forest types in Cerrado. A similar result was reported by Sano et al. [38], who combined optical and radar images to improve the classification of different vegetation types in the Cerrado biome. The study had a high overall classification accuracy, which used both sensors in regions of savanna and grasslands formations. Sano et al. [38] used data from the dry and rainy seasons and showed the importance of the time series in improving the classification of different types of vegetation. Additionally, Sano et al. [38] showed better performance of radar data (JERS-1 SAR) compared to optical data (Landsat). In contrast, our results showed that optical data performed better for classification, compared to the radar data. However, this study used a higher number of radar images using L Band compared to our study, which increased the efficiency of mapping vegetation, due to the sensitivity to identify the various structures of the forest, consequently better distinguishing the type of forest, as reported by Lucas et al. [69], Garestier [70], and Santoro [71]. Carvalho et al. [37] used images from ALOS-PALSAR and Landsat to map the different types of vegetation and the results agree on our findings. The highest overall accuracy and kappa values were from the S2 classification; therefore, in our results, the use of radar images did not reach the highest accuracy and kappa values. Carvalho et al. [37] showed that the use of radar data did not improve classification accuracy; however, the study used only one data from radar imaging. Concerning GLCM textures, the same study showed similar results. Grey Level Co-occurrence Matrix textures images had a high variable importance during the Random Forest classification, in particular for entropy, which showed the disorder of GLCM elements. This may be related to the differences in the backscattering of the vegetation type classes.

Regarding the user's accuracy, the secondary forest was better classified using optical images, whereas the other three classes were better classified using optical and radar images. The optical bands were the most important variables for the RF classifier. The texture images were the second most important ones. Several authors presented similar results achieved in this study [62,72,73]. All mentioned studies showed an improvement in the separability of land cover types employing texture images. The coherence image from TanDEM-X was the third most important variable. Schlund et al. [72] and Baron and Erasmi [62] showed an improvement in the discrimination of forest against other classes using coherence as well.

Other studies about classification of vegetation type in the Cerrado biome, such as Mesquita et al. [35], were in regions where the vegetation has a smaller gradient compared to regions within the Arc of Deforestation, such as Distrito Federal, Minas Gerais, and São Paulo. The IBGE and the MMA mapped vegetation types from the whole Cerrado biome. The studies used Landsat images from the year 2004 and scaling of 1:250,000, which is not enough to detect the gradients of the Cerrado biome. The mapping of vegetation types in transition zones is still a challenge, due to these not having a clear border [74]. However, these regions play an important role in the conservation of the Amazon and Cerrado biome, wherein 75% of the deforestation in Amazon occurs.

5. Conclusions

In this paper, we evaluated the use of optical and radar remote sensing for mapping different types of vegetation in the transitional area between the Cerrado and Amazon biomes. The method described in this study improved the mapping of vegetation type in the Arc of Deforestation in the Cerrado biome and can be applied to create accurate vegetation type maps. We evaluated the use of four different sensors, one optical sensor (Sentinel 2) and three radar sensors (Sentinel 1, ALOS, TanDEM-X), for better vegetation type identification and area discrimination, so that these can be used for better calculations of biomass loss and carbon storage in the high dynamic Arc of Deforestation in Brazil.

When applying a supervised random forest classification, the highest overall accuracy and kappa coefficient were obtained using only the Sentinel 2A for classification. However, of the three classifications that obtained the highest overall accuracy and kappa values, two used radar and optical images. Bands 5, 11, and 12 of Sentinel 2A, texture images from Sentinel 1A cross-polarization, and coherence of TanDEM-X were the most important images in order to separate each class, as calculated by the random forest variable importance. The combination of optical and radar sensor data usually improves the vegetation classification. Nevertheless, in our study, the single use of optical sensors was sufficient to discriminate the four forest classes in the study area: cerradão (Open Forest), cerrado denso (Dense Woodland), gallery forest, and secondary forest classes in a highly fragmented complex vegetation biome. Such information is relevant for the upcoming mapping of vegetation types in the endangered Cerrado/Amazon ecotone.

Author Contributions: Conceptualization, F.d.S.M., G.G. and S.E.; methodology, F.d.S.M. and S.E.; validation, F.d.S.M., S.E. and G.G.; formal analysis, F.d.S.M. and D.B.; investigation, F.d.S.M.; writing—original draft preparation, F.d.S.M.; writing—review and editing, F.d.S.M., D.B., G.G., V.L. and S.E.; visualization, F.d.S.M., D.B., G.G., V.L. and S.E.

Funding: This research was conducted during a scholarship financed by CAPES—Brazilian Federal Agency for Support and Evaluation of Graduate Education within the Ministry of Education of Brazil (99999.001387/2015-04). The funding of fieldwork was provided from Geo-Gender-Chancenfonds, and Georg-August University School of Science (GAUSS). V.L. was supported by FAPESC (2017TR1762) and CNPq (436863/2018-9; 313887/2018-7).

Acknowledgments: The authors would like to thank the Earth Observation Center (DLR) for providing TanDEM-X data. The Japan Aerospace Exploration Agency (JAXA) for providing ALOS/PALSAR-data, which were obtained under the 4th ALOS Research Announcement (RA, Process 1090), the European Space Agency (ESA) for providing free access to Sentinel 1A and Sentinel 2A data, and the Instituto Nacional de Meteorologia (INMET) for proving free access to precipitation data.

Conflicts of Interest: The authors declare no conflict of interest. The funders had no role in the design of the study; in the collection, analyses, or interpretation of data; in the writing of the manuscript, or in the decision to publish the results.

References

1. Myers, N.; Mittermeier, R.A.; Mittermeier, C.G.; da Fonseca, G.A.; Kent, J. Biodiversity hotspots for conservation priorities. *Nature* **2000**, *403*, 853. [CrossRef]
2. Klink, C.A.; Machado, R.B. A conservação do Cerrado brasileiro. *Megadiversidade* **2015**, *1*, 147–155.
3. IBGE. Mapa de Biomas do Brasil, Primeira Aproximação. 2004. Available online: https://ww2.ibge.gov.br/home/presidencia/noticias/21052004biomashtml.shtm (accessed on 10 December 2018).
4. Castro, A.A.J.F. Comparação florística de espécies do cerrado. *Silvicultura* **1994**, *15*, 16–18.
5. EMBRAPA. *Sistema Brasileiro de Classificação de Solos*; Embrapa: Brasilia, Brazil, 1999.
6. Sano, E.E.; Ferreira, L.G.; Asner, G.P.; Steinke, E.T. Spatial and temporal probabilities of obtaining cloud-free Landsat images over the Brazilian tropical savanna. *Int. J. Remote Sens.* **2007**, *28*, 2739–2752. [CrossRef]
7. Walter, B.H. Fitofisionomias do Bioma Cerrado: Síntese Terminológica e Relações Florísticas. Ph.D. Thesis, Universidade de Brasília, Brasília, Brazil, 2006.
8. Fearnside, P.M. Deforestation in Brazilian Amazonia: History, Rates, and Consequences. *Conserv. Biol.* **2005**, *19*, 680–688. [CrossRef]
9. Fearnside, P.M.; Righi, C.A.; Graça, P.M.L.D.A.; Keizer, E.W.H.; Cerri, C.C.; Nogueira, E.M.; Barbosa, R.I. Biomass and greenhouse-gas emissions from land-use change in Brazil's Amazonian "arc of deforestation": The states of Mato Grosso and Rondônia. *For. Ecol. Manag.* **2009**, *258*, 1968–1978. [CrossRef]
10. Felfili, M. Proposição de Critérios Florísticos, Estruturais e de Produção para o Manejo do Cerrado Sensu Stricto do Brasil Central. Ph.D. Thesis, Universidade de Brasília, Brasília, Brazil, 2008.
11. Silva Junior, C.; Aragão, L.; Fonseca, M.; Almeida, C.; Vedovato, L.; Anderson, L. Deforestation-Induced Fragmentation Increases Forest Fire Occurrence in Central Brazilian Amazonia. *Forests* **2018**, *9*, 305. [CrossRef]
12. Ramankutty, N.; Foley, J.A.; Olejniczak, N.J. People on the Land: Changes in Global Population and Croplands during the 20th Century. *AMBIO J. Hum. Environ.* **2002**, *31*, 251–257. [CrossRef]
13. Machado, R.B.; Ramos Neto, M.B.; Pereira, P.G.P.; Caldas, E.F.; Gonçalves, D.A.; Santos, N.S.; Tabor, K.; Steininger, M. *Estimativas de Perda da Área do Cerrado Brasileiro*; Relatório técnico; Conservação Internacional: Brasilia, Brazil, 2004.
14. Janssen, T.A.J.; Ametsitsi, G.K.D.; Collins, M.; Adu-Bredu, S.; Oliveras, I.; Mitchard, E.T.A.; Veenendaal, E.M. Extending the baseline of tropical dry forest loss in Ghana (1984–2015) reveals drivers of major deforestation inside a protected area. *Biol. Conserv.* **2018**, *218*, 163–172. [CrossRef]
15. Hirota, M.; Nobre, C.; Oyama, M.D.; Bustamante, M.M.C. The climatic sensitivity of the forest, savanna and forest-savanna transition in tropical South America. *New Phytol.* **2010**, *187*, 707–719. [CrossRef]

16. Macedo, M.N.; DeFries, R.S.; Morton, D.C.; Stickler, C.M.; Galford, G.L.; Shimabukuro, Y.E. Decoupling of deforestation and soy production in the southern Amazon during the late 2000s. *Proc. Natl. Acad. Sci. USA* **2012**, *109*, 1341–1346. [CrossRef]

17. INPE. Program for the Estimation of Amazon Deforestation (PRODES). 2011. Available online: http://www.dpi.inpe.br/prodesdigital/prodesmunicipal.php (accessed on 15 January 2019).

18. Soares-Filho, B.; Rajão, R.; Macedo, M.; Carneiro, A.; Costa, W.; Coe, M.; Rodrigues, H.; Alencar, A. Land use. Cracking Brazil's Forest Code. *Science* **2014**, *344*, 363–364. [CrossRef] [PubMed]

19. Marques, E.Q.; Marimon-Junior, B.H.; Marimon, B.S.; Matricardi, E.A.T.; Mews, H.A.; Colli, G.R. Redefining the Cerrado–Amazonia transition: Implications for conservation. *Biodivers. Conserv.* **2019**, 1–17. [CrossRef]

20. Nogueira, E.M.; Fearnside, P.M.; Nelson, B.W.; França, M.B. Wood density in forests of Brazil's 'arc of deforestation': Implications for biomass and flux of carbon from land-use change in Amazonia. *For. Ecol. Manag.* **2007**, *248*, 119–135. [CrossRef]

21. Bitencourt, M.D.; de Mesquita, H.N., Jr.; Mantovani, W.; Batalha, M.A.; Pivello, V.R. Identificação de fisionomias de cerrado com imagem índice de vegetação. In *Contribuição ao Conhecimento Ecológico do Cerrado*; Editora Universidade de Brasília: Brasilia, Brazil, 1997; pp. 316–320.

22. Liesenberg, V.; Galvão, L.S.; Ponzoni, F.J. Variations in reflectance with seasonality and viewing geometry: Implications for classification of Brazilian savanna physiognomies with MISR/Terra data. *Remote Sens. Environ.* **2007**, *107*, 276–286. [CrossRef]

23. Nascimento, E.R.P.; Sano, E.E. Identificação de Cerrado Rupestre por Meio de Imagens Multitemporais do Landsat: Proposta Metodológica. *Soc. Nat.* **2010**, *22*, 93–106. [CrossRef]

24. Ferreira, M.E.; Ferreira, L.G.; Sano, E.E.; Shimabukuro, Y.E. Spectral linear mixture modelling approaches for land cover mapping of tropical savanna areas in Brazil. *Int. J. Remote Sens.* **2007**, *28*, 413–429. [CrossRef]

25. Girolamo Neto, C.D.; Fonseca, L.M.G.; Körting, T.S. Assessment of texture features for Brazilian savanna classification: A case study in Brasília national park. *Braz. J. Cartogr.* **2017**, *69*, 891–901.

26. Schwieder, M.; Leitão, P.J.; da Cunha Bustamante, M.M.; Ferreira, L.G.; Rabe, A.; Hostert, P. Mapping Brazilian savanna vegetation gradients with Landsat time series. *Int. J. Appl. Earth Observ. Geoinf.* **2016**, *52*, 361–370. [CrossRef]

27. Müller, H.; Rufin, P.; Griffiths, P.; Barros Siqueira, A.J.; Hostert, P. Mining dense Landsat time series for separating cropland and pasture in a heterogeneous Brazilian savanna landscape. *Remote Sens. Environ.* **2015**, *156*, 490–499. [CrossRef]

28. Sano, E.E.; Rosa, R.; Brito, J.L.S.; Ferreira, L.G. Land cover mapping of the tropical savanna region in Brazil. *Environ. Monit. Assess.* **2010**, *166*, 113–124. [CrossRef]

29. MMA. Mapeamento do Uso e Cobertura da Terra do Cerrado. Projeto TerraClass Cerrado 2013; Brasília. 2015. Available online: http://www.mma.gov.br/images/arquivo/80049/Cerrado/publicacoes/Livro%20EMBRAPA-WEB-1-TerraClass%20Cerrado.pdf (accessed on 20 November 2018).

30. Almeida-Filho, R.; Shimabukuro, Y.E. Detecting areas disturbed by gold mining activities through JERS-1 SAR images, Roraima State, Brazilian Amazon. *Int. J. Remote Sens.* **2010**, *21*, 3357–3362. [CrossRef]

31. Evans, T.L.; Costa, M. Landcover classification of the Lower Nhecolândia subregion of the Brazilian Pantanal Wetlands using ALOS/PALSAR, RADARSAT-2 and ENVISAT/ASAR imagery. *Remote Sens. Environ.* **2013**, *128*, 118–137. [CrossRef]

32. Saatchi, S.S.; Nelson, B.; Podest, E.; Holt, J. Mapping land cover types in the Amazon Basin using 1 km JERS-1 mosaic. *Int. J. Remote Sens.* **2010**, *21*, 1201–1234. [CrossRef]

33. Santos, J.R.; Xaud, M.R.; Pardi Lacruz, M.S. Analysis of the backscattering signals of JERS-1 image from savanna and tropical rainforest biomass in Brazilian Amazonia. In Proceedings of the International Symposium on Resource and Environmental Monitoring, Budapest, Hungary, 1–4 September 1998.

34. Sano, E.E.; Pinheiro, G.C.C.; Meneses, P.R. Assessing JERS-1 synthetic aperture radar data for vegetation mapping in the Brazilian savanna. *J. Remote Sens. Soc. Jpn.* **2001**, *21*, 158–167.

35. Mesquita JR, H.N.; Bittencourt, M.D.; Kuntschik, G. Gradient analysis of cerrado vegetation physiognomies by SAR image processing (JERS-1). In Proceedings of the X Simpósio Brasileiro de Sensoriamento Remoto, Foz do Iguacú, Brasil, 21–26 Abril 2001.

36. Bitencourt, M.D.; de Mesquita, J.H.N.; Kuntschik, G.; da Rocha, H.R.; Furley, P.A. Cerrado vegetation study using optical and radar remote sensing: Two Brazilian case studies. *Can. J. Remote Sens.* **2007**, *33*, 468–480. [CrossRef]

37. Carvalho, L.; Rahman, M.; Hay, G.; Yackel, J. Optical and SAR imagery for mapping vegetation gradientsin Brazilian savannas: Synergy between pixel-based andobject-based approaches. In Proceedings of the International Conference of Geographic Object-Based Image, Ghent, Belgium, 29 June–2 July 2010.

38. Sano, E.E.; Ferreira, L.G.; Huete, A.R. Synthetic Aperture Radar (L band) and Optical Vegetation Indices for Discriminating the Brazilian Savanna Physiognomies: A Comparative Analysis. *Earth Interact.* **2005**, *9*, 1–15. [CrossRef]

39. Zaiatz, A.P.S.R.; Zolin, C.A.; Vendrusculo, L.G.; Lopes, T.R.; Paulino, J. Agricultural land use and cover change in the Cerrado/Amazon ecotone: A case study of the upper Teles Pires River basin. *Acta Amaz.* **2018**, *48*, 168–177. [CrossRef]

40. Goedert, W.J. *Solos dos Cerrados. Tecnologias e Estratégias de Manejo*; EMBRAPA, Centro de Pesquisa Agropecuária dos Cerrados; Nobel: Brasília, Brasil, 1986.

41. IBGE. Mapa de Solos do Brasil. 2001. Available online: ftp://geoftp.ibge.gov.br/informacoes_ambientais/pedologia/mapas/brasil/solos.pdf (accessed on 11 November 2018).

42. Rudorff, B.F.T.; Adami, M.; Risso, J.; de Aguiar, D.A.; Pires, B.; Amaral, D.; Fabiani, L.; Cecarelli, I. Remote Sensing Images to Detect Soy Plantations in the Amazon Biome—The Soy Moratorium Initiative. *Sustainability* **2012**, *4*, 1074–1088. [CrossRef]

43. Ribeiro, J.F.; Walter, B.M.T. As principais fitofisionomias do Bioma Cerrado. In *Cerrado: Ecologia e Flora*; Sano, S.M., Almeida, S.P., Ribeiro, J.F., Eds.; Embrapa Cerrados: Brasília, Brazil, 2008; pp. 151–212.

44. Ferreira, R.L.C. Estrutura e Dinâmica de uma Floresta Secundária de Transição, Rio Vermelho e Serra Azul de Minas, MG. Ph.D. Thesis, Universidade Federal de Viçosa, Viçosa, Brasil, 1997.

45. Louis, J.; Debaecker, V.; Pflug, B.; Main-Knorn, M.; Bieniarz, J.; Mueller-Wilm, U.; Cadau, E.; Gascon, F. Sentinel-2 Sen2Cor: L2A Processor for Users. In Proceedings of the ESA Living Planet Symposium 2016, Prague, Czech Republic, 9–13 May 2016.

46. Sabins, F.F. Remote sensing for mineral exploration. *Ore Geol. Rev.* **1999**, *14*, 157–183. [CrossRef]

47. Rosich, B.; Meadows, P. Absolute Calibration of ASAR Level 1 Products Generated with PF-ASAR. 2004. Available online: https://earth.esa.int/web/guest/-/absolute-calibration-of-asar-level-1-products-generated-with-pf-asar-4503 (accessed on 10 August 2018).

48. Small, D. Flattening Gamma: Radiometric Terrain Correction for SAR Imagery. *IEEE Trans. Geosci. Remote Sens.* **2011**, *49*, 3081–3093. [CrossRef]

49. Yague-Martinez, N.; Zan, F.D.; Prats-Iraola, P. Coregistration of Interferometric Stacks of Sentinel-1 TOPS Data. *IEEE Geosci. Remote Sens. Lett.* **2017**, *14*, 1002–1006. [CrossRef]

50. Haralick, R.M.; Shanmugam, K.; Dinstein, I.H. Textural Features for Image Classification. *IEEE Trans. Syst. Man Cybern.* **1973**, *SMC-3*, 610–621. [CrossRef]

51. Lee, J.-S.; Pottier, E. (Eds.) *Polarimetric Radar Imaging. From Basics to Applications*; CRC: London, UK; Taylor & Francis [Distributor]: Boca Raton, FL, USA, 2009; ISBN 142005497X.

52. Schreier, G. *SAR Geocoding. Data and Systems*; Wichmann: Karlsruhe, Germany, 1993; ISBN 3879072477.

53. Pottier, E.; Lee, J.; Ferro-Famil, L. PolSARPro_v4.2.0-Tutorial-Part1 Radar Polarimetry-4 Polarimetric Decompositions. 2007.

54. Pereira, L.; Furtado, L.; Novo, E.; Sant'Anna, S.; Liesenberg, V.; Silva, T. Multifrequency and Full-Polarimetric SAR Assessment for Estimating Above Ground Biomass and Leaf Area Index in the Amazon Várzea Wetlands. *Remote Sens.* **2018**, *10*, 1355. [CrossRef]

55. Cloude, S.R.; Pottier, E. A review of target decomposition theorems in radar polarimetry. *IEEE Trans. Geosci. Remote Sens.* **1996**, *34*, 498–518. [CrossRef]

56. Furtado, L.F.D.A.; Silva, T.S.F.; Novo, E.M.L.D.M. Dual-season and full-polarimetric C band SAR assessment for vegetation mapping in the Amazon várzea wetlands. *Remote Sens. Environ.* **2016**, *174*, 212–222. [CrossRef]

57. Krieger, G.; Moreira, A.; Fiedler, H.; Hajnsek, I.; Werner, M.; Younis, M.; Zink, M. TanDEM-X: A Satellite Formation for High-Resolution SAR Interferometry. *IEEE Trans. Geosci. Remote Sens.* **2007**, *45*, 3317–3341. [CrossRef]

58. Bamler, R.; Hartl, P. Synthetic aperture radar interferometry. *Inverse Probl.* **1998**, *14*, R1. [CrossRef]

59. Breiman, L. Random Forests. *Mach. Learn.* **2001**, *45*, 5–32. [CrossRef]

60. Bradter, U.; Thom, T.J.; Altringham, J.D.; Kunin, W.E.; Benton, T.G. Prediction of National Vegetation Classification communities in the British uplands using environmental data at multiple spatial scales, aerial images and the classifier random forest. *J. Appl. Ecol.* **2011**, *48*, 1057–1065. [CrossRef]

61. Baron, D.; Erasmi, S. High Resolution Forest Maps from Interferometric TanDEM-X and Multitemporal Sentinel-1 SAR Data. *PFG J. Photogramm. Remote Sens. Geoinf. Sci.* **2017**, *85*, 389–405. [CrossRef]
62. Congalton, R.G. A review of assessing the accuracy of classifications of remotely sensed data. *Remote Sens. Environ.* **1991**, *37*, 35–46. [CrossRef]
63. Olofsson, P.; Foody, G.M.; Herold, M.; Stehman, S.V.; Woodcock, C.E.; Wulder, M.A. Good practices for estimating area and assessing accuracy of land change. *Remote Sens. Environ.* **2014**, *148*, 42–57. [CrossRef]
64. Hermuche, P.M.; Sano, E.E. Identificação da floresta estacional decidual no Vão do Paraná, estado de Goiás, a partir da analise da reflectância acumulada de imagens do sensor ETM+/Landsat-7. *Revista Brasileira de Cartografia* **2011**, *63*, 415–425.
65. Barret, B.; Raab, C.; Cawkwell, F.; Grenn, S. Upland vegetation mapping using Random Forests with optical and radar satellite data. *Remote Sens. Ecol. Conserv.* **2016**, *2*, 212–231. [CrossRef]
66. Asner, G.P. Cloud cover in Landsat observations of the Brazilian Amazon. *Int. J. Remote Sens.* **2001**, 3855–3862. [CrossRef]
67. Yu, Y.; Saatchi, S. Sensitivity of L-Band SAR Backscatter to Aboveground Biomass of Global Forests. *Remote Sens.* **2016**, *8*, 522. [CrossRef]
68. Jensen, J.R. *Remote Sensing of the Environment. An Earth Resource Perspective*, 2nd ed.; Pearson Prentice Hall: Upper Saddle River, NJ, USA, 2007; ISBN 0131889508.
69. Lucas, R.M.; Milne, A.K.; Cronin, N.; Witte, C.; Denham, R. The potential of synthetic aperture radar (SAR) for quantifying the biomass of Australia's woodlands. *Rangel. J.* **2000**, *22*, 124–140. [CrossRef]
70. Garestier, F.; Dubois-Fernandez, P.C.; Guyon, D.; Le Toan, T. Forest Biophysical Parameter Estimation Using L- and P-Band Polarimetric SAR Data. *IEEE Trans. Geosci. Remote Sens.* **2009**, *47*, 3379–3388. [CrossRef]
71. Santoro, M.; Fransson, J.E.S.; Eriksson, L.E.B.; Magnusson, M.; Ulander, L.M.H.; Olsson, H. Signatures of ALOS PALSAR L-Band Backscatter in Swedish Forest. *IEEE Trans. Geosci. Remote Sens.* **2009**, *47*, 4001–4019. [CrossRef]
72. Schlund, M.; von Poncet, F.; Hoekman, D.H.; Kuntz, S.; Schmullius, C. Importance of bistatic SAR features from TanDEM-X for forest mapping and monitoring. *Remote Sens. Environ.* **2014**, *151*, 16–26. [CrossRef]
73. Inglada, J.; Vincent, A.; Arias, M.; Marais-Sicre, C. Improved Early Crop Type Identification By Joint Use of High Temporal Resolution SAR And Optical Image Time Series. *Remote Sens.* **2016**, *8*, 362. [CrossRef]
74. Ribeiro, J.F.; Walter, B.M.T. Fitofisionomias do Bioma Cerrado. 1998. Available online: http://www.bdpa.cnptia.embrapa.br/consulta/busca?b=ad&id=554094&biblioteca=vazio&busca=554094&qFacets=554094&sort=&paginacao=t&paginaAtual=1 (accessed on 10 August 2018).

Article

Woody Aboveground Biomass Mapping of the Brazilian Savanna with a Multi-Sensor and Machine Learning Approach

Polyanna da Conceição Bispo [1,2,*], Pedro Rodríguez-Veiga [2,3], Barbara Zimbres [4],
Sabrina do Couto de Miranda [5], Cassio Henrique Giusti Cezare [6], Sam Fleming [7],
Francesca Baldacchino [7], Valentin Louis [2], Dominik Rains [8,9], Mariano Garcia [10],
Fernando Del Bon Espírito-Santo [2], Iris Roitman [11], Ana María Pacheco-Pascagaza [2],
Yaqing Gou [2], John Roberts [2], Kirsten Barrett [2], Laerte Guimaraes Ferreira [6], Julia Zanin Shimbo [4],
Ane Alencar [4], Mercedes Bustamante [11], Iain Hector Woodhouse [7,12], Edson Eyji Sano [13],
Jean Pierre Ometto [14], Kevin Tansey [2] and Heiko Balzter [2,3]

[1] Department of Geography, School of Environment, Education and Development, University of Manchester, Oxford Road, Manchester M13 9PL, UK
[2] Centre for Landscape and Climate Research, School of Geography, Geology and the Environment, University of Leicester, Leicester LE1 7RH, UK; pedro.rodriguez@leicester.ac.uk (P.R.-V.); valentin.louis@leicester.ac.uk (V.L.); fdbes1@leicester.ac.uk (F.D.B.E.-S.); ampp2@leicester.ac.uk (A.M.P.-P.); yg137@leicester.ac.uk (Y.G.); jfr10@leicester.ac.uk (J.R.); kirsten.barrett@leicester.ac.uk (K.B.); kjt7@leicester.ac.uk (K.T.); hb91@leicester.ac.uk (H.B.)
[3] NERC National Centre for Earth Observation, University Road, Leicester LE1 7RH, UK
[4] Amazon Environmental Research Institute (IPAM), Brasília 71503-505, Brazil; barbara.zimbres@ipam.org.br (B.Z.); julia.shimbo@ipam.org.br (J.Z.S.); ane@ipam.org.br (A.A.)
[5] University of Goiás State (UEG), Palmeiras de Goiás 76190-000, Brazil; sabrina.couto@ueg.br
[6] Federal University of Goiás (UFG), Goiânia 74690-900, Brazil; cassio_cezare@hotmail.com (C.H.G.C.); laerte@ufg.br (L.G.F.)
[7] Carbomap Ltd., Edinburgh EH1 1LZ, UK; s.fleming@carbomap.com (S.F.); f.baldacchino@carbomap.com (F.B.); i.h.woodhouse@ed.ac.uk (I.H.W.)
[8] Department of Environment, Ghent University, 9000 Ghent, Belgium; dominik.rains@ugent.be
[9] Earth Observation Science, Department of Physics & Astronomy, University of Leicester, Leicester LE1 7RH, UK
[10] Department of Geology, Geography and Environment, University of Alcalá, 28801 Alcalá de Henares, Madrid, Spain; mariano.garcia@uah.es
[11] Department of Ecology, University of Brasília (UNB) and Brazilian Research Network on Global Climate Change—Rede Clima, Brasília 70910-900, Brazil; irisroitman01@gmail.com (I.R.); mercedes@unb.br (M.B.)
[12] School of Geosciences, University of Edinburgh, Edinburgh EH1 1LZ, UK
[13] Brazilian Agricultural Research Corporation (Embrapa Cerrados), Brasília 70770-901, Brazil; edson.sano@embrapa.br
[14] Earth System Science Center (CCST), National Institute for Space Research (INPE), Av dos Astronautas 1758, São José dos Campos 12227-010, Brazil; jean.ometto@inpe.br
* Correspondence: polyanna.bispo@manchester.ac.uk

Received: 16 July 2020; Accepted: 15 August 2020; Published: 19 August 2020

Abstract: The tropical savanna in Brazil known as the Cerrado covers circa 23% of the Brazilian territory, but only 3% of this area is protected. High rates of deforestation and degradation in the woodland and forest areas have made the Cerrado the second-largest source of carbon emissions in Brazil. However, data on these emissions are highly uncertain because of the spatial and temporal variability of the aboveground biomass (AGB) in this biome. Remote-sensing data combined with local vegetation inventories provide the means to quantify the AGB at large scales. Here, we quantify the spatial distribution of woody AGB in the Rio Vermelho watershed, located in the centre of the Cerrado, at a high spatial resolution of 30 metres, with a random forest (RF) machine-learning approach. We produced the first high-resolution map of the AGB for a region in the Brazilian Cerrado using a combination of vegetation inventory plots, airborne light detection and ranging (LiDAR) data, and multispectral and radar satellite images (Landsat 8 and ALOS-2/PALSAR-2). A combination of random forest (RF) models and jackknife analyses enabled us to select the best remote-sensing variables to quantify the AGB on a large scale. Overall, the relationship between the ground data from vegetation inventories and remote-sensing variables was strong ($R^2 = 0.89$), with a root-mean-square error (RMSE) of 7.58 Mg ha^{-1} and a bias of 0.43 Mg ha^{-1}.

Keywords: aboveground biomass; Cerrado ecosystem; random forest; SAR

---

1. Introduction

The tropical savanna in Brazil, known as the Cerrado, is the second-largest biome in South America, covering over 200 million ha or approximately 23% of the Brazilian territory [1]. It has the highest species richness and biodiversity among the world's savannas [2,3]. Gradients of tree density, height, canopy cover, and aboveground biomass (AGB) in the Cerrado vary according to the climate, fire regime, geomorphology, and soil nutrient availability, resulting in 19 distinctive ecoregions [4]. The Cerrado is characterised by a mosaic of grasslands, shrublands, and forestlands in varying proportions, depending on the location [5]. Its physiognomies range from *campo* (grasslands) to the typical Cerrado *stricto sensu* (trees and shrubs up to 8–10-m-high and with an understory dominated by grass) and the *cerradão* (forest formations with trees up to a height of 20-m-high) [6,7].

Although the aboveground carbon stock in the Cerrado is lower than that of the Brazilian Amazon, the conversion of the Cerrado biome to different types of land uses is occurring much faster than in the Brazilian Amazon, mainly because the Brazilian livestock and agricultural frontier has been expanding towards the northern parts of the Cerrado over the last decades [8,9]. This trend has been increasing due to the governmental policies put in place since 2018. The Cerrado land use and land cover (LULC) mapping project, conducted by the Brazilian Ministry of Environment (MMA), showed that, in 2013, approximately 43% (88 million ha) of the biome had already been converted to different land uses, with 55% (111 million ha) still covered by native vegetation [4]. The remaining 2% of the Cerrado were covered by water bodies and by the class "unidentified", which included areas covered by clouds and burned areas [10]. Most of the remaining natural areas have been undergoing degradation due to unsustainable selective logging and burning activities, often overlooked as threats to habitat integrity and connectivity [4,11]. According to MMA-2019 [12] and the MapBiomas alert platform (https://mapbiomas.org/en/project), the Cerrado was the Brazilian biome with the highest levels of deforestation between October 2018 and March 2019, losing 47,704 ha of native vegetation. In addition, about 95% of the deforestation alerts were in areas without any authorisation—that is, without a license for deforestation, which is issued by either the federal or state environmental agency. More than 1400 ha of deforestation took place in legal reserves (Brazil's environmental legislation obligates private property owners to retain a fixed proportion of their total area for native vegetation. These areas are called "legal reserves") [13].

Only 3% of the Cerrado is strictly protected by the law within conservation areas [12,14], and the high rates of vegetation loss and degradation have made the Cerrado the second-largest source of carbon emissions in Brazil [15]. In this context, it is essential to monitor AGB and carbon stocks effectively, and reliable maps are needed for climate change mitigation policies [16–18]. Uncertainties in current vegetation carbon stock estimates over the Cerrado are high, and biomass estimates vary by more than 50 Mg ha^{-1} within the same area [19–21]. This demands improvements in the accuracy and spatial resolution to estimate the AGB in this biome. The challenge here is to take the large latitudinal gradient and the high variation of the vegetation structure into consideration [15], as well as the paucity of field studies quantifying AGB over different regions of the biome [15,16].

Different types of sensors, such as satellite-based multispectral imagers and Synthetic Aperture RADAR (SAR) systems, as well as airborne light detection and ranging (LiDAR), have been successfully applied to estimate AGB in the tropics [21–26]. SAR and LiDAR have been increasingly used to estimate AGB in the last eight years [27–36]. The microwave pulses transmitted by a SAR system, especially at longer wavelengths such as the L- or P-band, interact with the branches and trunks, providing information about the forest structure, which is highly correlated to AGB [37–39]. Airborne LiDAR provides information on the canopy height and canopy cover [40–43], which are good proxies for woody AGB estimations [44]. The main limitations of LiDAR are that it is still an expensive technology and it is typically not available for large areas [45]. So far, only a few studies have explored the effectiveness of multi-sensor data synergy in tropical savannas [46–48]. In contrast, this approach has been intensively used to study other biomes [49–51].

Most studies focusing on AGB estimations in the Cerrado are based on establishing statistical models between remote-sensing measurements and field plots (i.e., allometric equations linking in situ measurements to parameters such as tree biomass) [5,16,52]. However, uncertainties remain high [53], and studies using remote sensing to estimate AGB in the Cerrado are still limited [53,54]. Bitencourt et al. [53] studied the Cerrado vegetation using optical and RADAR data, showing a strong relationship between these Earth Observation (EO) datasets and foliage biomass. The same authors also used Japanese Earth Resources Satellite (JERS-1) SAR observations to estimate the woody AGB of the savanna using a multiple regression analysis, resulting in $R^2 = 0.87$. Miguel et al. [54] used artificial neural networks to predict the wood volume and AGB of the savanna using satellite observations from the Linear Imaging Self-Scanner (LISS-III) sensor onboard the ResourceSat-1 satellite. The neural network approach showed a good accuracy for both the wood volume (R^2 ~0.98 and standard error of estimate (SEE) ~4.83%) and AGB (R^2 ~0.94 and SEE ~8.5%). Schwieder et al. [55] combined Landsat phenological metrics with aboveground carbon field samples of woodland savanna vegetation using random forest (RF) regression models to map the regional carbon distribution and to analyse the relationship between the phenological metrics and aboveground carbon stocks. The model performance varied among the three selected study areas, with root mean squared error (RMSE) values of 1.64 Mg ha^{-1} (mean relative RMSE 30%), 2.35 Mg ha^{-1} (mean relative RMSE 46%), and 2.18 Mg ha^{-1} (mean relative RMSE 45%), while the aboveground carbon distributions revealed characteristic spatial patterns. Biomass maps can be assessed through the biomass product accuracy requirements of satellite missions dedicated to the estimation of a biomass, such as those from the BIOMASS mission [56]. BIOMASS is aiming at errors smaller than ± 20% in terms of the relative RMSE (rel. RMSE) for AGB higher than 50 Mg ha^{-1} and ± 10 Mg ha^{-1} in terms of RMSE for AGB lower than 50 Mg ha^{-1}.

In this study, we quantified the spatial distribution of the AGB in the Rio Vermelho watershed, located in a central Cerrado region in Brazil. Using a machine-learning approach, we produced the first high-resolution AGB map (30 m) of a Brazilian Cerrado area based on a combination of vegetation inventory plots, airborne LiDAR data, and satellite images (Landsat 8 and ALOS–2/PALSAR-2). The two-stage upscaling approach (field to LIDAR and LIDAR to EO) was also applied for the first time in an area of this biome. We used a RF model and jackknife analyses to analyse the importance of the remote-sensing predictor variables, enabling us to select the best ones to quantify the AGB.

2. Data and Methodology

2.1. Study Area

The study site is located in the Rio Vermelho watershed, which is part of the Rio Araguaia watershed in the state of Goiás, Central Brazil (Figure 1), and it covers an area of 1,082,460 ha. The landscape topography is typically lowland, and the Cerrado biome's climate is semi-humid (*Aw* in the Köppen's climate classification system). It is a tropical savanna with two marked seasons: a dry winter (from May to September) and a rainy summer (from October to April). This watershed is characterised by a relatively constant air temperature throughout the year, with minimum and maximum temperatures of 20 °C and 32 °C, respectively) [57,58]. Intense and localised precipitation and frequent dry spells occur.

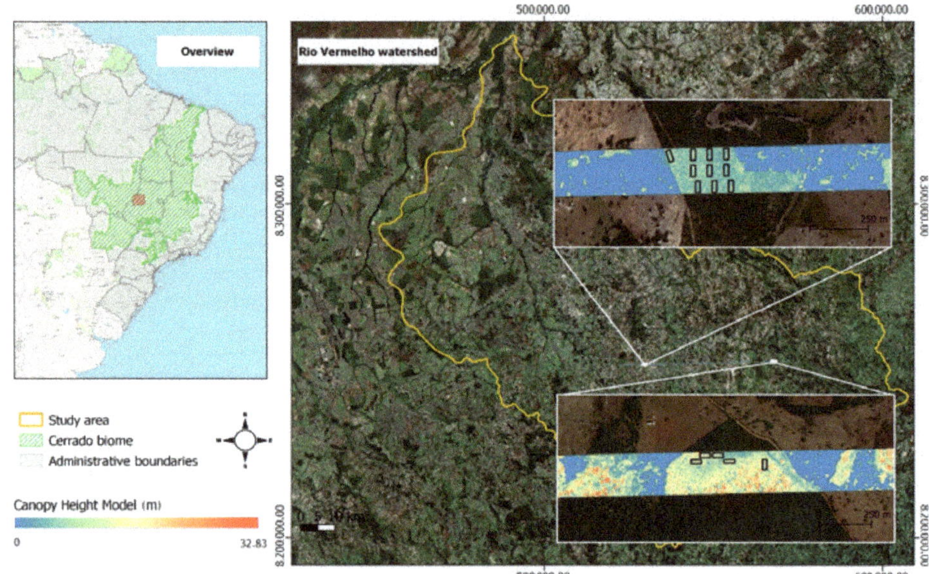

Figure 1. Location of the Rio Vermelho watershed study area (outlined in yellow in the right panel) located in the State of Goiás in Brazil and in the Cerrado biome (grey and green areas in the left overview panel). The location and a zoom of the canopy height light detection and ranging (LiDAR) footprints within the watershed (each of those covered an area of 16.82 ha) with the location of the field plots (black rectangles) are also shown. The entire right panel corresponds to the red square in the left overview panel. The background image corresponds to the © Bing™ aerial photo, a screenshot(s) reprinted with permission of Microsoft Corporation.

The Rio Vermelho watershed is part of a region in the Goiás State called "Mato Grosso Goiano". This denomination is based on the historical extensive and dense native forest cover (typical for the Mato Grosso State) in the region [59]. This part of the Brazilian Cerrado underwent intensive human settlement in the 18th century due to the Brazilian gold rush in Central Brazil [60]. Since forests in the Cerrado are supposedly associated with soils with good fertility, most of the forest-covered areas have already been converted to other uses, mainly croplands and pastures [59], sometimes resulting in land degradation. Today, forest formations are restricted to a small and few fragments. Pastures cover more than 57% of the study area [6], and livestock ranching is the main economic activity in the region. By 2018, only approximately 33% of the native vegetation persisted [12,61], and the remaining native vegetation is fragmented [62] and often located in legal reserves areas.

Vegetation

Most of the Cerrado natural vegetation is comprised of savanna formations (cerrado *stricto sensu*). However, forest formations such as woodlands, riparian forests, and seasonal forests play an important role in the carbon balance, because they have a higher carbon stock density [63]. In the Rio Vermelho watershed, woodlands or cerradão (in Portuguese) and seasonal forest formations predominate as the forest remnants. The forest inventory sites were located in two fragments of cerradão, and the plots follow a structural and biomass gradient from: (a) a savanna-cerradão transition zone (lower biomass), (b) cerradão, and (c) cerradão-seasonal forest transition zone (higher biomass).

Dystrophic cerradão is typical for deep, highly lixiviated soil (Oxisols and sandy soils) areas with a seasonal tropical climate, and their canopy height varies between 8 and 15 m. Canopy cover varies between 50% and 90%, with an understory of shrubs and grass [62]. They are structurally similar to seasonal forests and differ mainly in species composition, as they are comprised of seasonal forest as well as wooded savanna species [62,64].

Seasonal forests are often associated with more fertile soils (mesotrophic and eutrophic soils) and present different levels of deciduousness during the dry season. The canopy height varies between 15 and 25 m, and the forest cover is between 70% and 95% during the rainy season. In the dry season, the canopy cover can be lower than 50% in seasonal semideciduous forests and lower than 35% in seasonal deciduous forests [62].

In wooded savannas, most trees are shorter and sparser than in forest formations, allowing for a continuous herbaceous grassy layer. Wooded savannas are separated into subgroups according to their structural gradient: *cerrado ralo* (2–3-m-tall trees with 5–20% canopy cover), *cerrado sensu stricto* or *typical cerrado* (3–6-m-tall trees with 20–50% forest cover), and *cerrado denso* (8–15-m-tall trees with 50–70% forest cover) [65].

2.2. Methodology

Figure 2 shows the flowchart of the methodology used to produce the AGB map of the Rio Vermelho watershed. Our reference datasets consisted of ground measurements and airborne LiDAR data. We used a combination of Earth Observation (EO) datasets from multispectral passive optical and synthetic aperture radar (SAR) sensors as predictor variables to estimate the AGB of the study area. The description of each step of the methodology is provided in the following subsections.

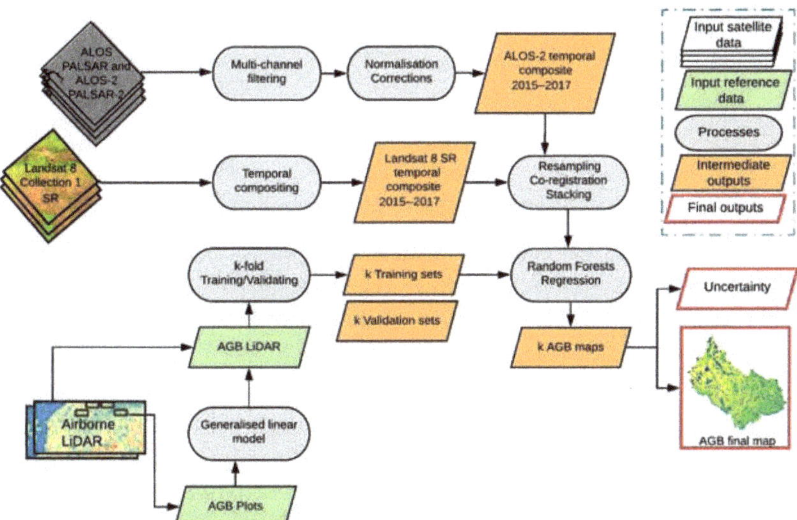

Figure 2. Flowchart showing the methodology used to produce the aboveground biomass (AGB) map of the Rio Vermelho watershed, Goiás State, Brazil. SR = surface reflectance.

2.2.1. Ground Truth Data

We used woody AGB data from 15 (20 m × 50 m) field plots located in cerradão vegetation to estimate the AGB from airborne LiDAR. These plots were established in 2014, and all trees with a diameter at breast height (dbh) ≥ 5 cm at 1.30 m above the ground were considered. The tree heights were measured with a digital clinometer (HAGLOF ECII-D). The basal area in m^2 was calculated from the dbh (basal area = $(\pi/4) * (dbh/100)^2$) (Table 1). Ten plots were located in the remnants of native vegetation in the municipality of Itapirapuã (Figure 1, right panel, top zoom), while five plots were located in the municipality of Goiás (Figure 1, right panel, lower zoom).

Table 1. Floristic and structural characterisations of the plots located in fragments of the native Cerrado vegetation in the Rio Vermelho watershed, Goiás State, Brazil. WS-FS = savanna-cerradão transition zone, TFS = cerradão, FS-SF = cerradão-seasonal forest transition zone; S = species richness, TD = tree density, DBH = diameter at breast height, H = height, TBA = tree basal area, AGB = aboveground biomass, and CV= coefficient of variation.

Plot ID	Vegetation Type	S (Species)	TD (Ind. ha^{-1})	DBH Range (cm) (Mean/CV%)	H (m) (Mean/CV%)	TBA (m^2/ha)	AGB (Mg ha^{-1})
Itapirapuã 1	WS-FS	38	990	5.0–36.7 (9.1/56.9)	1.7–11.2 (5.8/26.9)	13.5	19.3
Itapirapuã 2	WS-FS	32	920	5.0–45.5 (9.5/54.7)	1.6–13.4 (5.6/35.5)	10.8	21.2
Itapirapuã 3	WS-FS	45	1030	5.0–29.4 (10.4/54.5)	2.1–14.0 (5.8/35.9)	15.0	24.5
Itapirapuã 4	WS-FS	41	1040	5.3–52.0 (9.8/58.4)	1.3–12.6 (5.5/33.4)	14.4	28.2
Itapirapuã 5	TFS	36	1140	5.0–41.4 (10.3/60.4)	2.0–12.8 (6.0/31.9)	16.4	32.2
Itapirapuã 6	TFS	45	1570	5.0–34.7 (10.5/47.0)	1.5–12.9 (6.5/31.6)	22.6	35.3
Itapirapuã 7	TFS	50	1990	5.0–43.7 (9.1/52.9)	1.8–13.2 (6.5/33.6)	21.9	36.8
Itapirapuã 8	TFS	60	1440	5.0–48.0 (10.4/61.2)	2.6–13.2 (6.1/31.8)	20.1	40.2
Itapirapuã 9	TFS	35	1210	5.0–53.3 (10.1/62.9)	1.8–13.1 (6.5/30.8)	17.0	40.9
Goiás 10	TFS	41	1260	5.0–35.3 (10.9/57.3)	3.6–19.6 (8.7/38.6)	20.5	52.8
Itapirapuã 11	TFS	39	1260	5.0–42.3 (11.2/65.9)	1.7–14.9 (6.3/36.3)	24.3	54.3
Goiás 12	FS-SF	38	1310	5.0–49.3 (11.3/61.5)	2.5–19.3 (9.3/33.3)	24.6	70.4
Goiás 13	FS-SF	26	690	5.0–49.0 (13.0/73.6)	3.4–22.0 (10.2/43.5)	18.3	77.0
Goiás 14	FS-SF	27	820	5.0–44.2 (13.2/63.5)	3.0–38.0 (12.5/48.7)	22.0	98.3
Goiás 15	FS-SF	24	760	5.0–41.7 (14.0/67.3)	2.5–26 (10.9/52.5)	24.4	103.9

As mentioned, the two native vegetation fragments sampled are classified as cerradão. However, due to the high structural heterogeneity in the Cerrado biome, there were considerable variations in the structure of the woody vegetation within the sampled plots. The floristic and structural characteristics of the plots described in Table 1 (and from Figure S1 to Figure S20 in the Supplementary Materials) corroborate that the sampled plots adequately represent the gradient of floristic-structural variation inherent to forest formations from all of the Cerrado. Thus, four plots were classified as a savanna-cerradão transition zone, seven as cerradão, and four as a cerradão-seasonal forest transition zone [62]. The sampled plots showed a height gradient directly related to the aboveground biomass stocks of woody vegetation. In the savanna-cerradão transition zone, the mean height was below 6 m, in the cerradão, between 6 and 8 m, and, in the cerradão-seasonal forest transition zone, above 9 m (Table 1). These plots were representative of the structural variation found in the cerradão vegetation of the region, with AGB values ranging from 19 to 104 Mg ha^{-1} (Table 1).

In tropical ecosystems, species diversity is generally very high. Therefore, generalised (mixed-species) allometric models are more appropriate than species-specific equations [66,67]. We used the following mixed species allometric model [68] designed for cerradão (N = 87, R^2 = 0.94, residual standard error (RSE) = 0.13 Mg; RSE (%) = 49.83). The model estimates the woody AGB (only live trees were used) with a diameter at breast height $\geq$ 5 cm, excluding leaf biomass.

$$AGB = \exp\left[-12.29 + 2.69 * \ln(dbh) + 0.80 * \ln(h)\right] \tag{1}$$

where dbh is the diameter at breast height, and h is height.

In our study, these in situ woody AGB estimates were then used to establish an empirical model to estimate AGB as a function of the structural metrics derived from the airborne LiDAR.

2.2.2. LiDAR Data

We used 60 tiles of airborne LiDAR data (covering a total area of 1009.01 ha) collected by the Sustainable Landscapes project led by the Brazilian Enterprise for Agricultural Research Corporation (EMBRAPA). The LiDAR data covered the municipalities of Itapirapuã and Goiás, Goiás State, Brazil. The flights took place between 20 June 2015 and 7 July 2015. The LiDAR dataset has an average density of returns of ~45 ppm^2 (points per square meter). The average altitude of the flights was ~850 m, with a field of view of 12°. Two LiDAR sensors (Optech Orion M300 and Optech ALTM 09SEN243) were used in these campaigns, but the percentage of flight line overlaps was high (~65%). LiDAR returns of all 15 plots are shown from Figure S6 to Figure S20 in the Supplementary Materials.

The LASTools software [69] was used to process the LiDAR data and to generate the following variables derived from the LiDAR point cloud data: DSM (digital surface model), CHM (canopy height model), DTM (digital terrain model), CC (canopy cover coverage), CD (canopy density), Max-H (maximum height), Percentile_p15, Percentile_p10, Percentile_p20, Percentile_p30, Percentile_p35, Percentile_p55, Percentile_p40, Percentile_p45, Percentile_p50, Percentile_p60, and Percentile_p65. Highly correlated variables were excluded, and the ones containing the most unique information, CHM, CC, and CD, were used to establish statistical models to extrapolate the spatially limited LiDAR AGB estimates to both the optical and radar observations (Figure 2).

2.2.3. Optical Data

The United States Geological Survey (USGS) Landsat 8 Collection 1 Tier 1 data consists of surface reflectance products generated from the Landsat 8 Operational Land Imager (OLI), with 30-m spatial resolution. The USGS atmospherically corrects the scenes using the Landsat 8 Surface Reflectance Code (LaSRC), which also uses the C function of the mask (CFMASK) algorithm [70] to generate a cloud, shadow, water, and snow mask. We selected all the scenes acquired in the period of 2015–2017 (244 scenes) to generate a cloud and shadow-free median temporal composite (Figure 2). In addition to the spectral bands, we generated a range of vegetation indices that could potentially be representative for the vegetation canopy structure, seasonality, and a measure of vegetation greenness. These were the normalised difference vegetation index (NDVI), normalised burn ratio (NBR), normalised difference moisture index (NDMI), and the soil adjusted vegetation index (SAVI). These indices have previously been used in similar studies [21,23,71–73].

2.2.4. SAR Data

We used the global 25-m resolution ALOS PALSAR/PALSAR−2 annual mosaics, which are freely available at https://www.eorc.jaxa.jp/ALOS/en/palsar_fnf/fnf_index.htm. This dataset was pre-processed by the Japanese Aerospace Exploration Agency (JAXA) using L-band SAR images of the backscattering coefficient acquired by the Advanced Land Observing Satellite (ALOS) and Advanced Land Observing Satellite-2 (ALOS-2). The mosaics consist of 10 × 10-degree tiles pre-processed to correct for geometric distortions (ortho-rectification) and topographic effects [74]. The mosaics were calibrated to $\gamma0$ using the following equation:

$$\gamma^0 = 10 \; x \; log_{10}(DN)^2 + CF \tag{2}$$

where γ^0 is the gamma-naught in decibels (dB), DN is the digital number in unsigned 16 bit, and CF is a calibration constant of 83.0 dB.

We reduced the noise of the mosaics by applying a temporal multi-channel filter [75] with a 5 × 5-pixel moving average window. Then, we applied a temporal normalisation between PALSAR and PALSAR-2 images aiming to correct for soil moisture and sensor differences. For our final PALSAR-2 composite, we used the median value of the three mosaics (2015, 2016, and 2017) (Figure 2). We observed a geolocation error of ~80 m on average for the PALSAR-2 mosaics in comparison to HR (high-resolution) imagery and Landsat 8 scenes. We generated a Sentinel-1 SAR median γ^0 composite using 283 scenes from between 2016 and 2018, and we used this not as an input for our modelling but as a geolocation reference image. The PALSAR-2 composite was georeferenced against the Sentinel-1 composite using a SAR-to-SAR co-registration. This approach, instead of a SAR-to-optical approach, was chosen to avoid errors derived from the different viewing geometries of SAR and optical images. The variables used for this study were the SAR backscatter coefficients (γ^0_{HV} and γ^0_{HH}) and two SAR indices—namely, the radar forest degradation index (RFDI) and the cross-polarised ratio (CpR) [74].

2.3. Modelling Framework

We adopted a two-stage upscaling method from field measurements to airborne LiDAR point clouds and from the LiDAR-based estimates to satellite imagery. We estimated the woody AGB of the field plots and used these as a reference to estimate the woody AGB across the LiDAR footprints using LiDAR-derived structural variables as a predictor. Our field plots covered a representative range of flora and structure of the woody vegetation (Table 1) from 19 Mg ha^{-1} to 104 Mg ha^{-1}. As the LiDAR point clouds are sensitive to the forest structure, we assumed that these plots could express the physical and structural variations of the vegetation of the study site, and we allowed the AGB LiDAR model to extrapolate for values < 19 Mg ha^{-1}. This field-to-LiDAR procedure allowed us to increase our sampling from the 15 plots to thousands of observations derived from the LiDAR footprint covering a wide range of woody AGB and vegetation types. We then used these LiDAR AGB, representative for the vegetation in the region, in combination with EO datasets to estimate the woody AGB over the entire Rio Vermelho watershed.

The first step was to model the relationship between LiDAR woody vegetation structural variables and the AGB estimated from the field plots (Figure 2). We generated raster-based LiDAR vegetation structural variables with using a 1-m pixel size. We then used the average value of the pixels within the 20×50-m plots (0.10 ha) to develop our models. As we had a limited number of field plots, we had to limit the number of variables included in the empirical models as well. Based on a visual exploration of the relationships between the dependent and independent variables, as well as the distribution of each predictor, we identified three potential predictors: canopy height model (CHM), canopy density (CD), and canopy cover (CC). We then used all possible combinations of these variables to find the best generalised linear model to estimate the AGB based on the Akaike's information criterion corrected for a small sample size (AICc). This final model was used to estimate the AGB over the entire LiDAR footprint at a 30-m spatial resolution (0.09 ha). The next step was to model the relationship between the estimated LiDAR AGB and the EO datasets. These EO datasets were resampled, co-registered, and stacked at 30-m spatial resolutions (Figure 2). We used the LiDAR AGB pixels (N = 2,973) as training and test samples in a regression version of the random forest (RF) algorithm [76] in the Google Earth Engine [77], which was implemented within a k-fold cross-validation framework [73]. The following variables were used as predictors in the model: γ^0_{HV}; γ^0_{HH}; RFDI; CpR from the PALSAR-2 data; and blue, green, red, near infrared, and shortwave infrared bands, as well as NDVI, NBR, NBR2, NDMI, and SAVI from the Landsat 8 data. RF is a nonparametric machine-learning algorithm that uses a bootstrap technique to construct multiple decision trees. Jackknife tests were also run to compare the importance of the predictor variables based on a R^2 of the model. Due to the randomness (stochastic nature) of the RF algorithm, the importance may change for each run. Thus, the variable importance analysis was run for each k and then averaged to produce the overall importance figures. The test was performed running each single set of variables in isolation to assess the accuracy of each set. The higher the R^2, the higher the importance of the variable. Then, a set of variables was excluded each time from the whole set to assess the drop in the variance explained by the model.

The k-fold cross-validation framework was used to train and validate the algorithm (Figure 2), maximising the reference data available. Figure 2 shows the overview of the method used to produce the AGB map and the uncertainty characterisation in the Rio Vermelho watershed. The whole reference dataset was used for both training and validation purposes. The dataset was stratified by three AGB levels (low, medium, and high), and randomly sampled into the folds to ensure that all folds have similar probability distribution functions of the AGB. Then, k-1 folds were used to train the RF algorithm and to produce the AGB map, while the remaining fold was used for the validation. The process was then run k times, reusing the folds for training purposes but using them only once for validation purposes. The final outputs were k AGB maps over the study area. The mean value of the k AGB maps was used as the final AGB map, and we used the standard deviation (SD) of the k estimates for any given pixel to generate a prediction error ($\sigma_{prediction}$) map. This error also accounts for the representativeness of the sampling sites of the true distribution of AGB in the region [21]. The total SD (σ_{AGB}) was propagated, as explained in Rodriguez-Veiga et al. [29] and Saatchi et al. [21], as follows:

$$\sigma_{AGB} = \left(\sigma^2_{measurement} + \sigma^2_{LiDAR} + \sigma^2_{allometry} + \sigma^2_{sampling} + \sigma^2_{prediction} \right)^{\frac{1}{2}} \tag{3}$$

where the measurement error ($\sigma_{measurement}$) of the tree level parameters averaged at the plot scale was assumed as 10% [78], the LiDAR error (σ_{LiDAR}) was assumed as 1.5% (from a 0.11-m instrument error for an average 7.13 m canopy height), the allometric error ($\sigma_{allometry}$) was assumed to be 49.83% (Scolforo et al. [68]), and the sampling error ($\sigma_{sampling}$) from the variability of AGB within the pixels was estimated as 12.39% based on Réjou-Méchain et al. [79].

Our k-fold cross-validation assessment followed James et al. [80] and involved the calculation of the R^2, RMSE, rel. RMSE, and the mean bias error (MBE). Aside from the overall assessment, we also analysed the errors by biomass ranges. Finally, we also compared our AGB estimates to four other studies [21–23,81] (Table 2). Baccini et al. [23] used multi-sensor satellite data to estimate the aboveground live woody vegetation carbon density for pantropical ecosystems with a 500-m resolution. Saatchi et al. [21] mapped the total carbon stock in live biomasses (above- and belowground) in the tropics using a combination of data from in situ inventory plots and satellite LiDAR samples of the forest structure to estimate the carbon storage, plus optical and microwave imagery (1 km resolution), to extrapolate over the landscape. Santoro et al. [81] estimated AGB globally at a 100-m resolution by combining spaceborne SAR, LiDAR, and optical observations for the year 2010, with auxiliary datasets from forest inventories—namely, additional remote-sensing observations, climatological variables, and ecosystems classifications. Avitabile et al. [22] combined two existing datasets of vegetation aboveground biomasses (AGB) into a pantropical AGB map at a 1-km resolution using an independent reference dataset of field observations and locally calibrated it using high-resolution, harmonised, and upscaled biomass maps. For the comparison of our results to these maps, we aggregated all of them to the same spatial resolution (1 km).

Table 2. Summary of forest aboveground biomass (AGB) maps used for comparison with our study results. RMSE: root mean square error, rel. RMSE: relative root mean square error.

Study	Coverage	Year	Spatial Resolution	Methodology	Accuracy/Uncertainty
Saatchi et al. [21]	Pantropical	early 2000s	1 km	MaxEnt (field measurements, GLAS data, optical and microwave imagery)	Uncertainty from ±6% to ±53%
Baccini et al. [23]	Pantropical	2007–2008	500 m	Random Forest algorithm (field data, GLAS and MODIS data)	RMSE = 25 Mg C/ha for tropical America
Avitabile et al. [22]	Pantropical	2000–2008	1 km	Weighted linear averaging method (biomass reference datasets, Saatchi et al. [21] map and Baccini et al. [23] map)	RMSE = 87~98 Mg/ha; Mean error: almost null in most cases
Santoro et al. [81]	Global	2010	100 m	Water Cloud Model (ALOS PALSAR, Envisat, Landsat and field plots)	rel. RMSE = 57.1% and Bias = 10.6 Mg ha $^{-1}$ for tropical areas

3. Results

3.1. LiDAR-Derived AGB Map

Table 3 shows the models tested in this study. The AGB = f(CHM, CC) model was the one that best explained the field-based AGB variations (Table 3, adj R^2 = 0.93, RMSE = 6.74 Mg ha^{-1}, or 13% of the mean biomass values) and, therefore, was used to predict the AGB over the whole LiDAR flight footprint. This predicted LiDAR AGB was used to generate training and test samples to run the RF algorithm to extrapolate the AGB to the EO datasets covering the entire Rio Vermelho watershed. Although the AGB = f(CC, CD, CHM) model resulted in a slightly smaller RMSE and a higher AIC than AGB = f(CHM, CC), both models showed the same R^2 (= 0.93) and RMSE (6.74 Mg ha^{-1} or rel. RMSE = 13%). We therefore opted for the AGB = f(CHM, CC) model, as it is more parsimonious and only uses two independent variables (Table 3 and Figure S1 in the Supplementary Materials).

Table 3. Model comparison according to Akaike's information criterion corrected for a small sample size (AICc). Parameters for each model include: LogLik (log-likelihood), k (number of predictor variables), AICc (Akaike's information criterion corrected for a small sample size), ΔAICc (difference in AICc between the current and the best model), Adj R^2 (adjusted coefficient of determination), and RMSE (root mean square error). The intercept and coefficients, as well as the coefficient of determination (R^2) for each model, are also presented. CHM = canopy height model, CD = canopy density, and CC = canopy cover.

Model	LogLik	k	AICc	ΔAICc	Adj R^2	RMSE (Mg ha^{-1})
AGB = −61.92 + 4.88*CHM + 0.83*CC	−49.90	2	107.81	0.00	0.93	6.74
AGB = −107.18 + 2.21*CC − 1.26*CD + 6.34*CHM	−49.40	3	108.81	1.00	0.93	6.52
AGB = −30.03 + 4.30*CHM + 0.66*CD	−50.56	2	109.13	1.32	0.92	7.04
AGB = 2.44 + 6.25*CHM	−52.71	1	111.42	3.61	0.89	8.12
AGB = −27.17 − 2.31*CC + 3.22*CD	−54.83	2	117.66	9.85	0.86	9.36
AGB = −76.56 + 1.76*CD	−56.56	1	119.13	11.32	0.81	10.51
AGB = −173.08 + 2.48*CC	−60.72	1	127.44	19.63	0.69	13.86
AGB = 52.34	−69.39	0	142.78	34.97	0.00	24.70

3.2. AGB and Uncertainty Map

The AGB varied from 0 to 90 Mg ha^{-1} per pixel (Figure 3), whereas the pixel-scale uncertainty estimated by the error propagation approach ranged from 0 to 49 Mg ha^{-1} (Figure 4). Figure 5 shows the averaged variable importance analysis across the k-fold procedure for each set of variables derived from Landsat 8 (L8) and ALOS-2/PALSAR-2 (ALOS) included in the random forest (RF) model. These results indicated that the Landsat reflectances composite was the most important set of variables when predicting AGB. However, the ALOS indices contain more unique information not represented by the other variables, as shown by the largest decrease in accuracy when the variable was excluded. Figure 6 shows the overall accuracy assessment and Table 4 the assessment by AGB range.

The accuracy assessment between the AGB map predicted from EO and the AGB reference from LiDAR showed R^2 = 0.89, RMSE = 7.58 Mg ha^{-1}, rel. RMSE = 43%, and bias = 0.43 Mg ha^{-1} (Figure 6). Our map shows an underestimation of very high AGB (negative bias) and a slight overestimation of low AGB (positive bias) [28] (Table 4). We also found a RMSE of 6.39 Mg ha^{-1} (rel. RMSE = 61%) for AGB lower than 50 Mg ha^{-1} and a RMSE = 13.41 Mg ha^{-1} (rel. RMSE = 19%) for AGB higher than 50 Mg ha^{-1}. Details of the accuracy assessment by biomass range are shown in Table 4.

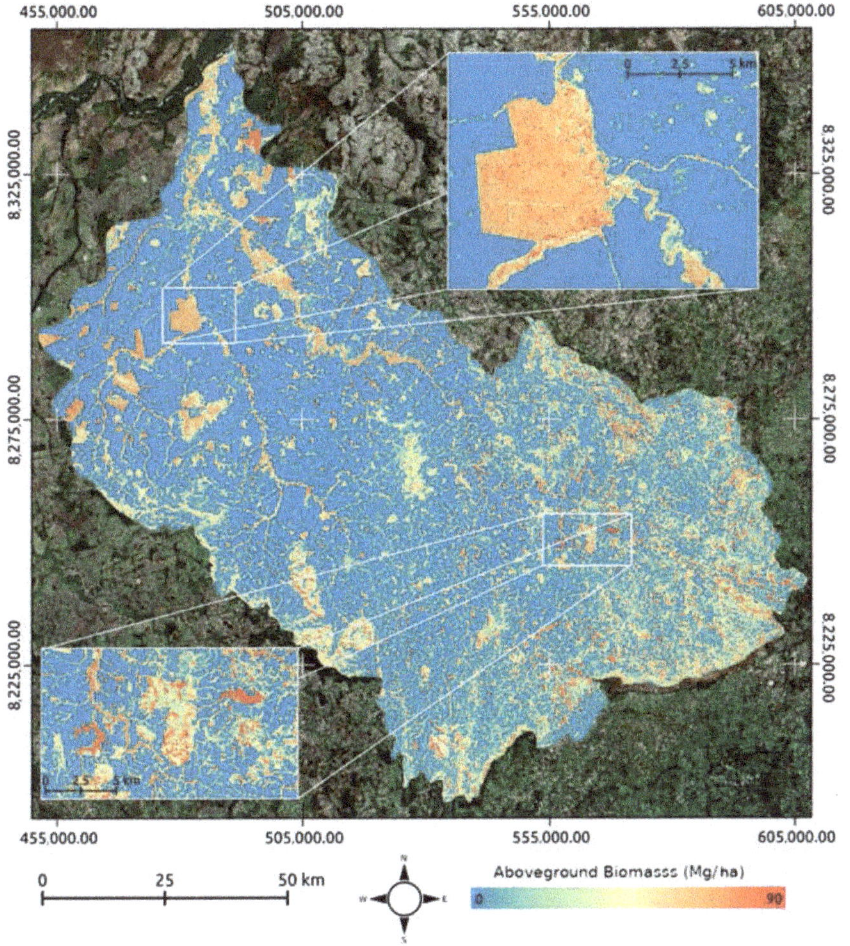

Figure 3. AGB map (30-m resolution) of the Rio Vermelho watershed in the Brazilian Cerrado, Goiás State. Zooms on the top right and bottom left of the figure show the details of the variations of the AGB between the different types of forest formations in the study area. Background image: © Bing™ aerial photo. Screenshot(s) reprinted with permission by Microsoft™ Corporation.

Table 4. Cross-validation assessment by the AGB range. MBE: mean bias error.

AGB Range (Mg ha^{-1})	Number of Pixels	Reference AGB (Mg ha^{-1})	Map AGB (Mg ha^{-1})	MBE (Mg ha^{-1})	RMSE (Mg ha^{-1})	Rel. RMSE
0–20	2084	4.4	5.7	1.3	4.8	109%
20–40	379	29.5	31.8	2.3	10.4	35%
40–60	287	49.3	48.8	−0.5	10.8	22%
60–80	152	69.2	62.6	−6.6	11.4	16%
>80	71	89.2	72.3	−17.0	20.0	22%

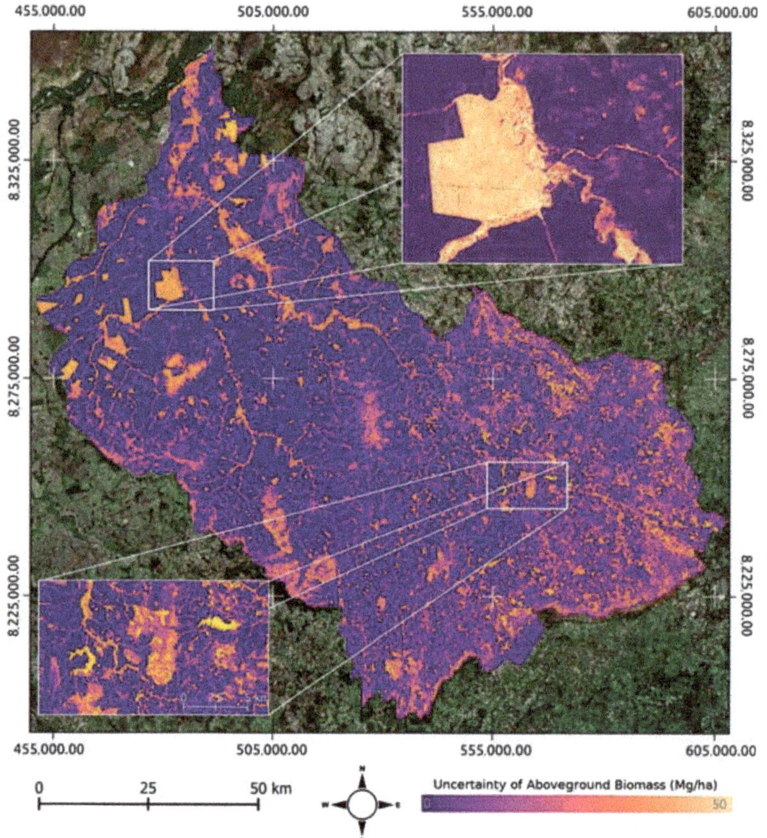

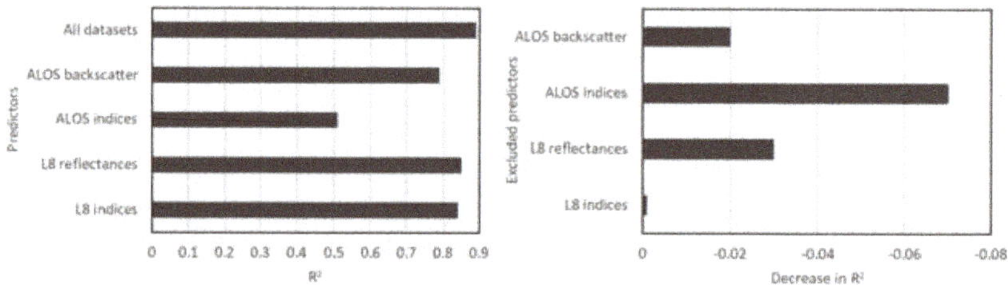

Figure 4. Error map (30-m resolution) for the Rio Vermelho watershed in the Brazilian Cerrado. Zooms on the top right and bottom left of the figure show the details of the variations of the AGB uncertainty between the different types of forest formations in the study area. Background image: © Bing™ aerial photo. Screenshot(s) reprinted with permission from Microsoft™ Corporation.

Figure 5. Averaged variable importance analysis across the k-fold procedure for each set of variables derived from Landsat 8 (L8) and ALOS-2/PALSAR-2 (ALOS) included in the random forest (RF) model. The R^2 for each single set of variables and all variables together (left) and decrease in R^2 for models excluding a single set of variables (right). ALOS backscatter: γ^0_{HV} and γ^0_{HH}. ALOS indices: radar forest degradation index (RFDI), cross-polarised ratio (CpR). L8 reflectances: blue, green, red, near infrared, shortwave infrared-1, and shortwave infrared-2. L8 indices: normalised difference vegetation index (NDVI), normalised burn ratio (NBR), NBR2, normalised difference moisture index (NDMI), and soil adjusted vegetation index (SAVI).

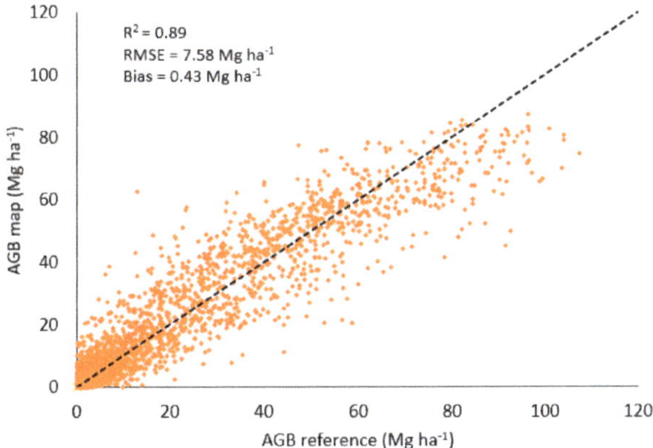

Figure 6. Cross-validation between the AGB map predictions and AGB reference data derived from the LiDAR point clouds. The black dash line corresponds to the y = x line. RMSE: root mean square error.

Figures 7 and 8 show the comparisons between the AGB estimates produced in our study and four other studies [21–23,81]. Although these maps were estimated using different methods and at different resolutions over the pantropical area, it is interesting to observe how the different data products compared to each other. While this study and Avitabile et al. [22] presented probability distribution functions skewed towards low biomass ranges, other studies [21,23,81] showed distributions tending towards much higher AGB levels, with averages two to three times larger than those found in this study and by Avitabile et al. [22] (Figure 7). Figure 8 shows that the datasets from [21,23,81] did not adequately represent all the variations that exist in the area due to the lower spatial resolution and, also, generally showed a greater overestimation of the AGB. Although Santoro et al. [81] provided more details in terms of the distribution of AGB in comparison with the other three maps, their map still overestimated the AGB compared to our results. In the map developed in this study, one can observe the AGB variation in detail, which is corroborated by the observed high coefficient of determination (R^2 = 0.89) between the AGB reference datasets and the final AGB map.

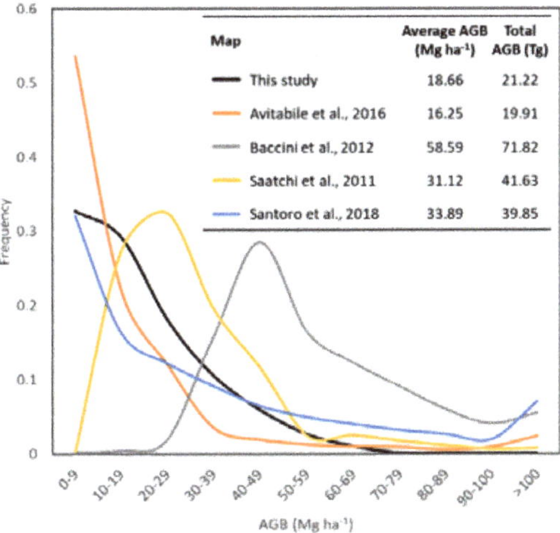

Figure 7. Distribution of the AGB intervals, average, and total AGB observed in this study and in four comparison maps for the Rio Vermelho watershed, Goiás State, Brazil. All the maps were resampled to 1-km resolutions.

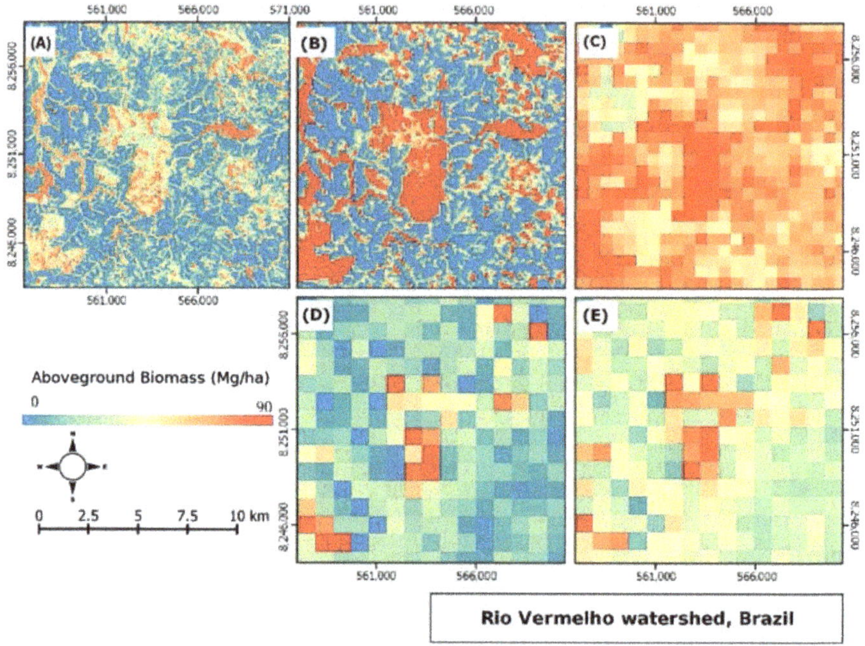

Figure 8. AGB maps over part of the Rio Vermelho watershed, Goiás State, Brazil produced by this study (30 m) (**A**) and by Santoro et al. [81] (100 m) (**B**), Baccini et al. [23] (500 m) (**C**), Avitabile et al. [22] (1 km) (**D**), and Saatchi et al. [21] (1 km) (**E**).

4. Discussion

In 2014, Brazil submitted a greenhouse gas (GHG) emission report (Forest Reference Emission Level (FREL) for Reducing Emissions from Deforestation for REDD+ under the United Nations Framework Convention on Climate Change (UNFCCC)) [82] as part of the country's commitment to reduce GHG emissions from deforestation in the Amazon (FREL Amazonia). Brazil indicated in that report that its national report would be the sum of the FRELs for each of the six Brazilian biomes. The FREL report for the Cerrado biome (FREL Cerrado), together with the FREL Amazonia, showed that emissions induced by land use changes accounted for approximately 73% of the emissions in Brazilian territory [82]. This shows the importance of the Cerrado biome for REDD+ policies and for the efforts of climate change mitigation in the country, such as the national inventories initiatives as part of the Third National Communication of Brazil to the UNFCCC [63].

The reduced number of studies on the quantification and spatial variability of the total Cerrado AGB stocks limits a full understanding of CO_2 emission patterns in this biome. These are especially critical for local and national climate governance strategies. Appropriate management and mitigation models will depend on the accuracy of this information. We produced the first high-resolution continuous map of AGB in an area of the Brazilian Cerrado, making use of a combination of vegetation inventory plots, airborne light detection and ranging (LiDAR) data, and satellite images (Landsat 8 and ALOS-2/PALSAR-2). Fieldwork in the Brazilian Cerrado is challenging, time-consuming, and expensive, and existing field datasets still do not entirely represent the extent of the biome. An alternative to increasing the sample size is using LiDAR by upscaling the information from field plots (which ideally should be representative of the analysed ecosystem). This strategy has been used in recent studies in other forest types [83,84]. We increased our sample size thousands of times, reaching a very high R^2 of 0.93 in the AGB LiDAR estimation. This consistent result was essential to allow us to use LiDAR as a reference for our AGB modelling process using satellite images and machine learning.

In our study, the RF algorithm proved to be a good strategy to estimate the AGB in the study area (Table 4 and Figure 6). This result can be attributed to the ability of these techniques to capture the nonlinearity present in the data, as they can approximate complex functions [85]. This is an important characteristic for the modelling of vegetation, such as ecological patterns governed by nonlinear and interactive processes at any ecological scale [85], since they usually present complex behaviours [28].

Current sources of AGB information for the Cerrado are the global and pantropical maps that have a limited spatial resolution (from 100 m to 1 km) and questionable high estimations for this biome (Figures 7 and 8) [2]. Figure 7 shows the distribution by AGB intervals, average, and total AGB observed from our study and the four global and pantropical maps over the Rio Vermelho watershed [21–23,81]. This study and the one conducted by Avitabile et al. [22] presented distributions with an average of 18.66 Mg ha^{-1} and 16.25 Mg ha^{-1}, respectively. On the other hand, Santoro et al. [81], Saatchi et al. [21], and Baccini et al. [23] showed distributions tending towards much higher AGB levels, with averages of 33.89 Mg ha $^{-1}$, 31.12 Mg ha $^{-1}$, and 58.59 Mg ha $^{-1}$, respectively. Saatchi et al. [21] and Baccini et al. [23] found the distributions of the AGB tending toward a normal distribution for the study site, which was not the case for the map produced by Avitabile et al. [22], Santoro et al. [81], and our study. Morandi et al. [2] recently calculated AGB values for the Cerrado biome using a large independent dataset of field plots. The study estimated an average of 20.4 ± 6.0 Mg ha^{-1} for the central areas of the Cerrado biome, where this watershed is located, which agrees with our study results. However, global and pantropical studies showed significantly higher AGB levels in comparison to Morandi et al. [2], our study (Figure 7), and to the AGB values measured in our field plots. We also observed the tendency of these products to overestimate the AGB on the lower AGB ranges, which results from the models attempting to compensate for the underestimation at the higher AGB ranges due to the potential signal saturation [28]. This agrees with previous studies that have indicated the inconsistency of these products when compared in several forest biomes [86,87]. Our regional AGB map (Figure 3) takes the local vegetation structure of the Brazilian Cerrado into account and has a spatial resolution of 30 m. Overall, the relationship between our reference data and the remote-sensing variables is relatively high (R^2 = 0.89), with a root mean square error (RMSE) of 7.58 Mg ha^{-1}. The accuracy assessment showed that our map underestimates high AGB levels and slightly overestimates low AGB levels (Table 4). This result fully meets the biomass product accuracy requirements set by the BIOMASS mission (RMSE < 10 Mg ha^{-1} for AGB < 50 Mg ha^{-1} and rel. RMSE < 20% for AGB > 50 Mg ha^{-1}) [56], with a RMSE of 6.39 Mg ha^{-1} for AGB < 50 Mg ha^{-1} and a rel. RMSE = 19% for values > 50 Mg ha^{-1}. It is important to highlight that we were interested in a representative range for cerradão AGB, including transitions to other vegetation types.

When analysing the statistical indicators of the goodness-of-fit (RMSE and R^2) (Table 4) using the RF technique, a higher accuracy was achieved for the AGB range > 40 Mg ha^{-1}. However, even though we are confident in our AGB estimations, we have to be cautious with the estimations of AGB values < 19 Mg ha^{-1}, areas that are not covered by woody vegetation (e.g. grasslands, agriculture, and bare soil), as our field-to-LiDAR AGB model was developed using field plots located only in woody vegetation areas. In Figure 6, we can observe an underestimation of the highest AGB ranges (AGB > 80 Mg ha^{-1}), which is also reflected in the negative bias for this high AGB level (Table 4). The range of predictions random forest regression can make is bound by the highest and lowest values in the training data. This can lead to overestimations in the lower value range and underestimations in the higher range [88]. The scatterplot between the satellite AGB estimates from our results and the LiDAR AGB showed this effect with the slight underestimation of high AGB (Table 4 and Figure 6). This effect was also observed in the studies of Nunes and Görgens [88] and Zhang and Lu [89].

It is reported that machine-learning techniques can provide more accurate estimates than classical regression models [88]. Silva et al. [90] evaluated the use of machine-learning techniques and mixed models to estimate the volume and AGB of individual trees in the Brazilian Cerrado. The machine-learning techniques presented, and mixed-effect models, showed similar and highly accurate results (R^2 = from 0.97 to 0.99 and RMSE = from 12% to 13%) [58]. According to Silva et al. [90], because the modelling of forest resources commonly presents complex relationships among the variables, nonlinear mixed-effects modelling (NLME) and machine-learning techniques such as RF, adaptive network-based fuzzy inference systems (ANFIS), and artificial neural networks (ANNs) may be good alternative modelling techniques. In our study, using RF, we produced an AGB map of an area in the Brazilian Cerrado with a R^2 of 0.89 and RMSE of 7.58 Mg ha^{-1}. Previous studies tried to estimate the AGB in savannas using satellite data and reached performances lower than the results from our work [54,91,92].

Over the last decades, several studies have used SAR, especially ALOS PALSAR L-band radar images, to estimate the AGB of tropical savannas. Mermoz et al. [93] used ALOS PALSAR data to map the AGB in savanna ecosystems in Cameroon. They argued that L-band PALSAR mosaics are suitable for the retrieval of savanna AGB (typically less than 100 Mg ha^{-1}) at the national and continental scales. In a similar study, Carreiras et al. [27] tested a combination of field data and ALOS PALSAR backscatter intensity to reduce the uncertainty in the estimation of vegetation AGB in the Miombo savanna woodlands of Mozambique (East Africa). They applied a machine-learning algorithm, resulting in a good fit and accurate model (R^2 = 0.95 and RMSE = 5.03 Mg ha^{-1}). However, since they used bootstrap samples in combination with a cross-validation procedure, the reported cross-validation statistics could be overoptimistic [22]. A combination of datasets from different sources, such as in this study, proved to be efficient when the goal was to reduce uncertainties in the AGB estimates. Recent studies have explored the combination of different data types. Braun et al. [91] used passive and active microwaves to estimate the AGB of savannas. They introduced the integration of passive brightness temperature as an additional variable for AGB estimation, based on the hypothesis that it contains information complementary to the microwave backscatter coefficient from active sensors.

As we found in our work, many studies have shown the advantages of combining optical and SAR datasets using machine-learning techniques such as RF for AGB estimations. The optical data are sensitive to the biophysical properties of the vegetation, and radar data are more sensitive to the electrical and geometrical information of the vegetation—more specifically, the moisture content and vegetation structure. Forkuor et al. [46] showed in a recent study that combining field inventory data with Sentinel 1 (SAR) and Sentinel 2 (optical) to estimate the AGB in the West African dryland forest (tropical savanna and woodlands) using a RF algorithm performed much better than either one of them alone, reaching a R^2 of 0.90. This corroborates our results, which have also shown an improvement in accuracy using the combination of SAR and optical datasets rather than using either one individually. Our study also showed the highest performance (R^2 = 0.89) when all variables were included. The variable importance analysis (Figure 5) showed a slightly superior R^2 for Landsat 8 reflectance data (R^2 = 0.84) in comparison to ALOS–2 backscatter data (R^2 = 0.79) when mapping the AGB using a single set of data. Landsat 8 indices seem to have an almost identical performance to the Landsat 8 reflectance data, and there is no significant drop in performance when the indices are excluded (Figure 6). Conversely, the exclusion of the ALOS indices from the whole set results in the largest drop of R^2 (i.e., −0.07), despite having the lowest performance when used alone (R^2 = 0.51). This agrees with previous studies that found a higher contribution of the optical reflectance data to estimate AGB, potentially due to the sensitivity of optical data to shadow and moisture [23,29,94], which can be a key factor in sparse vegetation such as Cerrado. Additionally, the relatively low AGB levels in the study area might provide a level playing field with L-band SAR data, which usually presents a sensitivity of the signal to higher AGB levels than optical data [95–99]. Our study highlighted the need to use remote sensing in combination with local vegetation inventories to effectively quantify the spatial variation of AGB in ecosystems of the Brazilian Cerrado. Future high-resolution maps of ABG will likely be more useful to quantify carbon emissions from Cerrado degradations at the local and regional scales. The methodology presented here has the potential to be used to generate the first of its kind AGB map of the entire Brazilian Cerrado, which is often neglected in carbon stock assessments of the South American continent.

5. Conclusions

We provided the first high-resolution map of the AGB (30-m resolution) of a Brazilian Cerrado area using a combination of field inventory plots, airborne light detection and ranging (LiDAR) data, and satellite images (Landsat 8 and ALOS−2/PALSAR−2). We used random forest (RF) models and jackknife analyses to study the importance of remote-sensing variables to quantify the AGB of cerradão at large scales. Overall, the relationship between the reference data and remote-sensing variables is strong (R^2 = 0.89), with a root mean square error (RMSE) of 7.58 Mg ha^{-1}. Spatially, our map slightly underestimated and overestimated the values of the AGB in few areas of the savanna (bias = 0.43 Mg ha^{-1}). However, this spatial bias is similar to other AGB maps. Our study highlights the need to use remote sensing in combination with local field inventories to effectively quantify the spatial variation of AGB in the ecosystems of the Brazilian Savanna.

Supplementary Materials: The following are available online at http://www.mdpi.com/2072-4292/12/17/2685/s1: Figure S1: One-on-one relationship between the selected airborne LiDAR predictors (CC (%): canopy cover; and CHM (m): canopy height model) and the mean AGB values measured in the field plots (Mg ha-1). Figure S2 to Figure S5: Field photos of natural vegetation found in the Rio Vermelho watershed. Figure S6 to Figure S20: LiDAR returns.

Author Contributions: Conceptualisation, P.d.C.B., P.R.-V., and H.B.; methodology, P.d.C.B. and P.R.-V.; formal analysis, P.d.C.B., P.R.-V., B.Z., S.d.C.d.M., S.F., F.B., M.G., D.R., and I.R.; validation, P.d.C.B., P.R.-V., S.d.C.d.M., C.H.G.C., B.Z., and L.G.F.; visualisation; P.d.C.B., P.R.-V., V.L., Y.G., A.M.P.-P., and D.R.; original draft preparation, P.d.C.B.; supervision, H.B.; writing—review and editing, P.d.C.B., P.R.-V., B.Z., S.d.C.d.M., C.H.G.C., S.F., F.B., V.L., D.R., M.G., F.D.B.E.-S., A.M.P.-P., Y.G., J.R., K.B., L.G.F., J.Z.S., A.A., I.R., M.B., I.H.W., E.E.S., J.P.O., K.T., and H.B. All authors have read and agreed to the published version of the manuscript.

Funding: The authors were supported by the Forests 2020 project from UK Space Agency's International Partnership Programme (IPP) under the Global Challenge Research Fund (GCRF). P. Rodriguez-Veiga and H. Balzter were also supported by the UK's Natural Environment Research Council, Agreement PR140015, between the NERC and National Centre for Earth Observation (NCEO) and ESA Biomass Climate Change Initiative (CCI+) project (4000123662/18/I-NB). S.C. Miranda was supported by the Edital Universal—CNPq (445420/2014-6). M. Bustamante was supported by the INCT MC Fase 2 CNPq (465501/2014-1).

Acknowledgments: The authors would like to acknowledge the Sustainable Landscapes Brazil project supported by the Brazilian Agricultural Research Corporation (EMBRAPA), the US Forest Service, USAID, and the US Department of State for the LiDAR data. The authors thank the MapBiomas team who provided the land cover map of Brazilian Cerrado with the different forest formations. The authors thank the editor and the anonymous reviewers for their valuable comments, which have improved the quality of this paper.

Conflicts of Interest: The authors declare no conflict of interest.

References

1. Instituto Brasileiro de Geografia e Estatistica (IBGE). Biomas E Sistema Costeiro-Marinho do Brasil. 2019. Available online: https://www.ibge.gov.br/apps/biomas/ (accessed on 18 August 2020).

2. Morandi, P.S.; Marimon, B.S.; Marimon-Júnior, B.H.; Ratter, J.A.; Feldpausch, T.R.; Colli, G.R.; Munhoz, C.B.R.; Júnior, M.C.D.S.; Lima, E.D.S.; Haidar, R.F.; et al. Tree diversity and above-ground biomass in the South America Cerrado biome and their conservation implications. *Biodivers. Conserv.* **2018**, *29*, 1519–1536. [CrossRef]

3. Klink, C.A.; Machado, R.B. Conservation of the Brazilian Cerrado. *Conserv. Boil.* **2005**, *19*, 707–713. [CrossRef]

4. Sano, E.E.; Rodrigues, A.A.; Martins, É.S.; Bettiol, G.M.; Bustamante, M.M.; Bezerra, A.S.; Couto, A.F.; Vasconcelos, V.; Schüler, J.; Bolfe, E.L. Cerrado ecoregions: A spatial framework to assess and prioritize Brazilian savanna environmental diversity for conservation. *J. Environ. Manag.* **2019**, *232*, 818–828. [CrossRef] [PubMed]

5. Ribeiro, S.C.; Fehrmann, L.; Soares, C.P.B.; Jacovine, L.A.G.; Kleinn, C.; Gaspar, R.D.O. Above- and belowground biomass in a Brazilian Cerrado. *For. Ecol. Manag.* **2011**, *262*, 491–499. [CrossRef]

6. Coutinho, L.M. O conceito de cerrado. *Rev. Bras. Bot.* **1978**, *1*, 17–23.

7. Delitti, W.B.C.; Meguro, M.; Pausas, J.G. Biomass and mineralmass estimates in a "cerrado" ecosystem. *Braz. J. Bot.* **2006**, *29*, 531–540. [CrossRef]

8. Brazil. Agricultural Development Plan Matopiba, Law no. 8.447–05/06/2015. 2015. Available online: http://www.planalto.gov.br/ccivil_03/_ato2015-2018/2015/decreto/d8447.htm (accessed on 20 December 2019).

9. Spera, S.A.; Galford, G.L.; Coe, M.T.; Macedo, M.N.; Mustard, J.F. Land-use change affects water recycling in Brazil's last agricultural frontier. *Glob. Chang. Boil.* **2016**, *22*, 3405–3413. [CrossRef]

10. MMA. *Mapeamento do Uso e Cobertura do Cerrado. Projeto TerraClass Cerrado 2013*; MMA: Brasília, Brazil, 2015.

11. Ferreira, J.N.; Pardini, R.; Metzger, J.P.; Fonseca, C.R.; Pompeu, P.S.; Sparovek, G.; Louzada, J. Towards environmentally sustainable agriculture in Brazil: Challenges and opportunities for applied ecological research. *J. Appl. Ecol.* **2012**, *49*, 535–541. [CrossRef]

12. Leite, J.P.R.; Araújo, D.L.S.; Duarte, M.D. Reflexos e considerações sobre a implementação do cadastro nacional de unidades de conservação do estado do piauí. *Sustentare* **2018**, *2*, 20–31. [CrossRef]

13. Metzger, J.P.; Bustamante, M.M.; Ferreira, J.; Fernandes, G.W.; Librán-Embid, F.; Pillar, V.D.; Prist, P.R.; Rodrigues, R.R.; Vieira, I.C.G.; Overbeck, G.E.; et al. Why Brazil needs its Legal Reserves. *Perspect. Ecol. Conserv.* **2019**, *17*, 91–103. [CrossRef]

14. ICMBio. Cerrado. Available online: http://www.icmbio.gov.br/portal/unidadesdeconservacao/biomas-brasileiros/cerrado (accessed on 10 February 2020).

15. De Miranda, S.D.C.; Bustamante, M.M.C.; Palace, M.; Hagen, S.; Keller, M.; Ferreira, L.G. Regional Variations in Biomass Distribution in Brazilian Savanna Woodland. *Biotropica* **2014**, *46*, 125–138. [CrossRef]

16. Roitman, I.; Bustamante, M.M.C.; Haidar, R.F.; Shimbo, J.Z.; Abdala, G.C.; Eiten, G.; Fagg, C.W.; Felfili, M.C.; Felfili, J.M.; Jacobson, T.K.B.; et al. Optimizing biomass estimates of savanna woodland at different spatial scales in the Brazilian Cerrado: Re-evaluating allometric equations and environmental influences. *PLoS ONE* **2018**, *13*, e0196742. [CrossRef] [PubMed]

17. Chave, J.; Réjou-Méchain, M.; Búrquez, A.; Chidumayo, E.; Colgan, M.S.; Delitti, W.B.; Duque, A.; Eid, T.; Fearnside, P.M.; Goodman, R.C.; et al. Improved allometric models to estimate the aboveground biomass of tropical trees. *Glob. Chang. Boil.* **2014**, *20*, 3177–3190. [CrossRef] [PubMed]

18. Temesgen, H.; Affleck, D.; Poudel, K.P.; Gray, A.; Sessions, J.; Hailemariam, T. A review of the challenges and opportunities in estimating above ground forest biomass using tree-level models. *Scand. J. For. Res.* **2015**, *30*, 1–10. [CrossRef]

19. Houghton, R.A. Aboveground Forest Biomass and the Global Carbon Balance. *Glob. Chang. Boil.* **2005**, *11*, 945–958. [CrossRef]

20. Sileshi, G.W. A critical review of forest biomass estimation models, common mistakes and corrective measures. *For. Ecol. Manag.* **2014**, *329*, 237–254. [CrossRef]

21. Saatchi, S.; Harris, N.L.; Brown, S.; Lefsky, M.; Mitchard, E.T.A.; Salas, W.; Zutta, B.R.; Buermann, W.; Lewis, S.L.; Hagen, S.; et al. Benchmark map of forest carbon stocks in tropical regions across three continents. *Proc. Natl. Acad. Sci. USA* **2011**, *108*, 9899–9904. [CrossRef]

22. Avitabile, V.; Herold, M.; Heuvelink, G.B.M.; Lewis, S.L.; Phillips, O.L.; Asner, G.P.; Armston, J.D.; Ashton, P.S.; Banin, L.; Bayol, N.; et al. An integrated pan-tropical biomass map using multiple reference datasets. *Glob. Chang. Boil.* **2016**, *22*, 1406–1420. [CrossRef]

23. Baccini, A.; Goetz, S.J.; Walker, W.S.; Laporte, N.T.; Sun, M.; Sulla-Menashe, D.; Hackler, J.; Beck, P.S.A.; Dubayah, R.; Friedl, M.A.; et al. Estimated carbon dioxide emissions from tropical deforestation improved by carbon-density maps. *Nat. Clim. Chang.* **2012**, *2*, 182–185. [CrossRef]

24. Baccini, A.; Walker, W.; Carvalho, L.; Farina, M.; Sulla-Menashe, D.; Houghton, R.A. Tropical forests are a net carbon source based on aboveground measurements of gain and loss. *Science* **2017**, *358*, 230–234. [CrossRef]

25. Marselis, S.M.; Tang, H.; Armston, J.D.; Calders, K.; Labrière, N.; Dubayah, R. Distinguishing vegetation types with airborne waveform lidar data in a tropical forest-savanna mosaic: A case study in Lopé National Park, Gabon. *Remote Sens. Environ.* **2018**, *216*, 626–634. [CrossRef]

26. Bispo, P.C.; Santos, J.R.; Valeriano, M.M.; Touzi, R.; Seifert, F.M. Integration of Polarimetric PALSAR Attributes and Local Geomorphometric Variables Derived from SRTM for Forest Biomass Modeling in Central Amazonia. *Can. J. Remote Sens.* **2014**, *40*, 26–42. [CrossRef]

27. Carreiras, J.; Vasconcelos, M.; Lucas, R.M. Understanding the relationship between aboveground biomass and ALOS PALSAR data in the forests of Guinea-Bissau (West Africa). *Remote Sens. Environ.* **2012**, *121*, 426–442. [CrossRef]

28. Rodríguez-Veiga, P.; Quegan, S.; Carreiras, J.; Persson, H.J.; Fransson, J.E.; Hoscilo, A.; Ziółkowski, D.; Stereńczak, K.; Lohberger, S.; Stängel, M.; et al. Forest biomass retrieval approaches from earth observation in different biomes. *Int. J. Appl. Earth Obs. Geoinf.* **2019**, *77*, 53–68. [CrossRef]

29. Rodríguez-Veiga, P.; Saatchi, S.; Tansey, K.; Balzter, H. Magnitude, spatial distribution and uncertainty of forest biomass stocks in Mexico. *Remote Sens. Environ.* **2016**, *183*, 265–281. [CrossRef]

30. Takagi, K.; Yone, Y.; Takahashi, H.; Sakai, R.; Hojyo, H.; Kamiura, T.; Nomura, M.; Liang, N.; Fukazawa, T.; Miya, H.; et al. Forest biomass and volume estimation using airborne LiDAR in a cool-temperate forest of northern Hokkaido, Japan. *Ecol. Inform.* **2015**, *26*, 54–60. [CrossRef]

31. Hernández-Stefanoni, J.L.; Castillo-Santiago, M.Á.; Mas, J.F.; Wheeler, C.E.; Andres-Mauricio, J.; Tun-Dzul, F.; George-Chacón, S.P.; Reyes-Palomeque, G.; Castellanos-Basto, B.; Vaca, R.; et al. Improving aboveground biomass maps of tropical dry forests by integrating LiDAR, ALOS PALSAR, climate and field data. *Carbon Balance Manag.* **2020**, *15*, 1–17. [CrossRef]

32. Cartus, O.; Santoro, M.; Kellndorfer, J. Mapping forest aboveground biomass in the Northeastern United States with ALOS PALSAR dual-polarization L-band. *Remote Sens. Environ.* **2012**, *124*, 466–478. [CrossRef]

33. Chen, Q. Modeling aboveground tree woody biomass using national-scale allometric methods and airborne lidar. *ISPRS J. Photogramm. Remote Sens.* **2015**, *106*, 95–106. [CrossRef]

34. Deng, S.; Katoh, M.; Guan, Q.; Yin, N.; Li, M. Estimating Forest Aboveground Biomass by Combining ALOS PALSAR and WorldView-2 Data: A Case Study at Purple Mountain National Park, Nanjing, China. *Remote Sens.* **2014**, *6*, 7878–7910. [CrossRef]

35. Li, A.; Glenn, N.F.; Olsoy, P.J.; Mitchell, J.J.; Shrestha, R. Aboveground biomass estimates of sagebrush using terrestrial and airborne LiDAR data in a dryland ecosystem. *Agric. For. Meteorol.* **2015**, *213*, 138–147. [CrossRef]

36. Puliti, S.; Saarela, S.; Gobakken, T.; Ståhl, G.; Næsset, E. Combining UAV and Sentinel-2 auxiliary data for forest growing stock volume estimation through hierarchical model-based inference. *Remote Sens. Environ.* **2018**, *204*, 485–497. [CrossRef]

37. Le Toan, T.; Beaudoin, A.; Riom, J.; Guyon, D. Relating forest biomass to SAR data. *IEEE Trans. Geosci. Remote Sens.* **1992**, *30*, 403–411. [CrossRef]

38. Joshi, N.; Mitchard, E.T.A.; Brolly, M.; Schumacher, J.; Fernández-Landa, A.; Johannsen, V.K.; Marchamalo, M.; Fensholt, R. Understanding 'saturation' of radar signals over forests. *Sci. Rep.* **2017**, *7*, 3505. [CrossRef]

39. Ouchi, K. Recent Trend and Advance of Synthetic Aperture Radar with Selected Topics. *Remote Sens.* **2013**, *5*, 716–807. [CrossRef]

40. Ghosh, S.M.; Behera, M.D. Forest canopy height estimation using satellite laser altimetry: A case study in the Western Ghats, India. *Appl. Geomat.* **2017**, *9*, 159–166. [CrossRef]

41. Minh, D.H.T.; Le Toan, T.; Rocca, F.; Tebaldini, S.; Villard, L.; Réjou-Méchain, M.; Phillips, O.L.; Feldpausch, T.R.; Dubois-Fernandez, P.; Scipal, K.; et al. SAR tomography for the retrieval of forest biomass and height: Cross-validation at two tropical forest sites in French Guiana. *Remote Sens. Environ.* **2016**, *175*, 138–147. [CrossRef]

42. Bispo, P.D.C.; Pardini, M.; Papathanassiou, K.P.; Kugler, F.; Balzter, H.; Rains, D.; Dos Santos, J.R.; Rizaev, I.G.; Tansey, K.; Dos Santos, M.N.; et al. Mapping forest successional stages in the Brazilian Amazon using forest heights derived from TanDEM-X SAR interferometry. *Remote Sens. Environ.* **2019**, *232*, 111194. [CrossRef]

43. Moura, Y.M.; Balzter, H.; Galvão, L.S.; Dalagnol, R.; Espírito-Santo, F.; Santos, E.G.; Garcia, M.; Bispo, P.D.C.; Junior, R.C.D.O.; Shimabukuro, Y.E. Carbon Dynamics in a Human-Modified Tropical Forest: A Case Study Using Multi-Temporal LiDAR Data. *Remote Sens.* **2020**, *12*, 430. [CrossRef]

44. Erten, E.; Lopez-Sanchez, J.M.; Yuzugullu, O.; Hajnsek, I. Retrieval of agricultural crop height from space: A comparison of SAR techniques. *Remote Sens. Environ.* **2016**, *187*, 130–144. [CrossRef]

45. Su, Y.; Guo, Q.; Xue, B.; Hu, T.; Alvarez, O.; Tao, S.; Fang, J. Spatial distribution of forest aboveground biomass in China: Estimation through combination of spaceborne lidar, optical imagery, and forest inventory data. *Remote Sens. Environ.* **2016**, *173*, 187–199. [CrossRef]

46. Forkuor, G.; Zoungrana, B.J.-B.; Dimobe, K.; Ouattara, B.; Vadrevu, K.P.; Tondoh, J.E. Above-ground biomass mapping in West African dryland forest using Sentinel-1 and 2 datasets - A case study. *Remote Sens. Environ.* **2020**, *236*, 111496. [CrossRef]

47. Heckel, K.; Urban, M.; Schratz, P.; Mahecha, M.D.; Schmullius, C. Predicting Forest Cover in Distinct Ecosystems: The Potential of Multi-Source Sentinel-1 and -2 Data Fusion. *Remote Sens.* **2020**, *12*, 302. [CrossRef]

48. Wessels, K.; Mathieu, R.; Knox, N.; Main, R.; Naidoo, L.; Steenkamp, K. Mapping and Monitoring Fractional Woody Vegetation Cover in the Arid Savannas of Namibia Using LiDAR Training Data, Machine Learning, and ALOS PALSAR Data. *Remote Sens.* **2019**, *11*, 2633. [CrossRef]

49. Chang, J.; Shoshany, M. Mediterranean shrublands biomass estimation using Sentinel-1 and Sentinel-2. In Proceedings of the 2016 IEEE International Geoscience and Remote Sensing Symposium (IGARSS), Beijing, China, 10–15 July 2016; pp. 5300–5303.

50. Omar, H.; Misman, M.A.; Kassim, A.R. Synergetic of PALSAR-2 and Sentinel-1A SAR Polarimetry for Retrieving Aboveground Biomass in Dipterocarp Forest of Malaysia. *Appl. Sci.* **2017**, *7*, 675. [CrossRef]

51. Sinha, S.; Jeganathan, C.; Sharma, L.K.; Nathawat, M.S.; Das, A.; Mohan, S. Developing synergy regression models with space-borne ALOS PALSAR and Landsat TM sensors for retrieving tropical forest biomass. *J. Earth Syst. Sci.* **2016**, *125*, 725–735. [CrossRef]

52. Fidelis, A.; Lyra, M.F.D.S.; Pivello, V.R. Above- and below-ground biomass and carbon dynamics in Brazilian Cerrado wet grasslands. *J. Veg. Sci.* **2012**, *24*, 356–364. [CrossRef]

53. Bitencourt, M.D.; De Mesquita, J.H.N.; Kuntschik, G.; Da Rocha, H.R.; Furley, P. Cerrado vegetation study using optical and radar remote sensing: Two Brazilian case studies. *Can. J. Remote Sens.* **2007**, *33*, 468–480. [CrossRef]

54. Miguel, E.P.; Rezende, A.V.; Leal, F.A.; Matricardi, E.A.T.; Vale, A.T.D.; Pereira, R.S. Redes neurais artificiais para a modelagem do volume de madeira e biomassa do cerradão com dados de satélite. *Pesqui Agropecu. Bras.* **2015**, *50*, 829–839. [CrossRef]

55. Schwieder, M.; Leitão, P.J.; Pinto, J.R.R.; Teixeira, A.M.C.; Pedroni, F.; Sanchez, M.; Bustamante, M.M.; Hostert, P. Landsat phenological metrics and their relation to aboveground carbon in the Brazilian Savanna. *Carbon Balance Manag.* **2018**, *13*, 7. [CrossRef]

56. Le Toan, T.; Quegan, S.; Davidson, M.; Balzter, H.; Paillou, P.; Papathanassiou, K.; Plummer, S.; Rocca, F.; Saatchi, S.; Shugart, H.; et al. The BIOMASS mission: Mapping global forest biomass to better understand the terrestrial carbon cycle. *Remote Sens. Environ.* **2011**, *115*, 2850–2860. [CrossRef]

57. Vieira, P.A.; Ferreira, N.C.; Ferreira, L.G. Análise da vulnerabilidade natural da paisagem em relação aos diferentes níveis de ocupação da bacia hidrográfica do Rio Vermelho, estado de Goiás. *Soc. Nat.* **2014**, *26*, 385–400. [CrossRef]

58. Vieira, P.A.; Ferreira, M.E.; Ferreira, L.G. Modelagem dinâmica da paisagem aplicada na análise de uso do solo na bacia hidrográfica do Rio Vermelho, Goiás, Brasil. *Rev. Bras. Cart.* **2015**, *67*, 1217–1230.

59. De Moura, J.U.; Bucci, R.L.F. Aspectos geográficos das micro-regiões do mato grosso goiano de goiás, meia ponte, sudeste goiano e planalto goiano. *Bol. Goiano Geogr.* **2009**, *1*, 60–94. [CrossRef]

60. Cavalcanti, M.A.; Lopes, L.M.; De Pontes, M.N.C. Contribuição Ao Entendimento Do Fenômeno Das Enchentes Do Rio Vermelho Na Cidade De Goiás, GO. *Bol. Goiano Geogr.* **2008**, *28*, 167–186. [CrossRef]

61. MapBiomas. Collection 4.0. 2020. Available online: https://mapbiomas.org (accessed on 20 February 2020).

62. Ribeiro, J.F.; Walter, B.M.T. As principais fitofisionomias do Bioma Cerrado. In *Ecologia e Flora*; Sano, S.M., Almeida, S.P., Ribeiro, J.F., Eds.; EMBRAPA: Brasília, Brazil, 2008; Volume 1, pp. 152–212.

63. Brazil. *Third National Communication of Brazil to the United Nations Framework Convention on Climate Change*; Ministry of Science, Technology and Innovation: Brasília, Brazil, 2016.

64. Instituto Brasileiro de Geografia e Estatistica (IBGE). Manual Técnico da Vegetação Brasileira. In *Manuais Técnicos em Geociências*, 2nd ed.; Ministério do Planejamento, Orçamento e Gestão, Instituto Brasileiro de Geografia e Estatística: Rio de Janeiro, Brazil, 2012; p. 275.

65. Ribeiro, J.F.; WAltER, B.M.T. Fitofisionomias do bioma Cerrado. In *Cerrado: Ambiente e Flora*; Sano, S.A.S., Ed.; Embrapa-CPAC: Planaltina, Brazil, 1998; pp. 89–166.

66. Brown, S. Measuring, monitoring, and verification of carbon benefits for forest–based projects. *Philos. Trans. R. Soc. A Math. Phys. Eng. Sci.* **2002**, *360*, 1669–1683. [CrossRef] [PubMed]

67. Chave, J.; Andalo, C.; Brown, S.; Cairns, M.A.; Chambers, J.Q.; Eamus, D.; Fölster, H.; Fromard, F.; Higuchi, N.; Kira, T.; et al. Tree allometry and improved estimation of carbon stocks and balance in tropical forests. *Oecologia* **2005**, *145*, 87–99. [CrossRef]

68. Scolforo, J.R.; Rufini, A.L.; Mello, J.; Trugilho, P.F.; Oliveira, A.; Silva, C. Equações para o peso de matéria seca das fisionomias, em Minas Gerais. In *Inventário Florestal de Minas Gerais: Equações de Volume, Peso de Matéria Seca e Carbono para Diferentes Fitofisionomias da Flora Nativa*; Scolforo, J.R.S., Oliveira, A.D., Acerbi, F.W., Jr., Eds.; UFLA: Lavras, Brazil, 2008.

69. LAStools. Efficient LiDAR Processing Software (Version 170323, Commercial). Available online: http://rapidlasso.com/LAStools (accessed on 16 July 2020).

70. Foga, S.; Scaramuzza, P.L.; Guo, S.; Zhu, Z.; Dilley, R.D.; Beckmann, T.; Schmidt, G.L.; Dwyer, J.; Hughes, M.J.; Laue, B. Cloud detection algorithm comparison and validation for operational Landsat data products. *Remote Sens. Environ.* **2017**, *194*, 379–390. [CrossRef]

71. Cartus, O.; Kellndorfer, J.; Walker, W.; Franco, C.; Bishop, J.; Santos, L.A.; Fuentes, J.M.M. A National, Detailed Map of Forest Aboveground Carbon Stocks in Mexico. *Remote Sens.* **2014**, *6*, 5559–5588. [CrossRef]

72. Halperin, J.; Lemay, V.; Chidumayo, E.; Verchot, L.; Marshall, P. Model-based estimation of above-ground biomass in the miombo ecoregion of Zambia. *For. Ecosyst.* **2016**, *3*, 14556. [CrossRef]

73. Rodríguez-Veiga, P.; Carreiras, J.; Smallman, T.L.; Exbrayat, J.-F.; Ndambiri, J.; Mutwiri, F.; Nyasaka, D.; Quegan, S.; Williams, M.; Balzter, H. Carbon Stocks and Fluxes in Kenyan Forests and Wooded Grasslands Derived from Earth Observation and Model-Data Fusion. *Remote Sens.* **2020**, *12*, 2380. [CrossRef]

74. Shimada, M.; Itoh, T.; Motooka, T.; Watanabe, M.; Shiraishi, T.; Thapa, R.B.; Lucas, R.M. New global forest/non-forest maps from ALOS PALSAR data (2007–2010). *Remote Sens. Environ.* **2014**, *155*, 13–31. [CrossRef]

75. Quegan, S.; Yu, J.J. Filtering of multichannel SAR images. *IEEE Trans. Geosci. Remote Sens.* **2001**, *39*, 2373–2379. [CrossRef]

76. Breiman, L. Random forests. *Mach. Learn.* **2001**, *45*, 5–32. [CrossRef]

77. Rodríguez-Veiga, P.; Barbosa-Herrera, A.P.; Barreto-Silva, J.S.; Bispo, P.C.; Cabrera, E.; Capachero, C.; Galindo, G.; Gou, Y.; Moreno, L.M.; Louis, V.; et al. Mapping the spatial distribution of colombia's forest aboveground biomass using sar and optical data. *ISPRS-Int. Arch. Photogramm. Remote Sens. Spat. Inf. Sci.* **2019**, *42*, 57–60. [CrossRef]

78. Mitchard, E.T.A.; Saatchi, S.; Lewis, S.L.; Feldpausch, T.R.; Woodhouse, I.; Sonke, B.; Rowland, C.; Meir, P. Measuring biomass changes due to woody encroachment and deforestation/degradation in a forest–savanna boundary region of central Africa using multi-temporal L-band radar backscatter. *Remote Sens. Environ.* **2011**, *115*, 2861–2873. [CrossRef]

79. Réjou-Méchain, M.; Muller-Landau, H.C.; Detto, M.; Thomas, S.C.; Le Toan, T.; Saatchi, S.S.; Barreto-Silva, J.S.; Bourg, N.A.; Bunyavejchewin, S.; Butt, N.; et al. Local spatial structure of forest biomass and its consequences for remote sensing of carbon stocks. *Biogeosciences* **2014**, *11*, 6827–6840. [CrossRef]

80. James, G.; Witten, D.; Hastie, T.; Tibshirani, R. *An Introduction to Statistical Learning*; Springer: New York, NY, USA, 2013; p. 103.

81. Santoro, M.; Cartus, O.; Carvalhais, N.; Rozendaal, D.; Avitabilie, V.; Araza, A.; De Bruin, S.; Herold, M.; Quegan, S.; Veiga, P.R.; et al. The global forest above-ground biomass pool for 2010 estimated from high-resolution satellite observations. *Earth Syst. Sci. Data Discuss.* **2020**. [CrossRef]

82. MMA. *Brazil's Forest Reference Emission Level for Reducing Emissions from Deforestation in the Cerrado Biome for Results-Based Payments for REDD+ under the United Nations Framework Convention on Climate Change*; MMA: Brasília, Brazil, 2017.

83. Urbazaev, M.; Thiel, C.; Cremer, F.; Dubayah, R.; Migliavacca, M.; Reichstein, M.; Schmullius, C. Estimation of forest aboveground biomass and uncertainties by integration of field measurements, airborne LiDAR, and SAR and optical satellite data in Mexico. *Carbon Balance Manag.* **2018**, *13*, 5–20. [CrossRef]

84. Wang, D.; Wan, B.; Liu, J.; Su, Y.; Guo, Q.; Qiu, P.; Wu, X. Estimating aboveground biomass of the mangrove forests on northeast Hainan Island in China using an upscaling method from field plots, UAV-LiDAR data and Sentinel-2 imagery. *Int. J. Appl. Earth Obs. Geoinf.* **2020**, *85*, 101986. [CrossRef]

85. Ryo, M.; Rillig, M.C. Statistically reinforced machine learning for nonlinear patterns and variable interactions. *Ecosphere* **2017**, *8*, e01976. [CrossRef]

86. Rodríguez-Veiga, P.; Wheeler, J.; Louis, V.; Tansey, K.; Balzter, H. Quantifying Forest Biomass Carbon Stocks from Space. *Curr. For. Rep.* **2017**, *3*, 1–18. [CrossRef]

87. Mitchard, E.T.A.; Saatchi, S.; Baccini, A.; Asner, G.P.; Goetz, S.J.; Harris, N.L.; Brown, S. Uncertainty in the spatial distribution of tropical forest biomass: A comparison of pan-tropical maps. *Carbon Balance Manag.* **2013**, *8*, 10. [CrossRef] [PubMed]

88. Nunes, M.H.; Gorgens, E.B. Artificial Intelligence Procedures for Tree Taper Estimation within a Complex Vegetation Mosaic in Brazil. *PLoS ONE* **2016**, *11*, e0154738. [CrossRef] [PubMed]

89. Zhang, G.; Lu, Y. Bias-corrected random forests in regression. *J. Appl. Stat.* **2012**, *39*, 151–160. [CrossRef]

90. Silva, J.P.M.; Da Silva, M.L.M.; Da Silva, E.F.; Da Silva, G.F.; De Mendonça, A.R.; Cabacinha, C.D.; Araújo, E.F.; Santos, J.S.; Vieira, G.C.; De Almeida, M.N.F.; et al. Computational techniques applied to volume and biomass estimation of trees in Brazilian savanna. *J. Environ. Manag.* **2019**, *249*, 109368. [CrossRef]

91. Braun, A.; Wagner, J.; Hochschild, V. Above-ground biomass estimates based on active and passive microwave sensor imagery in low-biomass savanna ecosystems. *J. Appl. Remote Sens.* **2018**, *12*, 046027. [CrossRef]

92. Lima-Bittencourt, C.; Astolfi-Filho, S.; Chartone-Souza, E.; Santos, F.R.; Nascimento, A.M.A. Analysis of Chromobacterium sp. natural isolates from different Brazilian ecosystems. *BMC Microbiol.* **2007**, *7*, 58. [CrossRef]

93. Mermoz, S.; Le Toan, T.; Villard, L.; Réjou-Méchain, M.; Seifert-Granzin, J. Biomass assessment in the Cameroon savanna using ALOS PALSAR data. *Remote Sens. Environ.* **2014**, *155*, 109–119. [CrossRef]

94. Avitabile, V.; Baccini, A.; Friedl, M.A.; Schmullius, C. Capabilities and limitations of Landsat and land cover data for aboveground woody biomass estimation of Uganda. *Remote Sens. Environ.* **2012**, *117*, 366–380. [CrossRef]

95. Wu, Y.; Strahler, A.H. Remote Estimation of Crown Size, Stand Density, and Biomass on the Oregon Transect. *Ecol. Appl.* **1994**, *4*, 299–312. [CrossRef]

96. Gibbs, H.K.; Brown, S.; O Niles, J.; Foley, J. Monitoring and estimating tropical forest carbon stocks: Making REDD a reality. *Environ. Res. Lett.* **2007**, *2*, 045023. [CrossRef]

97. Wagner, W.; Luckman, A.; Vietmeier, J.; Tansey, K.; Balzter, H.; Schmullius, C.; Davidson, M.; Gaveau, D.; Gluck, M.; Le Toan, T.; et al. Large-scale mapping of boreal forest in SIBERIA using ERS tandem coherence and JERS backscatter data. *Remote Sens. Environ.* **2003**, *85*, 125–144. [CrossRef]

98. Mitchard, E.T.A.; Saatchi, S.S.; Woodhouse, I.H.; Nangendo, G.; Ribeiro, N.S.; Williams, M.; Ryan, C.; Lewis, S.L.; Feldpausch, T.R.; Meir, P. Using satellite radar backscatter to predict above-ground woody biomass: A consistent relationship across four different African landscapes. *Geophys. Res. Lett.* **2009**, *36*. [CrossRef]

99. Lucas, R.M.; Armston, J.D.; Fairfax, R.; Fensham, R.; Accad, A.; Carreiras, J.; Kelley, J.; Bunting, P.; Clewley, D.; Bray, S.; et al. An Evaluation of the ALOS PALSAR L-Band Backscatter—Above Ground Biomass Relationship Queensland, Australia: Impacts of Surface Moisture Condition and Vegetation Structure. *IEEE J. Sel. Top. Appl. Earth Obs. Remote Sens.* **2010**, *3*, 576–593. [CrossRef]

Article

Mapping Three Decades of Changes in the Brazilian Savanna Native Vegetation Using Landsat Data Processed in the Google Earth Engine Platform

Ane Alencar [1,*], Julia Z. Shimbo [1], Felipe Lenti [1], Camila Balzani Marques [1], Bárbara Zimbres [1], Marcos Rosa [2], Vera Arruda [1], Isabel Castro [1], João Paulo Fernandes Márcico Ribeiro [1], Victória Varela [1], Isa Alencar [1], Valderli Piontekowski [1], Vivian Ribeiro [1,3], Mercedes M. C. Bustamante [4], Edson Eyji Sano [5] and Mario Barroso [6]

[1]   Amazon Environmental Research Institute (IPAM), SCN 211, Bloco B, Sala 201, Brasília 70836-520, Brazil; julia.shimbo@ipam.org.br (J.Z.S.); felipe.lenti@ipam.org.br (F.L.); camila.balzani@ipam.org.br (C.B.M.); barbara.zimbres@ipam.org.br (B.Z.); vera.arruda@ipam.org.br (V.A.); isabelcastro@ipam.org.br (I.C.); joao.ribeiro@ipam.org.br (J.P.F.M.R.); victoria.varela@ipam.org.br (V.V.); isa.alencar@ipam.org.br (I.A.); derlly@ipam.org.br (V.P.); vivian.ribeiro@ipam.org.br or vivian.ribeiro@sei.org (V.R.)

[2]   Programa de Pós-Graduação em Geografia Física, Faculdade de Filosofia, Letras e Ciências Humanas, Universidade de São Paulo, São Paulo 05508-080, Brazil; marcosrosa@usp.br

[3]   Stockholm Environment Institute (SEI), Linnégatan 87D, 115 23 Stockholm, Sweden

[4]   Departamento de Ecologia, Universidade de Brasília, Campus Darcy Ribeiro, Brasilia 70910-900, Brazil; mercedes@unb.br

[5]   Instituto Brasileiro do Meio Ambiente e dos Recursos Naturais Renováveis (IBAMA), SCEN Trecho 2, Edifício Sede, Brasília 70818-900, Brazil; edson.sano@ibama.gov.br

[6]   The Nature Conservancy Brasil (TNC), SCN Quadra 05 Bloco A Sala 1407—Torre Sul, Brasilia 70715-900, Brazil; mario.barroso@tnc.org

*   Correspondence: ane@ipam.org.br

Received: 11 February 2020; Accepted: 10 March 2020; Published: 13 March 2020

Abstract: Widespread in the subtropics and tropics of the Southern Hemisphere, savannas are highly heterogeneous and seasonal natural vegetation types, which makes change detection (natural vs. anthropogenic) a challenging task. The Brazilian Cerrado represents the largest savanna in South America, and the most threatened biome in Brazil owing to agricultural expansion. To assess the native Cerrado vegetation (NV) areas most susceptible to natural and anthropogenic change over time, we classified 33 years (1985–2017) of Landsat imagery available in the Google Earth Engine (GEE) platform. The classification strategy used combined empirical and statistical decision trees to generate reference maps for machine learning classification and a novel annual dataset of the predominant Cerrado NV types (forest, savanna, and grassland). We obtained annual NV maps with an average overall accuracy ranging from 87% (at level 1 NV classification) to 71% over the time series, distinguishing the three main NV types. This time series was then used to generate probability maps for each NV class. The native vegetation in the Cerrado biome declined at an average rate of 0.5% per year (748,687 ha yr^{-1}), mostly affecting forests and savannas. From 1985 to 2017, 24.7 million hectares of NV were lost, and now only 55% of the NV original distribution remains. Of the remnant NV in 2017 (112.6 million hectares), 65% has been stable over the years, while 12% changed among NV types, and 23% was converted to other land uses but is now in some level of secondary NV. Our results were fundamental in indicating areas with higher rates of change in a long time series in the Brazilian Cerrado and to highlight the challenges of mapping distinct NV types in a highly seasonal and heterogeneous savanna biome.

Keywords: Cerrado; land cover; grasslands; forests; monitoring; random forest; spectral indexes; vegetation seasonality

---

1. Introduction

Widespread in the subtropics and tropics of the Southern Hemisphere, savannas cover approximately 20% of the Earth's terrestrial surface—around 65% of Africa, 60% of Australia, and 45% of South America [1,2]. They are naturally heterogeneous in terms of climate, soil, biodiversity, and threats posed by human activities and land occupation [3]. This heterogeneity is a consequence of varying edaphic and climatic conditions, occurrence and frequency of fires and the impacts of herbivore populations, the latter mainly in the African continent [4,5]. These variations result in the provisioning of a myriad of ecosystem services, with benefits to human populations that depend directly and indirectly on the savannas as source of food, water, materials, and pollinators [5].

The Brazilian tropical savanna (Cerrado) is the second largest biome in South America, occupying approximately 2 million km^2. Although mainly distributed in the central part of Brazil, the Cerrado presents a large latitudinal and longitudinal variation, resulting in different ecoregions [6]. It is formed by a mosaic of open grasslands, shrublands, savanna woodlands, deciduous or semi-deciduous forests, and evergreen riparian forests [7]. It is the most biologically diverse savanna in the world and is considered one of the global hotspots for biodiversity conservation, as it is under severe human-induced threats [8,9]. The Brazilian Cerrado has already lost half of the biome to croplands and planted pastures [6,10,11], and the biome currently holds strategic importance for the Brazilian economy, owing to the production of agricultural commodities. The rate of conversion of native Cerrado vegetation (NV) is up to two times the conversion observed in the Amazon in the past five years [12], and most of the native vegetation conversion tends to occur in areas with dense vegetation (favorable climate and soil conditions) and flat terrains (suitable for mechanized farming) [13].

The conversion of NV has significant impacts on ecosystem functioning, such as regional climate regulation, hydrological stability, and biogeochemical cycles, associated with the loss of significant carbon stocks and the replacement of biodiverse ecosystems, presenting heterogeneous canopy and deep root systems with shallow-rooted monocultures [11,14].

It is, therefore, crucial to understand the spatial and temporal dynamics of land conversion taking into account the three main Cerrado vegetation types (grasslands, savanna, and forests), as differences in carbon stocks and water fluxes are related to the degree of woodiness of the vegetation. In that sense, a long time series of remotely sensed data with high to medium spatial and temporal resolutions provide the means to better understand the ongoing land cover processes in the biome in order to support decision making regarding its conservation [13,15].

The lack of systematic land use and land cover (LULC) maps for the Cerrado biome is partially because of the difficulty in classifying highly complex gradients of natural vegetation with important differences in woody and herbaceous layers [16]. The differentiation of anthropogenic land-use classes and the natural cover is not always straightforward. The Cerrado is also highly seasonal, with marked dry and rainy seasons, different vegetation strata, and different levels of deciduousness during the dry season [6,7], which makes change detection by remote sensing a challenge. The spectral responses of the vegetation change drastically from the rainy to dry season [17], and can be confused with land conversion processes. Moreover, disturbances such as fire are common in savannas [18,19], and pose other challenges to the discrimination of changes related to natural processes from those associated with anthropogenic conversion and degradation. Further, the spectral responses of savannas change drastically from the rainy to dry season (Jacon et al. 2017), often being confused with land conversion processes. Therefore, a method to effectively monitor land cover change over time, taking into consideration the effects of fire and seasonal variations, still needs to be developed.

There are several studies that have aimed to distinguish and map NV types in the Cerrado using moderate to low spatial resolution imagery [6,15,20–26]. Other initiatives have provided snapshots of Cerrado LULC at specific points in time. They include the following: Project of Conservation and Sustainable Use of the Brazilian Biological Diversity (PROBIO) from 2002 [26]; TerraClass Cerrado Project from 2013 [27]; Brazilian Foundation for Sustainable Development (FBDS) from 2013 [28]; National GHG Inventory Initiative, from 1994, 2002, and 2010 [29]; and the Brazilian Institute for Geography and Statistics (IBGE), from 2000, 2010, 2012, and 2014 [30] (Table S1). Therefore, the mapping efforts conducted so far for the biome have been done considering either a single year, a smaller area, or a short time interval, at low to moderate spatial resolution satellite imagery, and usually not distinguishing the three Cerrado NV types. Although quite challenging, an extended yearly time series of the spatio-temporal dynamics of the different Cerrado vegetation types using moderate spatial resolution satellite images still needs to be produced.

In this paper, we present a novel method for mapping the changes in Cerrado NV cover using Landsat imagery, which can be applied to other highly seasonal biomes. Landsat time series are the most suitable dataset for monitoring spatial and temporal dynamics because of their moderate spatial resolution (30 m), 16-day repeat pass, and the availability of historical data since 1985, with basically the same acquisition mode (similar spatial, spectral, and temporal resolutions, and same swath width) [31,32]. The drawback here is the number of scenes necessary to cover the entire biome (approximately 120 Landsat scenes) which is time-consuming in terms of downloading and image interpretation. With the recent development of cloud computing platforms (e.g., Google Earth Engine) [33,34], entire series of satellite images such as from the Landsat mission can be easily accessed, processed, and analyzed in a fully automated way. Within this context, the objective of this study was to map three decades of changes in native Brazilian Cerrado vegetation using a machine learning approach, based on Landsat time series obtained and processed in the Google Earth Engine platform. The image processing involved empirical and probabilistic decision tree classification and was based on spectral metrics that indicated seasonal and phenological variations, previously assessed using available field plot data. These maps were used to retrieve the temporal dynamics of the NV types, including loss and gain (NV net loss), to indicate the level of stability, and to identify the NV areas that have faced natural and anthropogenic changes over the last 33 years (1985–2017).

This effort is part of the MapBiomas project (https://mapbiomas.org), which aims to generate annual land use and land cover classification of Brazil using machine learning algorithms available in the Google Earth Platform. The project is a collaborative network in which different expert groups are responsible for the mapping of different Brazilian biomes and/or cross-cutting themes. The project is in constant development and methodological improvement, and has so far launched six collections. Collection 1.0 covered the period of 2008 to 2015 (published in 2016). Collections 2.0 and 2.3 covered the period of 2000 to 2016 (published 2018), while Collection 2.3 was the first time that the random forest algorithm was applied. Collections 3.0 and 3.1 expanded the period covered to 1985–2017. The work presented here consists of the methodology for mapping the native vegetation for the Cerrado biome in Collection 3.1, and our results comprise the final integration of our classification with other cross-cutting themes (e.g., pasture, agriculture, coastal zone, and urban infrastructure). All resulting maps and methodologies are freely available on the MapBiomas platform, to be downloaded and replicated to other regions in the world.

2. Materials and Methods

2.1. Study Area

The Cerrado biome occupies more than 2 million km^2, with ecotonal transitions with the Amazon (tropical rainforest), Caatinga (semi-arid), Atlantic Forest (coastal tropical forest), and Pantanal (wetland) [35]. It ranges over 10 Brazilian states (Bahia, Goiás, Maranhão, Mato Grosso, Mato Grosso do Sul, Minas Gerais, Paraná, Piauí, São Paulo, Tocantins), and the Federal District. The core area is concentrated in the Brazilian highlands, the central part of the country (Figure 1). This biome presents wide latitudinal variation (22.4 degrees), elevation ranging from sea level to 1800 m, and strong climatic seasonality (rainy season from October to March and dry season from May to September). The annual precipitation varies between 600 mm and 2000 mm, increasing from east (border with the semi-arid Caatinga) to west (boundary with the Brazilian Amazon rainforest). During the rainy season, short periods of drought (dry spells) can occur, while in the dry season, rainfall levels are deficient, and there is frequently no rain for three to five months. Relative air humidity can be lower than 15% in July and August [36]. The average annual temperature is approximately 22–23 °C, while the absolute maximum temperature does not vary much over the year, but can be higher than 40 °C [37]. The absolute minimum temperature, however, varies widely, reaching negative values in May to July, causing frosts in some regions in the southern part of the biome.

The Cerrado has three predominant NV types, which can be classified according to their degree of woodiness, from grasslands to savanna woodlands and forests. According to the classification system proposed by Ribeiro and Walter [7], grasslands can be classified into two types: *campo limpo* and *campo sujo*, characterized by the predominance of herbaceous-shrub species and, to a lesser extent, sparsely distributed small trees. Savanna vegetation has a variable tree-shrub stratum, with a canopy cover ranging from 50% to 90%, which allows the coexistence of a grass layer. Forests (forested savannas, also known as *cerradão*, and riparian forests) present dense vegetation, with relatively large trees and low cover of grasses. Total plant biomass and carbon stocks in the Cerrado vary according to the type of vegetation, with an average of 18.5 t C ha^{-1} in grasslands, 39.9 t C ha^{-1} in savannas, and 68.6 t C ha^{-1} in forests [29]. Data on biomass and vegetation structure from 262 field inventory plots in the Cerrado compiled by Roitman et al. [38] were used to assess the spectral metrics that defined the different NV types, as detailed in the next session. The biome was partitioned in a regular grid of tiles compatible with the 1:250,000 international cartographic charts, where individual classifications were run for each tile. Each tile covers an area of 1° of latitude by 1° 30′ of longitude [30], generating a total of 172 tiles.

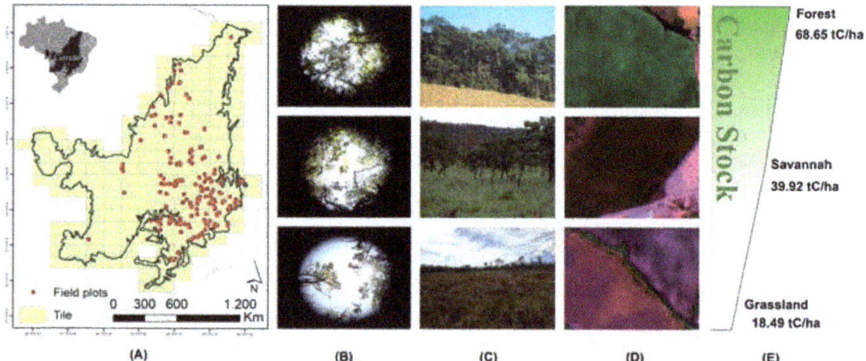

Figure 1. (**A**) Location of the Cerrado biome in Brazil and its subdivision into tiles of 1.5 degree by 1.0 degree, and field plots (in red), which were used to explore the spectral differences between vegetation types; (**B**) "fisheye" field photographs showing the canopy cover in a forest (top), savanna (middle), and grassland (bottom); (**C**) panoramic field photos of forest, savanna, and grassland; (**D**) R5G4B3 representative Landsat-8 color composites of forest, savanna, and grassland; and (**E**) average carbon stock estimates in each vegetation [29].

2.2. Classification Approach

The procedure to generate multi-temporal land cover maps in the Cerrado and to detect changes over time in the NV followed three main steps. The first step was the production of the year-based Landsat mosaics for the entire biome by defining the boundaries between the dry and wet seasons (Figure 2). These mosaics were used to generate spectral metrics, including sub-pixel fractions, indexes, and individual spectral bands. The second step included the selection of the best spectral variables for the definition of a preliminary empirical decision tree (EDT) classification, which in turn was used to derive a greater feature space to train the statistical decision tree (SDT) classifier. The last step was the classification of the annual mosaics using a machine learning approach, based on the class consistency map derived from the previous step.

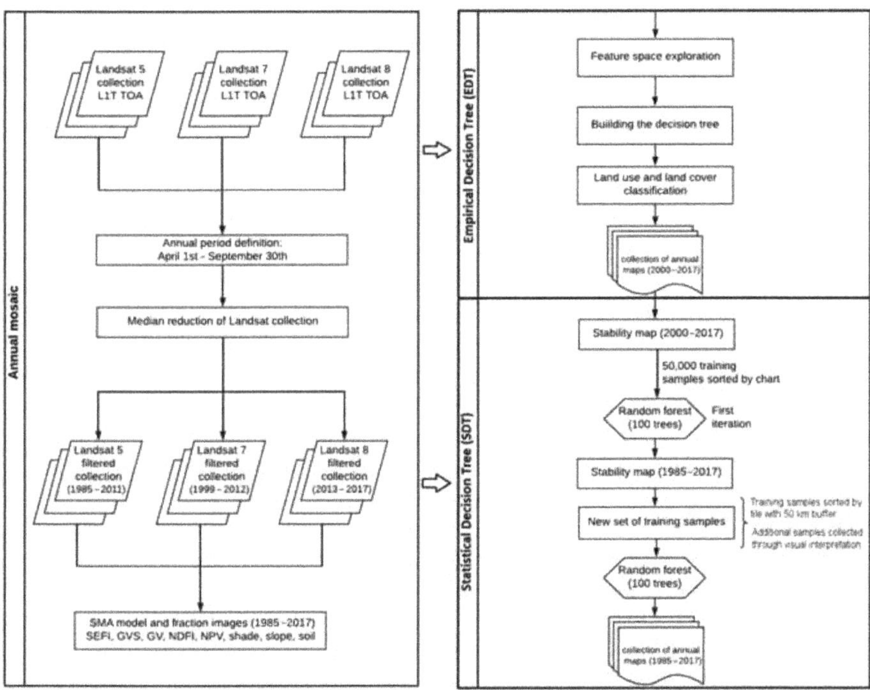

Figure 2. Workflow of the procedures conducted to map annual native vegetation cover in the Brazilian Cerrado biome, including the production of the annual mosaics, the definition of the empirical decision tree classification, and the statistical decision tree classification. L1T TOA = Collection 1, Tier 1 top-of-atmosphere reflectance; SMA = spectral mixing analysis; SEFI = savanna ecosystem fraction index; GVS = green vegetation and shade; GV = green vegetation; NDFI = normalized difference fraction index; and NPV = non-photosynthetic vegetation.

2.2.1. Annual Landsat Mosaics

The historical changes from 1985 to 2017 in the Cerrado NV were mapped based on the Landsat-5 Thematic Mapper (TM), Landsat-7 Enhanced Thematic Mapper Plus (ETM+), and Landsat-8 Operational Land Imager (OLI). The Landsat top-of-atmosphere reflectance collection (Collection 1 Tier 1 TOA reflectance), with a 30 m spatial resolution, was accessed via Google Earth Engine (GEE) platform. The entire image processing procedure was also conducted in this platform for cloud removal, classification, and post-classification routines (links to the scripts available in the Supplementary Materials—Table S2).

For each tile, best-pixel annual mosaics were produced using a combination of pixels from distinct Landsat scenes gathered during the year in consideration. The temporal window defining the seasonal limits was evaluated at the state level because of the high variations in annual precipitation over the extent of the Cerrado. On the basis of the annual precipitation patterns for each state, we identified the optimum period (OP) of images to compose the annual mosaics, by integrating the six months from the end of the rainy season to the end of the dry season for each tile (Figure S1). This procedure aimed to reduce NV commission errors provided by a greener mosaic (for example, by considering images from the end of the rainy season), or NV omission errors by considering a drier mosaic, composed by images from the end of the dry season (July to September).

After the definition of the initial and final dates of the optimum period of the mosaic, a cloud and cloud shadow masking procedure, as well as a data dimensionality reduction based on pixel median, were conducted. The cloud and cloud shadow pixels were removed by a mask that was produced with the temporal dark outlier mask (TDOM) (see details of the TDOM in [39]) and complemented using the quality assessment bitmask (BQA) (see details in https://www.usgs.gov/land-resources/nli/ landsat/landsat-collection-1-level-1-quality-assessment-band). The median algorithm was applied in each of the original bands, resulting in one final value per pixel and spectral band. The aggregation of this pixel-based composition of the annual mosaics was used for classification. To better represent seasonality, we used all normalized difference vegetation index (NDVI) values of the pixel in each year to produce dry and rainy median bands. The year-based NDVI values of each pixel were divided into quarters and the median pixels of the quarter with high NDVI were considered the rainy image and the median of the quarter with lower NDVI were considered the dry image.

2.2.2. Empirical Decision Tree

In order to derive a greater feature space used to train the statistical decision tree (SDT) classifier, an empirical decision tree was built with the most relevant parameters as inferred by the statistical evaluation of the input variables. This empirical decision tree (EDT) covered the period of 2000–2016, and sought to identify areas classified as the same NV over the entire period (stable samples). The factors selected for the EDT included the median of the original Landsat bands; spectral vegetation indices; and vegetation, non-photosynthetic vegetation, soil, and shade fractions derived from these bands (Table 1). A layer of slope data from the ALOS (Advanced Land Observing Satellite) global digital surface model with a 30 m resolution was also included (https://www.eorc.jaxa.jp/ALOS/en/aw3d30/index.htm).

This evaluation was conducted according to our feature space, which consisted of a sample of pixels selected based on 262 field inventory plots of the three NV types (see Figure 1 for the location of the plots within the biome) as well as samples from pasture and agriculture fields based on visual inspection. This procedure sought to identify the best metrics for highlighting the differences between NV types and major land-use classes such as pasture and agriculture fields (Figure S2). Among the metrics considered, those with the highest performances were the green vegetation (GV) index, non-photosynthetic vegetation (NPV) index, and soil and shade fractions, in addition to the green vegetation and shade (GVS) index and the savanna ecosystem fraction index (SEFI). While GVS highlights the difference between roughness and greenness of the vegetation, captured as the shade formed by the tree canopy [40], SEFI represents the combination of the amount of shade provided by the green vegetation with the non-photosynthetic materials and soils. Both GVS and SEFI were essential for separating forests and savannas from grasslands and other land use classes (pasture and agriculture) owing to the combination of vegetation heterogeneity and roughness, the amount of green material, and bare soil. Except for SEFI, which was developed for this study as an adaptation of the normalized difference fraction index (NDFI), the sub-pixel fraction images were based on the sub-pixel modeling and spectral library by Souza Jr. et al. [40].

Table 1. List of predictive input spectral variables considered to be included in the empirical decision tree classification. The median statistics were applied to all variables. L5 = Landsat-5; L7 = Landsat-7; L8 = Landsat-8; *Cloud* = fractional abundance of cloud within the pixel; OP = optimum period.

Input Variable	Meaning/Formula	Period	Reference
Blue band	Band 1 (L5 and L7); Band 2 (L8)	Annual, median OP	[41]
Green band	Band 2 (L5 and L7); Band 3 (L8)	Annual, median OP	[41]
Red band	Band 3 (L5 and L7); Band 4 (L8)	Annual, median OP	[41]
Near infrared (NIR) band	Band 4 (L5 and L7); Band 5 (L8)	Annual, median OP	[41]
Shortwave infrared (SWIR-1) band	Band 5 (L5 and L7); Band 6 (L8)	Annual, median OP	[41]
Shortwave infrared (SWIR-2) band	Band 7 (L5 and L7); Band 8 (L8)	Annual, median OP	[41]
NPV (non-photosynthetic vegetation fraction)	Fraction of abundance of non-photosynthetic vegetation within the pixel	Annual, median OP	[40]
GV (green vegetation fraction)	Fraction of abundance of green vegetation within the pixel	Annual, median OP	[40]
Soil fraction	Fraction of abundance of soil within the pixel	Annual, median OP	[40]
Shade fraction	$Shade = 100 - (GV + NPV + Soil + Cloud)$	Annual, median OP	[39]
GVS (green vegetation shade fraction)	$GVS = \frac{GV}{GV+NPV+Soil+Cloud}$	Annual, median OP	[39]
NDVI (normalized difference vegetation index)	$NDVI = \frac{NIR-Red}{NIR+Red}$	Annual, dry, and rainy	[42]
SAVI (soil-adjusted vegetation index)	$SAVI = 1.5\frac{NIR-Red}{NIR+Red+0.5}$	Annual, dry, and rainy	[43]
NDWI (normalized difference water index)	$NDWI = \frac{NIR-SWIR1}{NIR+SWIR1}$	Annual, dry, and rainy	[44]
EVI2 (enhanced vegetation index 2)	$EVI2 = 2.5\frac{NIR-Red}{NIR+2.4_{Red}+1}$	Annual, dry, and rainy	[45]
SEFI (savanna ecosystem fraction index)	$SEFI = \frac{GV+NPV-Soil}{GV+NPV+Soil}$	Annual, dry, and rainy	This study
DifNDFI	$DifNDFI = \frac{GVS-NPV-Soil}{GVS+NPV+Soil}$	Annual, dry, and rainy	[39]

Once the main input variables were selected, an EDT classification was semi-automatically created and calibrated with the variables for each tile and each year. A total of 2924 tiles (172 × 17 years) were calibrated based on the standard deviation values of each variable for each node, and manually adjusted using visual interpretation whenever necessary. The EDT is a conditional control statement algorithm used to support multistage decisions as a tree-like model consisting of a root node, several interior nodes, and several terminal nodes [42,43]. We used Fmask [45] as the first EDT node, which is a routine developed within the GEE platform to separate clouds, cloud shadow, and snow. Higher values of Fmask tend to be cloud and cloud shade, which, in this classification, were classified as non-observed areas. Non-observed areas on flat terrain were classified as water. Once the non-observed areas and water were classified, the remaining pixels were classified based on the V1 node defined by the SEFI variable. The combination of these fractions was used to distinguish the denser vegetation types from the sparser ones, capturing the differences between forests, savannas, and grasslands (Figure 3).

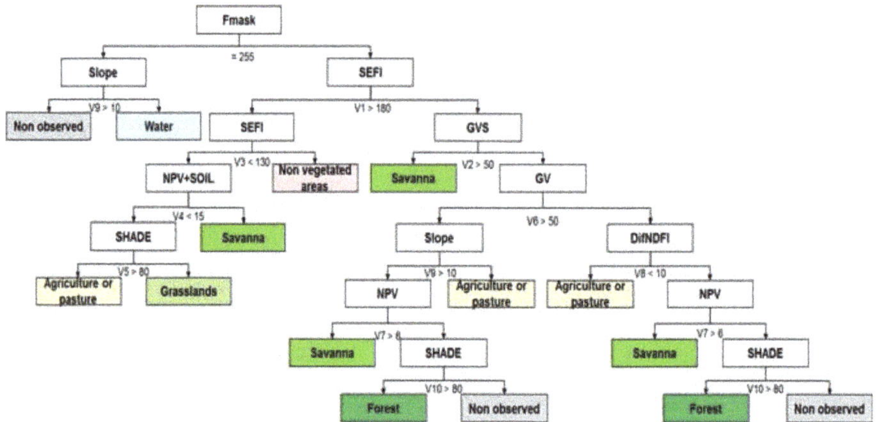

Figure 3. Empirical decision tree classification scheme based on the values of the spectral metrics in 262 known field inventory plots in the Cerrado biome, seeking to separate native vegetation classes (grassland, savanna, and forest) from land use-related classes (mostly agriculture and pastures) and other non-vegetated areas (e.g., urban areas, bare soils). Values below each split refer to the factors at each node above it.

The V1 node represented the main node of the EDT classification scheme, where areas with medium and high SEFI values were classified as forest or savanna with higher tree density. In the upper branch of this node, the GVS index (V2 node) was used to define the first separation of the savanna from other classes. This node is followed by the GV index (V6 node), which aimed to separate young crops, dense forests, and savannas, with higher levels of GV and GVS, from sparser forests and savannas, and drier, more mature crops. In these last two branches, NPV (V7 node) was used to separate savanna from forest, as the former has a higher content of non-photosynthetic material than the latter. Finally, shade (V10 node) was used to distinguish areas classified as forest, but that were, in fact, non-observed areas shaded by topography. The lower branch of the tree is related to areas with a more open vegetation structure. Therefore, nodes V3, V4, and V5 were used to separate open savanna vegetation types from grasslands, and grasslands from pasture and agriculture fields. This branch (specifically, node V3) also included the definition of non-vegetated areas (e.g., urban areas, bare soil). The main variables used to define the nodes in this last branch were the combination of NPV and soil, and the amount of shade, which represents the heterogeneity of surface cover.

Given the lack of robust temporal datasets of land cover maps for the entire Cerrado to be used as reference for machine learning classification (MLC), we used the resulting annual land cover maps from the EDT classification from 2000 to 2016 (Mapbiomas Collection 2.0 classification of Cerrado biome with distinct vegetation types, available in https://mapbiomas.org) as a reference for the distribution of the random training samples for the MLC. This land cover database also used the TerraClass LULC map for 2013 [12] as the reference for the spatial distribution of the three main native vegetation types.

2.2.3. Statistical Decision Tree Classification

On the basis of the time series maps produced with the preliminary EDT classification, frequency maps were built to demonstrate the spatial and temporal consistency of each class, indicating the number of times in which each pixel was consistently classified as a given class throughout the period from 2000 to 2016. This routine created a map with frequency values ranging from 0 to 17, which is the total number of years in that period. The resulting values of this procedure ended up representing the probability of each pixel belonging to a given class. This probability map was used as the reference map to collect training samples for the SDT routines.

The application of the SDT to map NV types over 33 years was based on a machine learning approach with two main steps. In the first step, a new reference dataset of annual maps was created for an updated period from 2000 to 2017 (MapBiomas Collection 2.3, available in https://mapbiomas.org). In this initial SDT classification, the random forest classification algorithm (Breiman, 2001) available in the GEE platform was trained using 50,000 samples (pixels) per tile selected randomly in the areas mapped with higher frequency/consistency for each one of the three main NV types. The minimum number of years in which a pixel would have to be classified to consider a given class as consistent varied between classes (Table 2). The final number of training samples, therefore, varied between tiles as a function of the consistent area availability. Each combination of tile and year had its training sample dataset for application in the random forest classification round (number of trees = 100). The results of this classification were visually assessed to identify areas of spatial inconsistency between adjacent tiles. Additional training samples were included from neighboring tiles to increase consistency in areas of spatial discontinuity.

In the second step, the NV maps resulting from the first random forest classification model were used to produce a new map of consistently classified NV classes. Over this map, new sampling points were selected to train the final map classification for the entire period (1985–2017) (MapBiomas Collection 3.1, available in https://mapbiomas.org). To address the issue of the spatial discontinuity between adjacent tiles in this final classification, the automatic sampling of points considered a buffer of 50 km around each tile. Such a procedure ensures that part of the training samples (about 10%) was representative of the variation found in the contact zone between tiles. Finally, additional samples were selected visually over the annual mosaics to complement the training samples in tiles with few consistently classified NV classes. The STD classifications were based on a larger number of spectral metrics including the ones extracted from the optimum mosaic and used for the EDT (Table 1), as well as other metrics retrieved for the entire annual dataset collection (Table S3).

Table 2. Probability thresholds were used to define areas of consistent classification for each land use and land cover (LULC) class. Values inside parentheses indicate the minimum number of years a pixel had to be classified as a given class to be considered consistent, considering a total of 17 years for the initial classification from 2000 to 2016, and 33 years for the final classification from 1985 to 2017. NA = not applicable. SDT, statistical decision tree.

LULC Class	Initial SDT Classification 2000–2016	Final SDT Classification 1985–2017
	Probability Threshold (Number of years)	**Probability Threshold (Number of years)**
Forest	88% (15)	100% (33)
Savanna	82% (14)	100% (33)
Grassland	88% (15)	100% (33)
Agriculture/pasture	71% (12)	94% (31)
Non-vegetated areas	71% (12)	NA
Water	71% (12)	100% (33)

2.3. Post Classification

A series of spatial and temporal filters were applied to the resulting maps in the GEE platform. The spatial filter segmented and indexed the classes of each collection into contiguous regions, which were subsequently identified and reclassified based on the following criterion: areas less than or equal to 0.5 ha (i.e., approximately 5 pixels) are reclassified based on the majority of the neighboring classes. A temporal filter was applied to identify and correct class transitions throughout the time series (i.e., 3 to 5 consecutive years), as well as to classify pixels with no data caused by cloud cover. For example, a pixel classified as non-forest in a given year t_i (where $i = 1985$–2017), and forest in year $t_i - 1$ and $t_i + 1$, was reclassified as forest for the year t_i. Several transition rules were defined and applied to be used in the temporal filter to deal with specific phenological and land use transitions (Table S4).

2.4. Integration with MapBiomas Cross-Cutting Themes

In order to implement the final MapBiomas land use and land cover maps for the Cerrado biome, the results of the Cerrado NV classification were integrated with other MapBiomas cross-cutting theme maps (e.g., pasture, agriculture, coastal zone, and urban infrastructure classes), which were independently developed [41,45]. This integration is hierarchical and follows prevalence rules to combine the classification results from all themes. For instance, urban areas and agriculture (i.e., crop fields) had prevalence over NV classes. More details are described in the ATBD (algorithm theoretical basis document) available at the MapBiomas website (http://mapbiomas.org/). After integration, the last step is spatial filtering to remove isolated class patches smaller than half a hectare as well as to remove noise caused by Landsat data misregistration. This spatial filter procedure is the same as described above in Section 2.3.

2.5. Accuracy Assessment

To assess the accuracy of each year and the NV classes, we used a set of 21,000 independent sampling points based on visual interpretation of Landsat data by three expert analysts (Figure S3). The sampling design and visual interpretation were performed by collaborators at the Image Processing and Geoprocessing Laboratory at the Federal University of Goiás, Brazil (LAPIG/UFG), using in the Temporal Visual Inspection web platform (tvi.lapig.iesa.ufg.br) [46]. The number of pixels collected to compose the reference dataset was pre-determined by statistical sampling techniques considering four tiles of the Cerrado subdivision (described in Section 2.1) as a minimum unit of analysis as well as six classes of slope, according to the shuttle radar topography mission data. The accuracy analysis was based on the approach proposed by Stehman et al. [47], using confusion matrix, overall accuracy, and omission and commission errors. The evaluation of a pixel in a given year was considered valid only when two or three interpreters agreed on the class observed in that pixel. The accuracy was then calculated using metrics that compare the class mapped with the class observed by the interpreters following the good practices proposed by Olofsson et al. [48]. We report year-based accuracy estimates with circa 5% error for each class in the mapping. The accuracy was reported for two levels of disaggregation: Level 1, which encompasses all three vegetation types as one class (NV); and Level 2, which separates all three types of NV.

2.6. Native Vegetation Net Loss

Once the annual NV maps were spatially and temporally filtered, NV net loss was calculated. The concept of NV net loss considered in this paper comprised the difference between the first and the last NV maps of the time series, representing the area difference of the Cerrado vegetation types in 1985 in comparison with 2017. Owing to the characteristics of the annual NV maps, which were built independently for each year, the net loss represented land cover change undergone by the NV over the past 33 years, accommodating both gains and losses over the whole time series.

2.7. Stability of Native Vegetation

The stability of the Cerrado NV was defined as the number of times in which a pixel was classified as the same given class (NV and non-NV) in the 17-year period initial SDT classification and in the 33-year period final SDT classification, indicating class consistency over time. The individual annual land cover maps were reclassified into four classes (forest; savanna; grassland; and non-NV land cover classes—agriculture, pasture, water, and non-vegetated areas). Next, they were overlaid and reduced to a single map to build this class consistency map. The NV stability map was classified into four classes for the final SDT classification (1985–2017): (i) stable NV areas classified as the same NV type in the 33 years of analysis; (ii) unstable NV areas, which had been classified as more than one vegetation type over time; (iii) stable non-NV areas, which had never been classified as NV over the time series; and (iv) NV areas converted to other non-NV land cover classes at some point in the time series, regardless of whether they subsequently recovered or not. The stability map was overlaid with the 2017 NV map indicating the current stable and unstable NV areas as well as the NV areas under recovery. The unstable NV areas and the NV areas converted to other land cover classes indicate natural and anthropogenic instability in the biome over the past three decades, respectively.

3. Results

3.1. Accuracy of Native Cerrado Vegetation Mapping

The strategy of using the semi-automatic calibrated EDT to retrieve the first set of annual consistent NV maps from 2000 to 2016 was fundamental for the improved performance of SDT classification. It guided the application of the SDT for the entire time series (1985 to 2017) in the highly complex Cerrado landscape, which lacked reference maps for gathering sets of training sampling points. The machine learning approach used here, which combined the SDT based on randomly selected samples in areas previously classified by the EDT as stable over the first time series (2000–2016), increased accuracy at Level 1 (where all NV types were aggregated as one NV class) by 4%: accuracy at Level 1 for the EDT alone was 83%, while accuracy for the final SDT classification (built upon the EDT-derived samples) was 87%.

The average overall accuracy of the final SDT classification at Level 2, analyzing the three NV classes separately, averaged 71%, ranging from 67% to 74% over the 33 years. The average balanced accuracy of forest (84%) was consistently higher than the accuracy of savanna (73%) and grassland (73%) throughout the time series (Figure S4). A confusion matrix was produced by aggregating the 33 annual confusion matrices into a single, average matrix (presented in Table S5). Also, omission and commission errors for all three classes throughout the time series are presented in Table S6. All classes presented high commission errors, with higher errors in the grasslands (Tables S5 and S6). These results suggest a relatively high degree of confusion between savanna and forests, and between savanna and grasslands, as expected in the naturally heterogeneous Cerrado. This is especially the case of *cerradão*, which, even in the field, can be alternately classified as a dense typical savanna or sparse forest. Confusion was higher with grasslands. Although the accuracy decreased from Level 1 to Level 2, as expected, it increased from the beginning to the end of the time series, and from the first to the final classifications.

3.2. Spatial and Temporal Patterns of the Cerrado Native Vegetation

The time series maps resulting from the SDT machine learning classification using the random forest algorithm indicated that, in 2017, 112 million hectares (Mha) (55%) of the original Cerrado range was still covered by native vegetation (Table 3). Among the three NV types, savanna is the most abundant. This class covered 52% (52.5 Mha) of the NV area in the biome in 2017 (i.e., 26% of the total area of the biome), followed by forest and grassland covering 36% (37.9 Mha) and 22% (22.2 Mha) of the total Cerrado NV, respectively (18% and 11% of the total area of the biome, respectively). Between 1985 and 2017, the Cerrado faced an overall net loss of 24.7 Mha, representing 10% of the original Cerrado distribution and 18% of the existing NV in 1985, resulting in 55% of the original NV distribution. Forests faced a net loss of almost half the total NV loss in the period, with 75% of this loss happening from 1985 to 2000. For savannas, a net loss of approximately 11 Mha could be observed, but, in contrast with the forests, the largest proportion of the net loss in the savanna happened after the year 2000 (52%) (Figure 4). Grasslands remained practically stable throughout this period, with an observed net loss of 1.9 Mha, representing 1% of the original range of the biome (Figure 5).

In the past 33 years (1985 to 2017), the native vegetation in the Cerrado biome has been declining at an average rate of 0.5% year (748,687 ha yr^{-1}), particularly affecting forests and savannas (Figure 5, Table 3). Although savannas and forests presented similar amounts of area converted to other land use classes (around 11.3 million hectares), proportionally, forests lost more NV area than savannas (23% of forest loss in relation to 18% of savanna loss by 2017). Moreover, the forests presented a higher conversion rate from 1985 to 2017 (0.7% yr^{-1}) compared with savannas (0.5% yr^{-1}) and grasslands (0.2% yr^{-1}), as the original area of forest is almost one-fourth of the original area of savanna (Table 3).

Table 3. Native vegetation area in the Brazilian Cerrado at the beginning (1985) and the end (2017) of the time series, and net loss of native vegetation over the whole period between 1985 and 2017. Absolute and proportional net loss rates per year were also calculated. NV = native Cerrado vegetation.

NV Type	NV (ha)		NV Net Loss (ha)	NV Net Loss (%)	Annual Net Loss Rate (ha yr^{-1})	Annual Net Loss Rate (%)
	1985	2017				
Forest	49,265,967	37,908,046	11,357,921	23	344,179	0.7
Savanna	63,810,140	52,456,330	11,353,810	18	344,055	0.5
Grasslands	24,204,833	22,209,899	1,994,934	8	60,453	0.2
Total NV	137,280,941	112,574,275	24,706,666	18	748,687	0.5

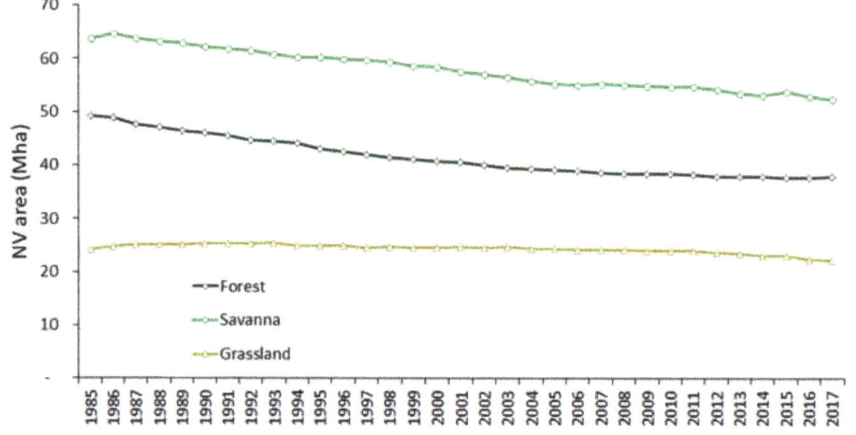

Figure 4. Area (hectare) occupied by the three native vegetation types (forest in dark green, savanna in light green, and grassland in beige) in the Brazilian Cerrado over the time period of 1985–2017. Mha = million hectares; NV = native Cerrado vegetation.

The remnant NV is located mostly in the center and northwestern part of the biome, more specifically in Mato Grosso (19%), Tocantins (18%), Maranhão (15%), Minas Gerais (18%), and Goiás (12%) (Table S7; Figure S6). These states account for almost 80% of the remaining Cerrado NV in 2017. Other states such as Bahia, Mato Grosso do Sul, and Piauí accounted for 19% of the current NV area in the biome (Table S7). Mato Grosso was the state that presented the highest levels of NV lost in the past 33 years, accounting for one-fourth of the whole NV net loss in the Cerrado from 1985 to 2017 (this state contains approximately 17% of the Cerrado area). The NV type lost that lost the most in this period in the Mato Grosso was savanna, followed by forest (Figure S7). Other states with important net loss of NV were Goiás and Mato Grosso do Sul (17% each), followed by Tocantins (13%), Maranhão (10%), Minas Gerais (8%), Bahia (7%), and Piauí (2%) (Table S7). Most of the NV net loss in Mato Grosso occurred between 1995 and 2005 (Figure S8), a period with large scale expansion of soybean in this state [49]. Mato Grosso do Sul and Goiás faced considerable losses in the 1985–1995 time period, being considered older agriculture frontier. The most recent NV net loss is happening in the states of the Matopiba region (Maranhão, Tocantins, Piauí, and Bahia). These four states together represent 55% of the loss of NV that happened recently from 2005 to 2017, representing the new region of Cerrado where agriculture is growing at the expense of NV loss [11,50].

Only 14% of the remaining Cerrado NV is protected in indigenous lands and conservation units. Grasslands are the most protected NV types (22% of their total area), followed by forests (17%) and savannas (9%) (Table S8). In the states where the majority of the NV conversion has already taken place, the remnant NV is mostly located within protected areas (states of São Paulo, Mato Grosso do Sul, and Mato Grosso) (Figure 5).

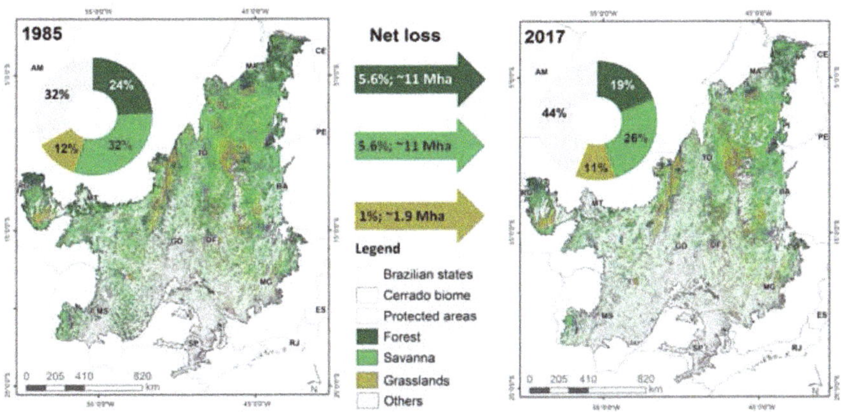

Figure 5. Comparison between the first (1985) and last (2017) maps of the time series, showing the net loss of the three predominant native vegetation classes in relation to the total area of the Brazilian Cerrado. Mha = million hectares.

The distribution of the three NV classes and their net losses also varied depending on the regions of the Cerrado (Figure 6). Forests are dominant in the transition of the Cerrado to the Amazon biome, for example, in the northwestern portion of the states of Maranhão and western Mato Grosso. These two areas concentrated the largest NV losses. Savannas are currently concentrated mainly in the northeast of the biome (states of Tocantins, Bahia, and Piauí). The states of Mato Grosso, Mato Grosso do Sul, and Goiás presented the largest savanna losses along the time series. Grasslands are mainly clustered in the state of Tocantins and surrounding areas. Only a few tiles presented more than 50% of grassland loss over time.

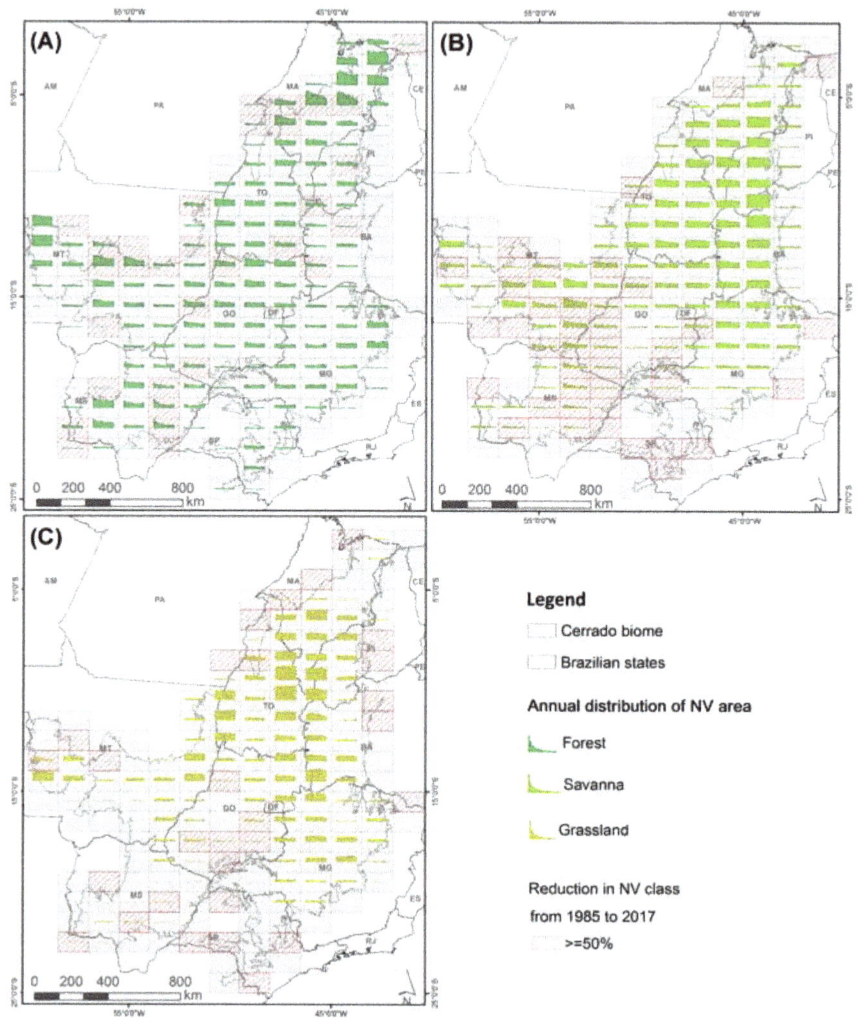

Figure 6. Temporal distribution of (**A**) forest, (**B**) savanna, and (**C**) grassland per year and per tile from 1985 to 2017 in the Brazilian Cerrado. Tiles with a reduction of over 50% in each NV class over the time series are highlighted in red.

3.3. Stability of the Cerrado NV Classes

Forests were the vegetation type that presented the highest intra-annual map consistency, mostly concentrated in the northern portion of Maranhão state and western portion of Mato Grosso along the border with the Amazon biome (Figure 7; Table 4). Approximately 43% of NV areas mapped as forest from 1985 to 2017 were consistently mapped as forest from 27 to 33 times. Savanna was the second NV class in terms of temporal consistency, with 36% mapped consistently for more than 27 years over the biome. Concentration of savanna is found especially in Bahia and Piauí states. For grasslands, consistently mapped areas comprised 32% of the total grassland mapped, especially located in Tocantins and in the bordering Mato Grosso and Bahia states. Stable areas are primarily located in large fragments of NV (Figures 7 and 8).

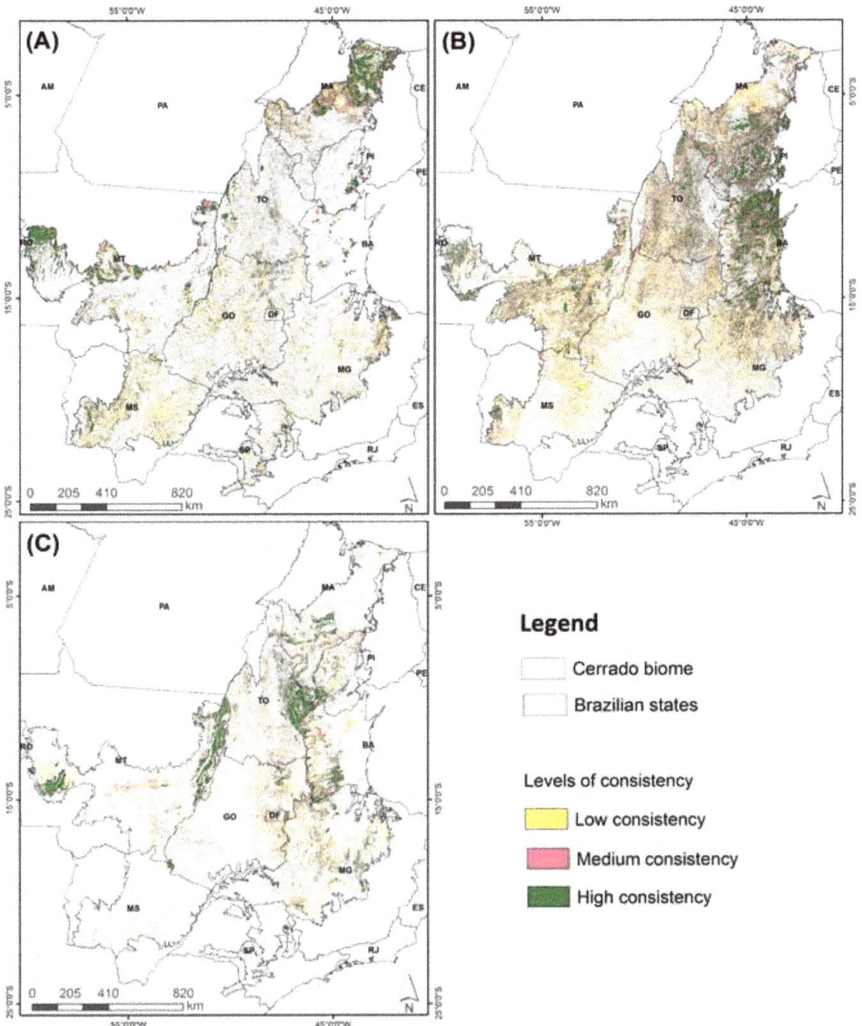

Figure 7. Classification probability maps of three native vegetation classes (**A**—forest, **B**—savanna, and **C**—grassland) in the Brazilian Cerrado, indicating the levels of consistency (low: 1–11 years; medium: 12–26 years; and high: 27–33 years) observed for each class in the 1985–2017 period.

Table 4. Consistency of the native vegetation classes in terms of the number of years in which the pixels were classified as the same class. Three consistency categories were considered: low: 1–11 years; medium: 12–26 years; and high: 27–33 years.

Consistency	Forest		Savanna		Grassland	
	Area (ha)	Percent (%)	Area (ha)	Percent (%)	Area (ha)	Percent (%)
Low	24,371,387	34	31,295,167	36	21,230,505	48
Medium	16,595,370	23	24,016,090	28	8,947,041	20
High	30,357,159	43	30,689,708	36	14,146,397	32

The stability map of the final SDT classification indicated that 36% of the Cerrado is composed of stable NV (classified as the same NV class over the entire time series) (Figure 8). Unstable areas of

NV, which varied among different types of NV class, occupy 7% of the biome. Meanwhile, 38% of the biome was covered by NV that was converted to other land cover classes (e.g., pasture and crop fields), and 19% of areas that were classified as non-NV throughout the time series. These results suggest that the majority of the standing vegetation in 2017 is stable in terms of NV type (65%), while 12% varied among NV classes (natural instability) and 23% was converted at some point in time to other non-NV classes and recovered to NV by the end of the period (anthropogenic instability; Table 5).

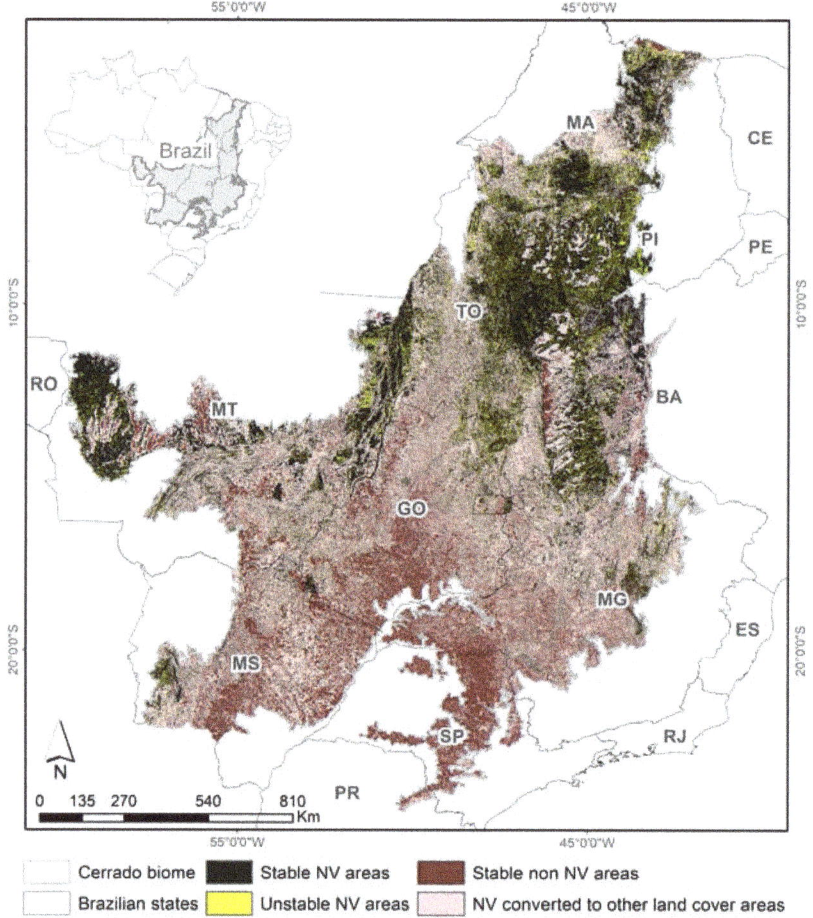

Figure 8. Analysis of stability of native vegetation (NV) and other non-NV classes in the Brazilian Cerrado from 1985 to 2017, presenting stable NV and non-NV areas, as well as the unstable areas that either changed to other NV classes or to other non-NV land cover classes.

Table 5. Out of the area classified as native vegetation (NV) in 2017, the total area that did not change class (stable), the total area that changed among other NV classes (natural instability), and the total area that had been previously converted (anthropogenic instability followed by NV recovery).

Native Vegetation in 2017	Area (ha)	Percentage (%)
Stable NV	75,678,855	65
NV that changed among NV	14,474,258	12
NV that had been previously converted (NV recovery)	26,357,350	23

4. Discussion

4.1. Innovative Machine Learning Approach to Map Temporal Dynamics of Cerrado Native Vegetation

The combination of a priori EDT with STD classification was able to adequately discriminate the three main NV types from a heterogeneous and seasonal biome, which is the case of the Brazilian Cerrado, over the entire time series of the Landsat mission (1985–2017). The use of cloud computing and the GEE platform has allowed the generation and calibration of automatic or semi-automatic methods for continuously monitoring land use and land cover with a medium to high spatial and temporal resolution over decades [51,52]. This procedure is a milestone in terms of creating robust baseline reference maps that can be used to train machine learning algorithms for monitoring and detecting changes in NV classes in complex and highly seasonal landscapes. This is especially relevant in regions where reference maps for guiding times series classification are insufficient or completely lacking. This is especially important for savanna ecosystems such as the Cerrado, which are currently undergoing changes owing to severe pressure for conversion, as well as facing degradation related to climate change [6,10,11].

Even though the Landsat image collection does not represent the most adequate imagery for capturing intra-annual or phenological variability of the NV because of the 16-day interval of image acquisition, it has been commonly used to map changes in NV in highly complex and variable ecosystems [22,26,53]. In addition, the Landsat mission is the only one that currently provides a historical perspective of change within a period of at least three decades of imagery with medium to high spatial resolutions [31,32]. It can also provide a consistent measure of the inter-annual variability of NV over the entire Earth's surface. Some other initiatives have generated classification maps for the entire Brazilian Cerrado biome using Landsat free-archives (Table S1). The most widely used approaches in Brazil are the visual interpretation and traditional supervised classifications. These processes are very slow and costly, and classification is possible only in specific time periods [6,12,29,30] and was never applied to the entire Landsat series. Therefore, the generation of a semi-automatic method applied to long time series to discriminate the main types of NV present in the Cerrado biome with levels of accuracy compatible with existing initiatives was innovative and our main goal.

The results of this effort offer, to the scientific community, the first long-term annual dataset of NV (forest, savanna, and grassland) compatible with some of the few existing single-year NV maps for the biome. The average overall accuracy at Level 1 of 87% from 1985 to 2017 was similar to that obtained by Sano et al. [15] in their land cover map of the Cerrado for 2002 (90% overall accuracy), and by the TerraClass Cerrado Initiative for 2013 (80% overall accuracy). The distinction between NV classes from other land use types (Level 1 aggregation) presented an overall accuracy of 71%, which was the same as that reported by Sano et al. [15] (71% overall accuracy), but using much fewer points (315 field data validation points compared with 21,062 validation points used in this study) and only for a single year (2002). The accuracy increased from the beginning to the end of the time series, representing a gain of stability when we have a greater quality and quantity of images available.

The main challenges in distinguishing NV classes in the Cerrado include the misclassification between grasslands and planted pasture (ca. 8%), as well as the confusion between savannas and the other two NV types (forest and grasslands) [15]. Savannas are often misclassified as forests (omission error, ca. 39%), and grasslands are misclassified as savannas (commission error; ca. 46%) (Table S5). About 20% of pixels classified as forest were, in fact, savanna, while approximately 75% of the pixels classified as forest in the reference data were correctly classified (Table S5). Even though we had considerable errors of omission and commission at the edge of the NV types, this long time series allowed us to identify those areas where the stable vegetation is concentrated, and those where confusion among vegetation types is expected. This contributed greatly to the definition of the more unstable areas in the biome, and highlights where the map can and must be improved.

Moreover, the confusion among NV types occurs because they are distributed along a natural gradient of vegetation structure, with the savannas ranging from dense to very sparse woodlands dominated by grasses [7]. The study conducted by Ribeiro and Walter [7], or even the system of classification of vegetation proposed by the Brazilian Institute of Geography and Statistics (IBGE), subdivided the Cerrado into at least seven more detailed vegetation types, known as phytophysiognomies. The classification of these phytophysiognomies, which are defined by specific ranges of vegetation structure, is even more challenging. One way to improve this classification is by adding structural characteristics to the spectral components of the NV, such as height. Some studies have demonstrated the ability of LiDAR (Light Detection and Ranging) technology to differentiate between Cerrado vegetation types [54,55]. However, this time series represents the best comprehensive temporal data set on NV distribution in Cerrado.

4.2. Temporal and Spatial Dynamics of the Cerrado Native Vegetation

Up until 2017, remnant NV covered about 55% of the Brazilian Cerrado, although it was very unevenly distributed over the biome. While NV comprises 90% in the northern part of the biome, only 15% is left in its southern portions. The area classified as NV in 2017 included some areas of secondary NV. Although the approach used in this study generates temporally independent annual land cover maps that do not differentiate primary NV areas from regenerating areas, this proportion of remnant NV (55%) is compatible with those found in other studies [6,12].

Savannas still represent the predominant NV type in the Cerrado, occupying 52% of the NV area in 2017. Forests and grasslands occupy 36% and 22%, respectively. The prevalence of savannas is consistent with the results obtained by other authors [6,12]. This distribution is explained by the spatial configuration of the Cerrado NV, forming a gradient of mosaics related to a series of continuous environmental characteristics such as topography, soil, and climate [7].

Over the last three decades, the annual net loss rate of the Cerrado NV was 0.50% per year or 748,687 ha per year, which is 20% lower than the official annual gross deforestation estimates for the past 10 years (944,521 ha yr^{-1}) [12]. This difference was expected as our concept of NV net loss includes NV areas that recovered after conversion, while gross deforestation estimates by the National Institute of Spatial Research (INPE) [12] do not include them. Beuchle et al. [23] reported a 0.60% net loss rate from 2000 to 2010 for the Cerrado biome, which is comparable to our observed average rate. Most of the net loss occurring in the first half of the time series (1985 to 2000) was in forests owing to cropland expansion.

The majority of the NV net loss occurred in the southern part of the Cerrado, in the states of Goiás, Mato Grosso do Sul, and Mato Grosso (Table S7), and mainly in the early stages of the time series, so that they are considered as older frontiers of agricultural expansion in the country. In these regions, most of the remaining NV mapped in 2017 is located within protected areas. The newest agriculture frontier in the Cerrado, responsible for large-scale deforestation, is located in the northern part of Cerrado, in a region known as MATOPIBA (southern portion of states of Maranhão and Piauí, north of Tocantins and west of Bahia) [11,56]. Native vegetation conversion to agricultural areas in the Cerrado biome is related to biophysical characteristics and infrastructure improvements [57].

Savannas and forests were the most affected NV types in terms of net loss due to the increase of pastures and agriculture in the biome along with the time series. Even though savannas are the most abundant vegetation type in the biome, forests faced most of the net loss over the time series. While savannas lost 11 Mha in 33 years, forests accounted for a comparable net loss in absolute terms, even though they represent two-thirds of the area of savanna. Grasslands correspond to the areas mostly unsuitable for agriculture usually owing to them being located over shallow and poor soils, and despite being mostly unprotected by law [57]. Detecting conversion of grasslands was quite difficult because of the spectral confusion between native grasslands and planted pastures, usually requiring observations of phenological timings [26]. These timings are easily missed by monitoring approaches that do not capture intra-annual changes, which is corroborated by the lower accuracies observed for this class.

4.3. Stability of the Cerrado Native Vegetation Classes

In a highly seasonal and heterogeneous biome, vegetation stability evaluation is a challenge that can be overcome with consistent long-term information. We managed to identify areas where specific types of NV are dominant, as well as areas with high instability in terms of NV mapping. This strategy demonstrated that the forests presented the highest consistency over time compared with savannas and grasslands in the Brazilian Cerrado. On the other hand, grasslands were the NV with the lowest mapping consistency over time. These patterns are expected and can be explained by the effects of seasonality on the spectral response of NV types dominated by grasses, which are more susceptible to drought [15,20]. These results also help to guide further improvements in classifying the unstable NV areas using either higher temporal and/or spatial resolution imagery.

The consistency maps can be also used to suggest whether the main drivers of instability are either natural (e.g., seasonality, fire) or anthropogenic (e.g., NV conversion to agriculture). In this way, we identified that 36% of the Cerrado is covered by NV that was mapped consistently over time as a specific NV class, so it can be considered as stable. Only 7% of the biome presented NV classes that were highly unstable in terms of mapping, probably associated with natural disturbances. Other areas of high NV instability were related to areas that, at some point in time, were converted to other land uses (e.g., agriculture and pasture), representing 38% of the biome, and were considered anthropogenic. This analysis suggests that, from 55% of the Cerrado area with NV, only 43% remained as NV during at least 33 years. When the stability map was masked by the 2017 NV map, it revealed that the majority (62%) of the NV types were stable over time, while 23% were in the process of recovery. This indicates that around 12% of the Cerrado biome is currently in the process of regeneration. These numbers are fundamental for assisting public policies towards Cerrado conservation.

5. Conclusions

This study was the first to apply a combination of EDT and SDT classification techniques for mapping different types of native vegetation in the Cerrado biome, using a long time series of satellite data (1985–2017). The use of Landsat archive and cloud computing freely available in the Google Earth Engine platform allowed the generation and calibration of a semi-automatic method for monitoring the spatiotemporal land cover dynamics over three decades. We managed to obtain annual-based estimates of the remnant NV in this biome dominated by a very heterogeneous landscape, marked seasonality, and intensive human land occupation. The results showed that approximately 45% of the Cerrado NV was converted into some type of land use and land occupation by 2017. The conversion is occurring rapidly, especially in the northern part of the biome where most of the remnant NV is found nowadays. This study also showed some areas where the NV appeared as a mosaic of grasslands, savannas, and forests, indicating a limitation of Landsat data to discriminate these three vegetation types properly, perhaps because of its moderate spatial resolution (30 m). Despite our promising results, discriminating different NV types in highly seasonal and heterogeneous ecosystems such as the Cerrado remains challenging, especially if the goal is to discriminate between phytophysiognomies rather than general classes. As a future investigation, we suggest including other satellite data such as the Sentinel-1 (radar) and Sentinel-2 (optical) images obtained with 10 m spatial resolution and repeat pass of five days, which are available on the Internet without cost.

Supplementary Materials: The following supplementary materials are available online at http://www.mdpi.com/2072-4292/12/6/924/s1, **Table S1.** Characteristics of the existing native vegetation maps covering the entire Brazilian Cerrado and produced by semi-automatic methods and visual interpretation of varying remote sensing data; **Table S2.** Scripts used in the initial and final statistical decision tree classification of the native Cerrado vegetation; **Table S3.** Spectral metrics, indexes, and other variables used to train the random forest SDT classification; **Table S4.** Rules of the temporal filters applied to classify native Cerrado vegetation. RG = General Rule; RP = First Year Rule; RU = Last Year Rule; FF = Forest; SV = Savanna; GL = Grassland; AG = Mosaic of Agriculture and Pasture; AR = Rocky Outcrop; WB = Water; and NO = Non-Observed; **Table S5.** Confusion matrix of NV classes at level 2. Other land cover classes include agriculture and pasture fields and non-vegetated areas. UA = user´s accuracy; PA = producer´s accuracy; **Table S6.** Omission and commission errors of the 33 classified maps; **Table S7.** State-level distribution of native vegetation (NV) and NV net loss from 1985 to 2017; **Table S8.** Distribution of the native Cerrado vegetation (NV) inside protected areas (indigenous lands and conservation units) from 1985 to 2017; **Figure S1.** Mean monthly precipitation and normalized difference vegetation index (NDVI) data from forest and savanna vegetation. Gray rectangle corresponds to the six-month optimum period for deriving the Landsat mosaic of the Brazilian Cerrado; **Figure S2.** Boxplots representing the distribution of the dominant land cover classes of the Brazilian Cerrado over six spectral metrics included in the empirical decision tree classification. GVS = green vegetation and shade; GV = green vegetation; NPV = non-photosynthetic vegetation; SEFI = savanna ecosystem fraction index; **Figure S3.** Location of 21,000 sampling points used for accuracy assessment. Other = water, non-vegetated, agriculture, and pasture fields; **Figure S4.** Overall accuracy for the three classifications (empirical decision tree (EDT), initial statistical decision tree (SDT), and final SDT), considering the aggregated legend; **Figure S5.** State-level proportion of native vegetation (NV) in 2017, other non-NV land cover types in 2017, and NV net loss from 1985 to 2017. MS—Mato Grosso do Sul, GO—Goiás, MT—Mato Grosso, TO—Tocantins, MA—Maranhão, MG—Minas Gerais, BA—Bahia, PI—Piauí, SP—São Paulo, and DF—Federal District; **Figure S6.** State-level net loss of native vegetation (NV) from 1985 to 2017. Mha = million hectares. See Figure S6 for state identification; **Figure S7.** State-level native vegetation (NV) net loss in three different periods: 1985–1995, 1995–2005, and 2005–2017. M ha = million hectares. See Figure S6 for state identification.

Author Contributions: Conceptualization, A.A. and J.Z.S.; Methodology, A.A., M.R., F.L., C.B.M., and J.Z.S.; Validation, A.A., F.L., C.B.M., I.C., I.A., and J.Z.S.; Formal analysis, A.A., F.L., C.B.M., B.Z., I.C., V.A., and J.Z.S.; Investigation, A.A., F.L., C.B.M., V.A., J.P.F.M.R., J.Z.S., I.C., I.A., M.M.C.B., E.E.S., M.B., V.R., and V.P.; Writing—original draft preparation, A.A., J.Z.S., and B.Z.; Writing—review and editing, A.A., J.Z.S., F.L., B.Z., C.B.M., M.M.C.B., M.R., E.E.S., and M.B.; Visualization, A.A., B.Z., F.L., C.B.M., V.A., J.P.F., M.R., V.V., and I.C.; Supervision, A.A.; Project administration, A.A. and J.Z.S.; Funding acquisition, A.A., J.Z.S., and M.B. All authors have read and agreed with the published version of the manuscript.

Funding: This research was conducted by the MapBiomas Initiative. It was financed by the Moore Foundation, with the support of The Nature Conservancy, as well as the UK Space Agency's International Partnership Programme (IPP) under the Global Challenge Research Fund (GCRF), with the support of Ecometrica.

Acknowledgments: The authors would like to thank the MapBiomas team and Google Earth Engine for providing access to the Landsat cloud processing. The authors are grateful for the field plot locations and accuracy analysis from University of Brasília and Federal University of Goiás/LAPIG, respectively.

Conflicts of Interest: The authors declare no conflict of interest. The funders had no role in the design of the study, in the collection, analysis, or interpretation; in the writing of the manuscript; or in the decision to publish the results.

References

1. Bourlière, F.; Hadley, M. Present-day savannas: An overview. In *Tropical Savannas*; Bourlière, F., Ed.; Elsevier: Amsterdam, The Netherlands, 1983; pp. 1–17.
2. Scholes, R.J.; Hall, D.O. The carbon budget of tropical savannas, woodlands and grasslands. *Sci. Comm. Probl. Environ. Int. Counc. Sci. Unions* **1996**, *56*, 69–100.
3. Solbrig, O.T. The diversity of the savanna ecosystem. In *Biodiversity and Savanna Ecosystem Processes*; Springer: Berlin/Heidelberg, Germany, 1996; pp. 1–27.
4. Gillson, L. Evidence of Hierarchical Patch Dynamics in an East African Savanna? *Landsc. Ecol.* **2004**, *19*, 883–894. [CrossRef]
5. Marchant, R. Understanding complexity in savannas: Climate, biodiversity and people. *Curr. Opin. Environ. Sustain.* **2010**, *2*, 101–108. [CrossRef]
6. Sano, E.E.; Rosa, R.; Scaramuzza, C.A.M.; Adami, M.; Bolfe, E.L.; Coutinho, A.C.; Esquerdo, J.C.D.M.; Maurano, L.E.P.; da Narvaes, I.S.; de Oliveira Filho, F.J.B.; et al. Land use dynamics in the Brazilian Cerrado in the period from 2002 to 2013. *Pesqui. Agropecuária Bras.* **2019**, *54*. [CrossRef]
7. Ribeiro, J.F.; Walter, B.M.T. Fitofisionomia do Bioma Cerrado. In *Cerrado: Ambiente e Flora*; Sano, S.M., Almeida, S.P., Eds.; Embrapa: Brasilia, Brazil, 1998; pp. 89–166.
8. Mittermeier, R.A.; Turner, W.R.; Larsen, F.W.; Brooks, T.M.; Gascon, C. Global Biodiversity Conservation: The Critical Role of Hotspots. In *Biodiversity Hotspots*; Springer: Berlin/Heidelberg, Germany, 2011; pp. 3–22.
9. Strassburg, B.B.N.; Brooks, T.; Feltran-Barbieri, R.; Iribarrem, A.; Crouzeilles, R.; Loyola, R.; Latawiec, A.E.; Oliveira Filho, F.J.B.; Scaramuzza, C.A.M.; Scarano, F.R.; et al. Moment of truth for the Cerrado hotspot. *Nat. Ecol. Evol.* **2017**, *1*, 99. [CrossRef] [PubMed]
10. Bustamante, M.; Nardoto, G.; Pinto, A.; Resende, J.; Takahashi, F.; Vieira, L. Potential impacts of climate change on biogeochemical functioning of Cerrado ecosystems. *Braz. J. Biol.* **2012**, *72*, 655–671. [CrossRef]

11. Spera, S.A.; Galford, G.L.; Coe, M.T.; Macedo, M.N.; Mustard, J.F. Land-use change affects water recycling in Brazil's last agricultural frontier. *Glob. Chang. Biol.* **2016**, *22*, 3405–3413. [CrossRef]

12. INPE (Instituto Nacional de Pesquisas Espaciais). Programa de Monitoramento da Amazônia e Demais Biomas—Bioma Cerrado. Available online: http://terrabrasilis.dpi.inpe.br/downloads/ (accessed on 10 November 2018).

13. Rocha, G.F.; Ferreira, L.G.; Ferreira, N.C.; Ferreira, M.E. Detecção de desmatamentos no bioma Cerrado entre 2002 e 2009: Padrões, tendências e impactos. *Rev. Bras. Cart.* **2011**, *3*, 341–349.

14. Dias, L.C.P.; Macedo, M.N.; Costa, M.H.; Coe, M.T.; Neill, C. Effects of land cover change on evapotranspiration and streamflow of small catchments in the Upper Xingu River Basin, Central Brazil. *J. Hydrol. Reg. Stud.* **2015**, *4*, 108–122. [CrossRef]

15. Sano, E.E.; Rosa, R.; Brito, J.L.S.; Ferreira, L.G. Land cover mapping of the tropical savanna region in Brazil. *Environ. Monit. Assess.* **2010**, *166*, 113–124. [CrossRef]

16. Glenn, E.; Huete, A.; Nagler, P.; Nelson, S. Relationship Between Remotely-sensed Vegetation Indices, Canopy Attributes and Plant Physiological Processes: What Vegetation Indices Can and Cannot Tell Us about the Landscape. *Sensors* **2008**, *8*, 2136–2160. [CrossRef] [PubMed]

17. Jacon, A.D.; Galvão, L.S.; dos Santos, J.R.; Sano, E.E. Seasonal characterization and discrimination of savannah physiognomies in Brazil using hyperspectral metrics from Hyperion/EO-1. *Int. J. Remote Sens.* **2017**, *38*, 4494–4516. [CrossRef]

18. Roy, D.P.; Boschetti, L.; Giglio, L. Remote Sensing of Global Savana Fire Occurrence, Extent, and Properties. In *Ecosystem Function in Savannas: Measurement and Modeling at Landscape to Global Scales*; Hill, M.J., Hanan, N.P., Eds.; CRC Press: Boca Raton, FL, USA, 2011; pp. 239–254.

19. Gomes, L.; Miranda, H.S.; Silvério, D.V.; Bustamante, M.M.C. Effects and behaviour of experimental fires in grasslands, savannas, and forests of the Brazilian Cerrado. *For. Ecol. Manage.* **2020**, *458*, 117804. [CrossRef]

20. Ferreira, L.G.; Huete, A.R. Assessing the seasonal dynamics of the Brazilian Cerrado vegetation through the use of spectral vegetation indices. *Int. J. Remote Sens.* **2004**, *25*, 1837–1860. [CrossRef]

21. Ratana, P.; Huete, A.R.; Ferreira, L. Analysis of Cerrado Physiognomies and Conversion in the MODIS Seasonal–Temporal Domain. *Earth Interact.* **2005**, *9*, 1–22. [CrossRef]

22. Ferreira, M.E.; Ferreira, L.G.; Sano, E.E.; Shimabukuro, Y.E. Spectral linear mixture modelling approaches for land cover mapping of tropical savanna areas in Brazil. *Int. J. Remote Sens.* **2007**, *28*, 413–429. [CrossRef]

23. Beuchle, R.; Grecchi, R.C.; Shimabukuro, Y.E.; Seliger, R.; Eva, H.D.; Sano, E.; Achard, F. Land cover changes in the Brazilian Cerrado and Caatinga biomes from 1990 to 2010 based on a systematic remote sensing sampling approach. *Appl. Geogr.* **2015**, *58*, 116–127. [CrossRef]

24. Schwieder, M.; Leitão, P.J.; da Cunha Bustamante, M.M.; Ferreira, L.G.; Rabe, A.; Hostert, P. Mapping Brazilian savanna vegetation gradients with Landsat time series. *Int. J. Appl. Earth Obs. Geoinf.* **2016**, *52*, 361–370. [CrossRef]

25. Hill, M.J.; Zhou, Q.; Sun, Q.; Schaaf, C.B.; Palace, M. Relationships between vegetation indices, fractional cover retrievals and the structure and composition of Brazilian Cerrado natural vegetation. *Int. J. Remote Sens.* **2017**, *38*, 874–905. [CrossRef]

26. Müller, H.; Rufin, P.; Griffiths, P.; Barros Siqueira, A.J.; Hostert, P. Mining dense Landsat time series for separating cropland and pasture in a heterogeneous Brazilian savanna landscape. *Remote Sens. Environ.* **2015**, *156*, 490–499. [CrossRef]

27. Brazil, M. *TerraClass: Mapeamento do Uso e Cobertura do Cerrado: Projeto TerraClass Cerrado 2013*; Ministério do Meio Ambiente (MMA): Brasilia, Brazil, 2015.

28. Fbds, F.B.; Para, D.S. Projeto de Mapeamento em Alta Resolução dos Biomas Brasileiros. Available online: http://geo.fbds.org.br/ (accessed on 1 November 2019).

29. Ministério da Ciência, Tecnologia e Inovação. *III Inventário Brasileiro de Emissões e Remoções Antrópicas de Gases de Efeito Estufa não Controlados pelo Protocolo de Montreal*; MCTIC: Brasilia, Brazil, 2015.

30. IBGE. *Monitoramento da Cobertura e uso da Terra—2000, 2010, 2012, 2014, 2015—Em Grade Territorial Estatística*; IBGE: Rio de Janeiro, Brazil, 2017.

31. Wulder, M.A.; White, J.C.; Loveland, T.R.; Woodcock, C.E.; Belward, A.S.; Cohen, W.B.; Fosnight, E.A.; Shaw, J.; Masek, J.G.; Roy, D.P. The global Landsat archive: Status, consolidation, and direction. *Remote Sens. Environ.* **2016**, *185*, 271–283. [CrossRef]

32. Phiri, D.; Morgenroth, J. Developments in Landsat Land Cover Classification Methods: A Review. *Remote Sens.* **2017**, *9*, 967. [CrossRef]

33. Gorelick, N.; Hancher, M.; Dixon, M.; Ilyushchenko, S.; Thau, D.; Moore, R. Google Earth Engine: Planetary-scale geospatial analysis for everyone. *Remote Sens. Environ.* **2017**, *202*, 18–27. [CrossRef]

34. Parente, L.; Mesquita, V.; Miziara, F.; Baumann, L.; Ferreira, L. Assessing the pasturelands and livestock dynamics in Brazil, from 1985 to 2017: A novel approach based on high spatial resolution imagery and Google Earth Engine cloud computing. *Remote Sens. Environ.* **2019**, *232*, 111301. [CrossRef]

35. IBGE. *Biomas e Sistema Costeiro-Marinho do Brasil: Compatível com a Escala 1:250.000*; IBGE: Rio de Janeiro, Brazil, 2019.

36. Assad, E.D. *Chuva nos Cerrados: Análise e Espacialização*; Embrapa-CPAC: Brasilia, Brazil, 1994.

37. Coutinho, L.M. O bioma do cerrado. In *Eugen Warming e o Cerrado Brasileiro: Um Século Depois*; UNESP: São Paulo, Brazil, 2002; pp. 77–91.

38. Roitman, I.; Bustamante, M.M.C.; Haidar, R.F.; Shimbo, J.Z.; Abdala, G.C.; Eiten, G.; Fagg, C.W.; Felfili, M.C.; Felfili, J.M.; Jacobson, T.K.B.; et al. Optimizing biomass estimates of savanna woodland at different spatial scales in the Brazilian Cerrado: Re-evaluating allometric equations and environmental influences. *PLoS ONE* **2018**, *13*, e0196742. [CrossRef]

39. Housman, I.; Chastain, R.; Finco, M. An Evaluation of Forest Health Insect and Disease Survey Data and Satellite-Based Remote Sensing Forest Change Detection Methods: Case Studies in the United States. *Remote Sens.* **2018**, *10*, 1184. [CrossRef]

40. Souza, C.M.; Roberts, D.A.; Cochrane, M.A. Combining spectral and spatial information to map canopy damage from selective logging and forest fires. *Remote Sens. Environ.* **2005**, *98*, 329–343. [CrossRef]

41. Diniz, C.; Cortinhas, L.; Nerino, G.; Rodrigues, J.; Sadeck, L.; Adami, M.; Souza-Filho, P. Brazilian Mangrove Status: Three Decades of Satellite Data Analysis. *Remote Sens.* **2019**, *11*, 808. [CrossRef]

42. Swain, P.H.; Hauska, H. The decision tree classifier: Design and potential. *IEEE Trans. Geosci. Electron.* **1977**, *15*, 142–147. [CrossRef]

43. Safavian, S.R.; Landgrebe, D. Separating and tracking multiple beacon sources for deep space optical communications. *Free. Laser Commun. Technol. XXII* **1991**, *21*, 660–674.

44. Zhu, Z.; Woodcock, C.E. Continuous change detection and classification of land cover using all available Landsat data. *Remote Sens. Environ.* **2014**, *144*, 152–171. [CrossRef]

45. Parente, L.; Ferreira, L. Assessing the Spatial and Occupation Dynamics of the Brazilian Pasturelands Based on the Automated Classification of MODIS Images from 2000 to 2016. *Remote Sens.* **2018**, *10*, 606. [CrossRef]

46. Nogueira, S.H.M.; Parente, L.L.; Ferreira, L.G. Temporal Visual Inspection: Uma Ferramenta Destinada À Inspeção Visual De Pontos Em Séries Históricas De Imagens De Sensoriamento Remoto. In Proceedings of the Anais do XXVII Congresso Brasileiro de Cartografia e XXVI Exposicarta, Rio de Janeiro, Brazil, 6–9 November 2017; pp. 624–628.

47. Stehman, S.V. Estimating area and map accuracy for stratified random sampling when the strata are different from the map classes. *Int. J. Remote Sens.* **2014**, *35*, 4923–4939. [CrossRef]

48. Olofsson, P.; Foody, G.M.; Herold, M.; Stehman, S.V.; Woodcock, C.E.; Wulder, M.A. Good practices for estimating area and assessing accuracy of land change. *Remote Sens. Environ.* **2014**, *148*, 42–57. [CrossRef]

49. Macedo, M.N.; DeFries, R.S.; Morton, D.C.; Stickler, C.M.; Galford, G.L.; Shimabukuro, Y.E. Decoupling of deforestation and soy production in the southern Amazon during the late 2000s. *Proc. Natl. Acad. Sci. USA* **2012**, *109*, 1341–1346. [CrossRef]

50. Zalles, V.; Hansen, M.C.; Potapov, P.V.; Stehman, S.V.; Tyukavina, A.; Pickens, A.; Song, X.P.; Adusei, B.; Okpa, C.; Aguilar, R.; et al. Near doubling of Brazil's intensive row crop area since 2000. *Proc. Natl. Acad. Sci. USA* **2019**, *116*, 428–435. [CrossRef]

51. Azzari, G.; Lobell, D.B. Landsat-based classification in the cloud: An opportunity for a paradigm shift in land cover monitoring. *Remote Sens. Environ.* **2017**, *202*, 64–74. [CrossRef]

52. Deines, J.M.; Kendall, A.D.; Crowley, M.A.; Rapp, J.; Cardille, J.A.; Hyndman, D.W. Mapping three decades of annual irrigation across the US High Plains Aquifer using Landsat and Google Earth Engine. *Remote Sens. Environ.* **2019**, *233*, 111400. [CrossRef]

53. Grecchi, R.C.; Gwyn, Q.H.J.; Bénié, G.B.; Formaggio, A.R. Assessing the spatio-temporal rates and patterns of land-use and land-cover changes in the Cerrados of southeastern Mato Grosso, Brazil. *Int. J. Remote Sens.* **2013**, *34*, 5369–5392. [CrossRef]

54. Ferreira, L.G.; Urban, T.J.; Neuenschawander, A.; de Araújo, F.M. Use of Orbital LIDAR in the Brazilian Cerrado Biome: Potential Applications and Data Availability. *Remote Sens.* **2011**, *3*, 2187–2206. [CrossRef]
55. Zimbres, B.; Shimbo, J.; Bustamante, M.; Levick, S.; Miranda, S.; Roitman, I.; Silvério, D.; Gomes, L.; Fagg, C.; Alencar, A. Savanna vegetation structure in the Brazilian Cerrado allows for the accurate estimation of aboveground biomass using terrestrial laser scanning. *For. Ecol. Manage.* **2020**, *458*, 117798. [CrossRef]
56. Noojipady, P.; Morton, C.D.; Macedo, N.M.; Victoria, C.D.; Huang, C.; Gibbs, K.H.; Bolfe, L.E. Forest carbon emissions from cropland expansion in the Brazilian Cerrado biome. *Environ. Res. Lett.* **2017**, *12*, 025004. [CrossRef]
57. Garcia, A.S.; Ballester, M.V.R. Land cover and land use changes in a Brazilian Cerrado landscape: Drivers, processes, and patterns. *J. Land Use Sci.* **2016**, *11*, 538–559. [CrossRef]

Article

Environmental Drivers of Water Use for Caatinga Woody Plant Species: Combining Remote Sensing Phenology and Sap Flow Measurements

Rennan A. Paloschi [1,*], Desirée Marques Ramos [2], Dione J. Ventura [1], Rodolfo Souza [3], Eduardo Souza [4], Leonor Patrícia Cerdeira Morellato [2], Rodolfo L. B. Nóbrega [5], Ítalo Antônio Cotta Coutinho [6], Anne Verhoef [7], Thales Sehn Körting [1] and Laura De Simone Borma [1]

1   National Institute for Space Research—INPE, São José dos Campos, SP 12227-010, Brazil; dione.silva@inpe.br (D.J.V.); thales.korting@inpe.br (T.S.K.); laura.borma@inpe.br (L.D.S.B.)
2   Department of Biodiversity, São Paulo State University—UNESP, Rio Claro, Jaboticabal, SP 14884-900, Brazil; dm.ramos@unesp.br (D.M.R.); patricia.morellato@unesp.br (L.P.C.M.)
3   Department of Biological and Agricultural Engineering, Texas A&M University, College Station, TX 77843, USA; rodolfo.souza@tamu.edu
4   Academic Unit of Serra Talhada, Federal Rural University of Pernambuco, Serra Talhada, PE 52171-900, Brazil; eduardo.ssouza@ufrpe.br
5   Department of Life Sciences, Imperial College London, Buckhurst Road, Ascot SL5 7PY, UK; r.nobrega@imperial.ac.uk
6   Department of Biology, Federal University of Ceara, Fortaleza, FL 33612, Brazil; italo.coutinho@ufc.br
7   Department of Geography and Environmental Science, The University of Reading, Reading RG6 6AR, UK; a.verhoef@reading.ac.uk
*   Correspondence: rennan.paloschi@inpe.br

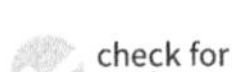
check for **updates**

Citation: Paloschi, R.A.; Ramos, D.M.; Ventura, D.J.; Souza, R.; Souza, E.; Morellato, L.P.C., Nóbrega, R.L.B.; Coutinho, Í.A.C.; Verhoef, A.; Körting, S.; et al. Environmental Drivers of Water Use for Caatinga Woody Plant Species: Combining Remote Sensing Phenology and Sap Flow Measurements. *Remote Sens.* **2021**, *13*, 75. https://doi.org/10.3390/rs13010075

Received: 31 October 2020
Accepted: 16 December 2020
Published: 28 December 2020

Publisher's Note: MDPI stays neutral with regard to jurisdictional claims in published maps and institutional affiliations.

Abstract: We investigated the water use of Caatinga vegetation, the largest seasonally dry forest in South America. We identified and analysed the environmental phenological drivers in woody species and their relationship with transpiration. To monitor the phenological evolution, we used remote sensing indices at different spatial and temporal scales: normalized difference vegetation index (NDVI), soil adjusted vegetation index (SAVI), and green chromatic coordinate (GCC). To represent the phenology, we used the GCC extracted from in-situ automated digital camera images; indices calculated based on sensors included NDVI, SAVI and GCC from Sentinel-2A and B satellites images, and NDVI products MYD13Q1 and MOD13Q1 from a moderate-resolution imaging spectroradiometer (MODIS). Environmental drivers included continuously monitored rainfall, air temperature, soil moisture, net radiation, and vapour pressure deficit. To monitor soil water status and vegetation water use, we installed soil moisture sensors along three soil profiles and sap flow sensors for five plant species. Our study demonstrated that the near-surface GCC data played an important role in permitting individual monitoring of species, whereas the species' sap flow data correlated better with NDVI, SAVI, and GCC than with species' near-surface GCC. The wood density appeared to affect the transpiration cessation times in the dry season, given that species with the lowest wood density reach negligible values of transpiration earlier in the season than those with high woody density. Our results show that soil water availability was the main limiting factor for transpiration during more than 80% of the year, and that both the phenological response and water use are directly related to water availability when relative saturation of the soil profile fell below 0.25.

Keywords: plant water availability; tree phenology; phenocams; Sentinel-2; MODIS

1. Introduction

A better understanding of plant water availability and water use is of great importance for reliable assessments of ecosystem's resilience to droughts [1]. Water availability is critical for plant growth, inducing phenological transitions and, ultimately, plant survival.

Additionally, information on plant water balance also allows the development of more realistic soil water assessment tools and land surface models, which is a widely acknowledged requirement for research related to the plant–soil–atmosphere continuum [2].

Plant species have their own specific adaptive mechanisms to cope with droughts [3,4], which are particularly important in water-limited ecosystems such as seasonally dry tropical forests (SDTFs).

The SDTF is a unique biome that occurs in low-latitude areas of fertile soils with a low annual precipitation (ranging from about 250 to 1500 mm year^{-1}) and a prolonged dry season that extends for five to six months [5–7]. The vegetation ranges from tall forests with closed canopies to scrublands rich in succulents and thorn-bearing plants [7,8]. The long water-limited period has been shown to be selective for deciduous, thorny, and succulent plant species that show a marked leaf senescence during the dry season followed by synchronous leaf flushing at the beginning of the rainy season [3,4,7–9]. The deciduity and absence of a grassy layer are important characteristics that distinguish SDTFs from other mild seasonally dry tropical biomes such as the savannas [8].

In South America, SDTFs are largely represented by the Caatinga vegetation, the predominant vegetation in northeast Brazil and located in a semi-arid region that extends over 800.000 km^2 [10–12]. The climate is predominantly hot and dry, with temperatures around 26 °C and high evapotranspiration (1500 to 2000 mm year^{-1}) coupled with low annual rainfall (400 to 800 mm year^{-1}) [11,13]. This results in high water deficits, which are aggravated by the short rainy season that usually lasts three to five months and show erratic rain episodes [10,13]. The water deficit is even more severe during catastrophic severe droughts that may last three to five years [11].

For most Caatinga plant species, vegetative and reproductive structures develop exclusively during the rainy season [3,14]. In fact, the fast metabolic response, which is unique to the Caatinga, allows species to produce a synchronous leaf flush with the onset of the rainy season [3,14].

Caatinga tree species are able to cope with long periods of drought by exhibiting different transpiration-reducing leaf morphological and photosynthetic characteristics. They also display a considerable variability in wood density and related water storage properties [3,14]. Lima et al (2012) show a high correlation between wood density and water storage in Caatinga vegetation and relate these characteristics to a seasonal phenology. Several studies have shown that high wood density and thick cell walls tend to protect plants from cavitation but, in general, both are associated with low water storage in stem tissues [15–19]. On the other hand, low wood density and thin cell walls increase the chances of cavitation but water storage may be higher and the maintenance of a high water potential may be prolonged [15–17]. The combination of these diverse plant traits produce different phenological evolutions throughout the growing season that depend on different plant water use strategies [18–20].

Eddy covariance (EC) and remote sensing (RS) techniques have been used to evaluate terrestrial environments because they allow vegetation systems to be studied at different scales: from local to global scales. The use of RS is gaining importance as satellites provide low-cost products and images with increasing spatial and temporal resolutions. The advances in technology have allowed several RS vegetation indices (VIs) to be used, improving our understanding of temporal and spatial changes in plant communities [21–23]. The direct relationship between water use and phenological response observed by RS has been used in models that seek to estimate evapotranspiration (ET) and gross primary production [24], mainly through the use of empirical or semi-empirical parameterisations, taking advantage of vegetation indices such as normalized difference vegetation index (NDVI) [25]. This approach is particularly successful in the Brazilian semi-arid regions due to the strong synchronicity between phenology and plant-water availability [3].

An alternative approach to monitor vegetation phenology is near-surface RS, which uses in-situ automated cameras (phenocams) [26]. This technique, in combination with EC measurements and remote sensing, has been applied in Brazilian SDTFs [4] to investigate the drivers that regulate phenological patterns and vegetation responses to seasonal and severe droughts.

Regarding water use, EC and RS can be used to estimate ET [25], a flux that encompasses processes other than transpiration, such as evaporation of soil water, canopy-intercepted water, and other free water surfaces. However, RS generally provides discontinuous time series of ET (e.g., due to cloudy conditions) [27]. Moreover, none of these tools allow direct measurements of transpiration at the individual tree or species level. Hence, these techniques cannot explain species-specific strategies that are strongly dependent on individual structural, phenological, and physiological traits. For such, in-situ or near-surface methods make monitoring possible.

Continuous in-situ estimation of transpiration is possible with sap flow sensors. Among the methods used to estimate sap flow, thermal dissipation probes (TDP) have been widely used and improved throughout the years [28]. Information on transpiration is essential for studies related to plant water use, including those in SDTFs [29,30]. Sap flow measurement techniques can provide time series of diurnal and seasonal water fluxes to estimate plant water use. The TDP method, suggested by Granier [31,32], is widely applied to estimate tree transpiration because the sensor system is simple to construct, easy to install, and relatively inexpensive [33,34].

When analysing time series and patterns of water use, vapor pressure deficit (VPD) has been considered as the main environmental driver of plant transpiration under most conditions, representing the atmospheric demand Grossiord 2019 [35]. performed continuously-logged sap flow monitoring experiments in humid tropical forests and observed that sap flow was highly dependent on VPD. However, for the less humid regions studied, very high VPD led to a reduction in transpiration as a result of the strong negative dependence of stomatal opening on VPD Grossiord et al. Similarly, Butz et al (2018) monitored sap flow of both deciduous and non-deciduous species and showed that evergreen species have a stronger relationship with seasonal VPD variation than deciduous ones, and a slightly higher dependence of sap flow on soil moisture than on VPD. Overall, these studies indicate that in more water-limited systems, species will be more dependent on soil water availability.

Here, we combine several datasets and approaches, as well as ground truth data, in order to test the efficiency of remote sensing data to monitor a significant physiological process such as water use. We aim to investigate the relationship between plant water use and phenology for Caatinga woody plant species at both the species and community level using different scales: moderate RS spatial resolution, high RS spatial resolution, and near-surface cameras. In addition, we intend to evaluate the environmental drivers and factors that limit plant phenology (leaf flush and senescence) and transpiration. We hypothesize that: (1) phenology at the community level (represented by RS data) and at the species level (represented by near surface remote sensing data) are fully explained by soil water availability; (2) RS data for the Caatinga vegetation are strongly correlated to variations in water use strategies among species; (3) variations in the sap flow signal are fully explained by changes in water availability.

2. Materials and Methods

2.1. Study Site

The study site (Figure 1) is located in central-northern part of the Brazilian state of Pernambuco, in the municipality of Serra Talhada (7°58′12″ S, 38°23′06″ W, 455 m a.s.l.). The soils are predominantly Aridisols Argid and Entisols orthents [36]. The area has a relatively smooth relief and comprises various xerophilous vegetation types [37]. The local climate is Bsh (arid-steppe-hot arid) according to the Köppen system in a transition region to the Aw (winter dry season) [38–40], with a mean annual rainfall of 680 mm yr^{-1} and a mean temperature of 23.8 °C [40]. The experimental site includes a 50 × 100 m monitoring plot (Figure 1) which is part of the Nordeste project plot network [41], and a 10 m tall eddy covariance (EC) flux tower (site ID BR-CST in AmeriFlux network; 7.9682° S, 38.3842° E) located 200 m away from the monitoring plot.

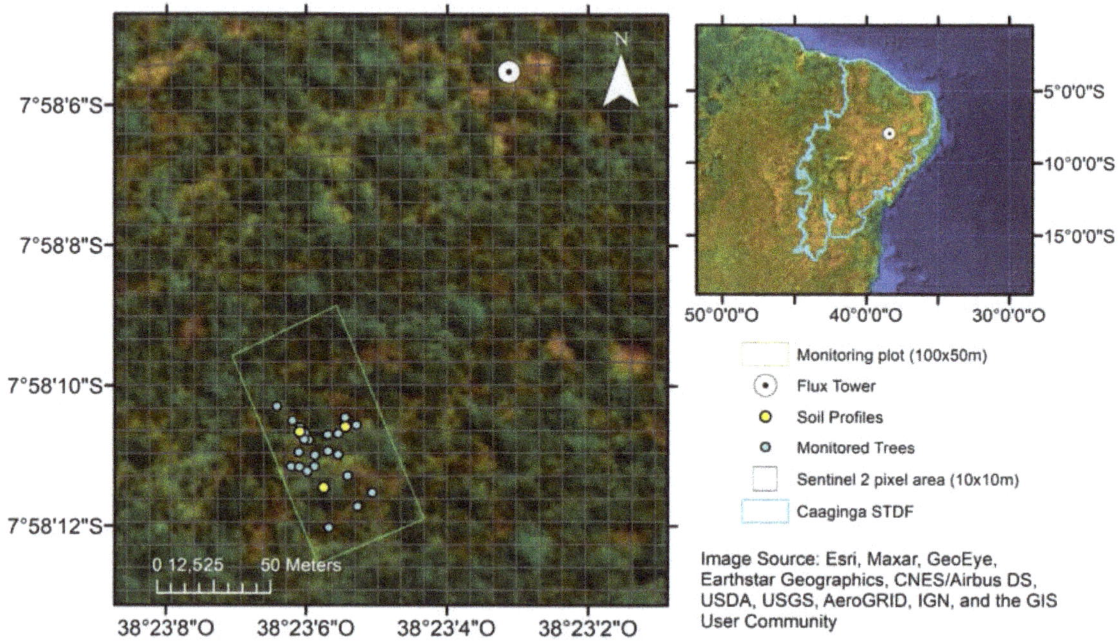

Figure 1. Study site: Map of the study area showing eddy covariance (EC) flux tower location (where the phenocam is installed), and the monitoring plot (where sap flow and soil moisture sensors were installed); location of the study site in the context of the Brazilian Caatinga. Spatial resolution: 15 cm.

Our study was carried out from April 2018 to March 2019, during which the in-situ pluviometer registered a pronounced rainy season from December to April (total precipitation of 499 mm), whereas the dry period lasted from May to November (total precipitation of 40 mm). Regarding the total rainfall in April, which marked the beginning of our experiment, that month was considered the wet-to-dry transition period. These precipitation values and their timing are consistent with the normal climate found for this region [40]. Therefore, the results found in this work can be extended to other years with normal rainfall.

2.2. Experimental Strategy

The experimental design is shown in the flowchart (Figure 2). In terms of ground data, the experimental strategy consists in: (i) measuring the sap flux density (F_d) on individual tree species found within the monitoring plot and normalizing it among the sampled species (F_{dn}); (ii) upscaling the normalized flux density (F_{dn}) to the level of the plot (or community; F_{dnc}); (iii) comparing the F_{dn} with the soil water availability. The purpose of these measurements was to evaluate the main drivers affecting the physiological response of Caatinga trees (i.e., the plant-water availability and VPD). To test the relationship between plant physiology patterns (sap flow density) and leaf phenology, we used VIs obtained from different sources. Besides being used as a proxy of leaf phenology, VIs from different sources were also evaluated in terms of their ability to represent certain physiological process, such as intra-annual variability of sap flow.

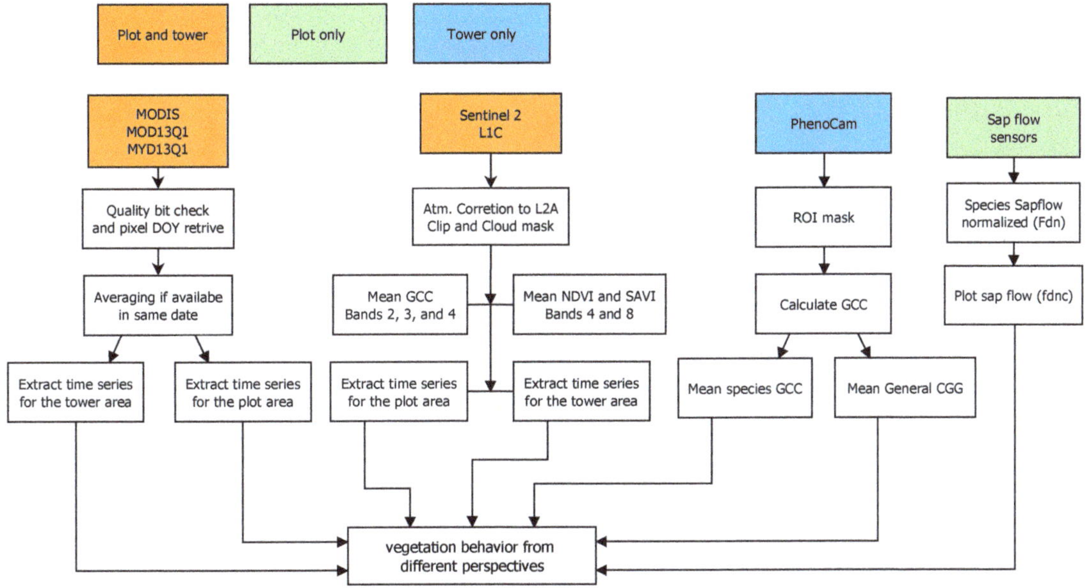

Figure 2. Flowchart of the experimental design.

2.2.1. Plant Species

A total of 24 species were found in the plot, of which 11 showed a dominance over 1.0: *Anadenanthera colubrina* (Vell.) Brenan, *Aspidosperma pyrifolium* Mart. & Zucc, *Astronium urundeuva* Engl., *Cenostigma nordestinum* Gagnon & G.P.Lewis, *Cereus jamacaru* DC., *Commiphora leptophloeos* (Mart.) J.B. Gillett, *Croton echioideus* Baill., *Erythroxylum pungens* O.E.Schulz, *Mimosa acutistipula* Benth. *Piptadenia flava* Benth. and *Senegalia polyphylla* (DC.) Britton & Rose.

We selected five species to monitor sap flow density, which accounted for 80% of the total dominance: *C. leptophloeos*; *A. colubrina*; *S. polyphylla* and; *A. pyrifolium*; *C. nordestinum*. All selected species are deciduous and were selected based on their abundance registered in the inventory of the monitoring plot and the Nordeste project plot network [41], and based on their relative abundance, as indicated in Table 1.

Table 1. Tree species monitored, wood density (W_d, g·cm^{-3}), mean diameter at breast height (DBH) mean species height (H.), dominance in the plot (D.), estimated sum of the sapwood area per species per species, percentage of the sapwood area in relation to the total sapwood area in the plot.

Species	W_d	DBH (cm)	H. (m)	D. (%)	Sap. Area (cm^2)	% of Total Sap.
Anadenanthera colubrina	0.59	13.4	7.2	4	175	6.2
Aspidosperma pyrifolium	0.54	8.1	4.9	30	772	27.3
Cenostigma nordestinum	0.66	13.8	5.4	30	1751	62.0
Commiphora leptophloeos	0.28	20.0	5.2	1	78	2.8
Senegalia polyphylla	— —	15	7.7	5.4	49	1.7
Total	— —	80	— —	— —	2824	100

W_d values obtained from [3].

All plant specimens were collected and identified by botanists at the Herbarium of the State University of Feira de Santana (HUEFS) Bahia, Brazil. The inventory included all plant individuals with a diameter at breast height (DBH) over 5 cm; the average height of trees in the plot was 5.2 m. Abundance of each species was determined by establishing the frequency of individuals of a given species within the monitoring plot.

We estimated sapwood depth using the linear regression given by 0.377 * DBH -1.075, R^2 of 0.67 ($p < 0.05$ for both coefficients). Additionally, we estimated the total sapwood area for all individuals of a given species, calculated the sapwood area of each species, and estimated the percentage of the total active xylem area for each species Table 1.

2.2.2. Ground Measurements

Sap flux density (F_d, g m^{-2} min^{-1}) for the selected individuals was measured using Granier's TDP method (see Supplementary Materials, Section S1C). The sap flux density at the individual level (F_d) was first normalized, yielding F_{dn}, and then scaled-up to the community level, F_{dnc}, considering the relative sapwood area of each species (Table 1). At the community level, F_{dnc} was calculated as the weighted average considering the relative contribution of the sapwood area for each species. The environmental drivers and plant traits used in the sap flow and phenology analysis are described in Table 2. The environmental drivers and plant traits used in the sap flow and phenology analysis were: vapour-pressure deficit (VPD) and plant water availability—expressed in terms of soil water content (S_w) and relative soil water saturation ($S_{e,prof}$; Table 2). S_w is the amount of water for the total soil profile of 77.5 cm. $S_{e,prof}$ is a fraction between; dependent on the actual profile soil moisture, θ_{prof}, and the difference between the saturated profile moisture content, $\theta_{s,prof}$, and residual profile moisture content, $\theta_{r,prof}$ (see Supplementary Materials Equation (S2)). For more details, see Supplementary Materials, as indicated in Table 2.

Table 2. Summary of measured or derived environmental variables: category of the variable; name; description; units; section number. See the Supplementary Material for more details.

Category	Variable	Description	Unit	SM s nr.
Atmospheric driving	R_n	Net Radiation	W m^{-2}	1.A
	T_{air}	Air temperature	°C	1.A
	VPD	Vapour-pressure deficit	hPa	1.A
	P	Rainfall	mm	1.A
Soil moisture status	$S_{e,prof}$	Relative saturation	-	1.B
	W_s	Soil Water Storage	mm	1.B
Transpiration and evapotranspiration	F_{dn}	Sap flow density normalized	-	1.C
	F_{dnc}	Community sap flow normalized	-	1.C
	ET	EC evapotranspiration	mm month^{-1}	1.A
Plant trait	W_d	Wood density	g·cm^{-3}	*
	Dominance	Relative frequency abundance	%	1.C
	Sap. area	Sum of estimated species sapwood area	cm^2	1.C

2.2.3. Remote Sensing Data

Information on land surface phenology and vegetation indexes (VIs) was obtained from remote sensing (RS) sensors which operate at different spatial and temporal scales: remote sensors MODIS/Terra + Aqua (250 m products MYD13Q1 and MOD13Q1) and MSI/Sentinel-2A + B (10 m), and near-surface remotely sensed data, obtained by one phenocam (Mobotix AG-Germany) installed at the top of the EC flux tower (see Supplementary Materials, Section S1A). The VIs derived from these RS products are NDVI (MODIS and Sentinel-2A), GCC (Sentinel-2A and phenocam) and SAVI (Sentinel-2A), as presented in Table 3.

Because of the rapid dynamics of the Caatinga vegetation, which responds to rain and soil water content within a few days, a short revisit time satellite was necessary.

The Sentinel-2 satellite was chosen because of its spatial resolution (10 m) and mainly because of its revisit time (3–10 days). The MODIS sensor was chosen for comparison purposes. Despite its moderate spatial resolution (250 m) considering the dimensions of our plot, the high revisit frequency and well known product quality provides a simple and effective way to compare the performance of the Sentinel-2 index. Since our objective is to analyze forest change over time, comparing the variables with a product resulting from a very low revisit time is essential (Terra and Aqua satellites cross every day, morning and afternoon, providing temporal VI products with much less cloud probability). The phenocam provided the field truth for the region and also enabled us to evaluate the phenology of each species separately.

Table 3. Remote Sensing data: Variable; Sensor/Satellite, spatial resolution; temporal resolution; respective section. Please see the Supplementary Materials for more details.

Variable	Sensor/Satellite	Spatial Resolution	Temporal Resolution	Section nr.
$NDVI_{modis}$	MODIS/TERRA + AQUA	250 m	8 days *	1.D
$NDVI_{S2}$	MSI/Sentinel 2A + 2B	10 m	3–10 days	1.D
GCC_{S2}	MSI/Sentinel 2A + 2B	10 m	3–10 days	1.D
GCC_{ns}	Near surface camera/NA	**	daily	1.D
$SAVI_{S2}$	MSI/Sentinel 2A + 2B	10 m	3–10 days	1.D

* pixel composition from daily image acquisitions; ** camera spatial resolution depends on the distance from camera's nadir.

The proportion of canopy coverage in relation to soil exposure is also seen by Sentinel-2 and MODIS. The coverage was considered spatially homogeneous since no significant differences ($p > 0.5$) were found between the temporal behavior of the sentinel individual pixels and individual species. This was due to the proximity of trees from different species, sometimes less than two meters, but also due to the spatial autocorrelation of the pixels. Each pixel influences its neighborhood [42], reducing the effective pixel resolution. It is worth mentioning that the spatial miss-registration in the Sentinel-2 images series vary by around 12 m (more than one pixel) but can be greater than 3 pixels according to [43]. For all these reasons, it was not possible to analyze the temporal behavior of Sentinel-2 pixels individually. Thus, a temporal profile was extracted considering the pixel mean of the monitoring plot area.

Regarding the phenocam images, the regions of interest (ROIs) were selected at the species and at the community level. At the species level, we selected pixels corresponding to the tree crowns of the four species found within the phenocam's field of view (Figure 3). *S. polyphylla*, which represents only 1.7% of the total sapwood area, was not present within the camera's field of view. We then extracted the GCC for each ROI and calculated the mean GCC for each species ($GCC_{ns,e}$. At the community level, we selected the ROI pixels corresponding to the entire vegetated area in the image, excluding bare soil and the tower (Figure 4). The community ROI includes trees, shrubs, and eventual forbs if present. The GCC was extracted from the community's ROI. Conservatively, the same ROI was used for the community during the entire study period, including the dry and rainy seasons. Figure 4 shows the contrast between the seasons: the fully developed tree crowns in the rainy season and the leafless crowns during the dry season.

2.3. Data Analysis

In order to establish relationships between plant water use and environmental drivers, we compared the temporal variability of F_{dnc} with the environmental data obtained from the micrometeorological flux tower (i.e., VPD, R_n and T_{air}) and with the relative saturation, $S_{e,prof}$. To identify the relationship at the community level, we used the same approach to compare F_{dnc} with NDVI (MODIS and Sentinel-2) and GCC_{S2}.

The VIs (GCC, SAVI, and NDVI) used to compare with the results obtained from the phenocams (GCC_{ns}) were calculated for the tower region; we assumed that the tower region consisted of a circle around the flux tower with a radius of 50 m, which is compatible with the camera field of view. The VIs (Table 3) used to compare with the sap flow (F_{dn} and F_{dnc}) data and with $S_{e,prof}$ were calculated using the pixels that covered the monitoring plot area (Figure 1).

To compare the VIs obtained from different scales, we used the transition dates of start (SOS) and end (EOS) of the growing season. The length of the growing season and associated phenological transitions, SOS and EOS [21], were calculated from the phenocam GCC and the RS vegetation indices GCC, SAVI, and NDVI based on Alberton et al. 2019.

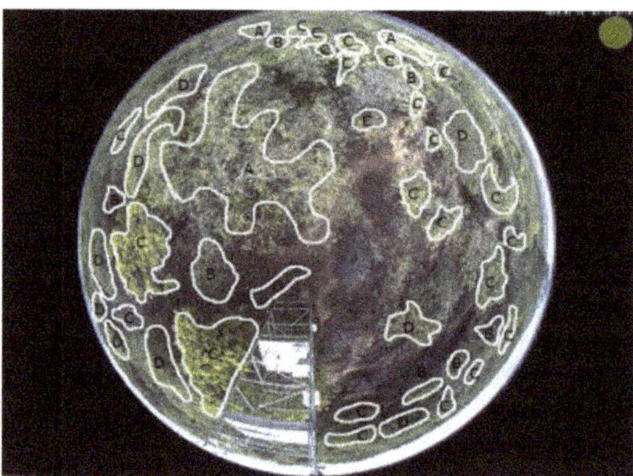

Figure 3. Example of a hemispherical image used for the proximal remote sensing phenological analyses at the study site, PE-Brazil. The areas selected in white represent tree crowns of individuals from four species: A, *Anadenanthera colubrina*; B, *Aspidosperma pyrifolium*; C, *Cenostigma nordestinum* and D, *Commiphora leptophloeos*.

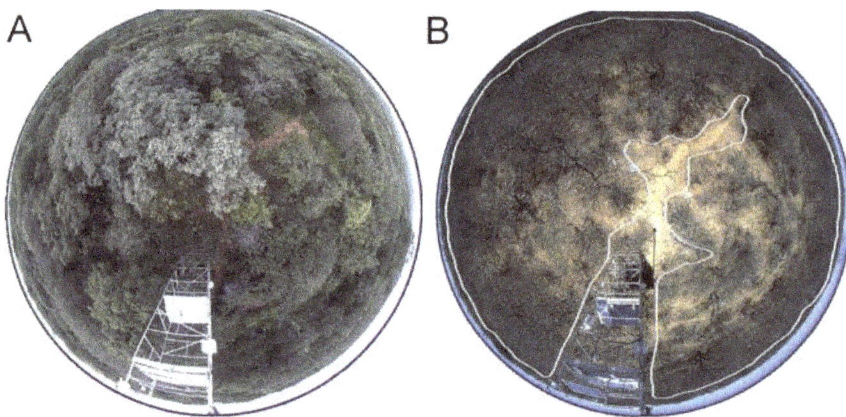

Figure 4. Hemispherical image used for the proximal remote sensing phenological analyses at the study site, PE-Brazil; (**A**) In the rainy season day of year (DOY) 107, 2018. (**B**) In the dry season DOY 256, 2018; the area selected in white represents the entire vegetation sampled as the region of interest (ROI) of the community for the extraction of phenological indices.

This method uses the confidence intervals of curve derivatives to identify changes in phenology [4]. First, we fitted a generalized additive model (GAM) using the species and community GCC as the response variable fitted by the sequence of consecutive days (time) of the phenological observations, which was used as an independent smoother variable to produce the phenological curves. We then calculated which regions of the curve had derivatives and detected when the derivatives were increasing (to determine SOS) or decreasing (to determine EOS) [4]. We considered SOS as the first day detected on the derivative at the left side of the GCC curve and EOS as the last day of the derivative at the right side of the curve. The transition dates calculated from the phenocam GCC were verified by visual inspection of the digital images and adjusted when necessary (for more details see Supplementary Materials Section S1D).

In addition, to evaluate whether variations in wood density could explain differences in water use and transition dates, we assessed species' wood density (W_d, g·cm^{-3}) and the difference between F_{dn} time series and near-surface GCC for each species.

Finally, the F_{dnc} data were compared with the VI that showed the greatest correlation with F_{dnc}. Thus, we compared NDVI indices ($NDVI_{modis}$ and $NDVI_{S2}$) with $S_{e,prof}$ using scatter plots in order to explicitly assess the dependence between variables across their data range.

3. Results

3.1. Hydroclimatological Drivers of Sap Flux Density

Figure 5a shows that the net radiation, R_n, was relatively low between April and August. This is part of the dry and cooler period where shortwave and longwave incoming radiation are lower than during the wet period, while albedo and longwave outgoing radiation are relatively more frequent, as previously observed in the Caatinga [22,44]. R_n remained approximately constant after September. The ET, shown in Figure 5b, varies between 0 and 125 mm month^{-1}, and the lowest values were found between August and November, when rainfall, (P) was negligible (see Figure 5a). Despite the low P in May and June, ET is still considerable because of the total water stored (S_w) in the soil profile (see Figure 5c), which allows transpiration to continue. It is worth noting that for leaves that are well-coupled to the atmosphere (i.e., with open stomata), the transpiration at the leaf level can be approximated by the product of the leaf-to-air VPD and the stomatal conductance according to Fick's law of diffusion [45]. However, in this region, our results show that intra-annual ET was inversely related to VPD ($p < 0.05$) (Figure 5b), most likely because, despite the higher VPD, plant species are mostly leafless from September to November.

The ET is in phase with S_w. The latter varied from a maximum of 184 mm at the beginning of April to a minimum of 74 mm at the end of November. However, after September S_w remained fairly constant (approximately 75 mm) until November when the rains started (Figure 5c). The highest air temperatures, T_{air}, coincided with the lowest values of ET, underlining the harsh climatic conditions during the dry season. T_{air} varied between 24 °C in July and 28.5 °C in November.

The community-level sap flow, F_{dnc} (Figure 5d), is in phase with S_w and ET. However, ET increased rapidly during the first rainy month, while F_{dnc} increased gradually, suggesting that a large part of the ET recorded during the first rainy month may be due to the evaporation of intercepted water and/or to soil water evaporation. Furthermore, the F_{dnc} varies in synchrony with the vegetation indices GCC_{S2}, $NDVI_{S2}$, and $NDVI_{modis}$ (Supplementary Materials, Figure S2).

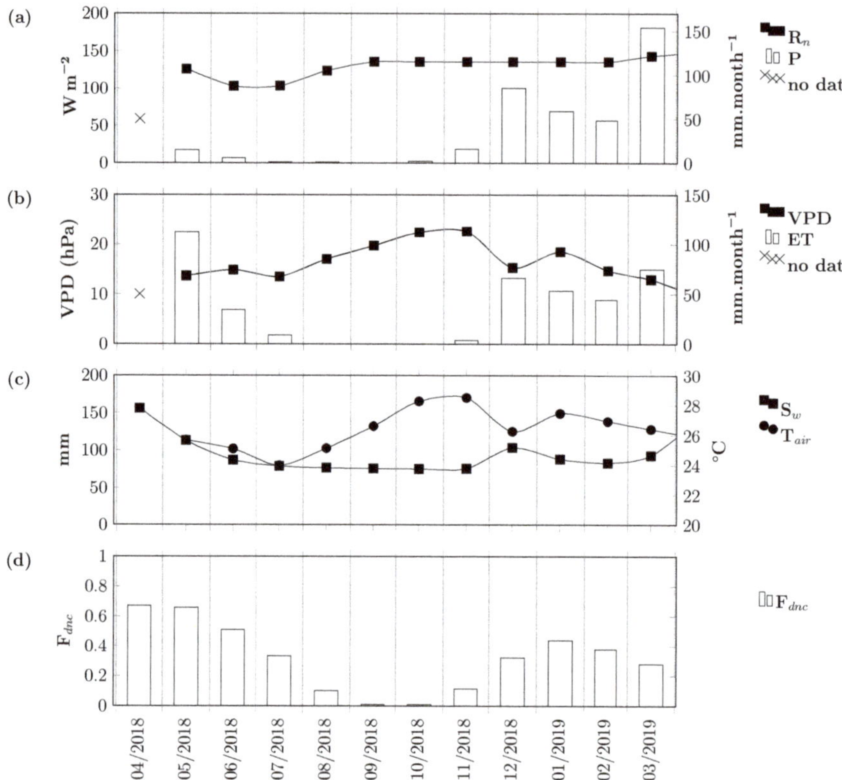

Figure 5. Climate variables, soil water availability, and community-level sap flow: (**a**) Net Radiation (R_n, W m^{-2}) and rainfall (P, mm month^{-1}); (**b**) vapour pressure deficit (VPD, hPa) and EC evapotranspiration (ET, mm month^{-1}); (**c**) soil water storage (S_w, mm) and monthly mean air temperature (T_{air}; °C); (**d**) Normalized sap flow for the plot community (F_{dnc}, dimensionless).

3.2. Transpiration, Soil Water Condition, and Remote Sensing Data

The normalized sap flux density at the species level, F_{dn}, was strongly affected by soil conditions when $S_{e,prof}$ was below 0.25 (Figure 6), indicating the high dependency of transpiration on stored water and demonstrating the threshold when plant water stress starts to limit transpiration. GCC and F_{dn} increased rapidly after the first rain, even with low water availability ($S_{e,prof}$ around 0.18), for all species except *A. pyrifolium*. After December 2019, the sap flow signal varied reasonably, but still followed the evolution of $S_{e,prof}$. This is due in part to climate variations, but also to the fact that a number of sensors failed during this period. Thus, the average F_{dn} for certain months was not based on all four sap flow sensors. All individuals from the studied species had a fully developed canopy at the beginning of the sap flow measurements. Therefore, our study period included two transition dates, the EOS of the ongoing growing season (2017–2018) and the SOS of the second growing season (2018–2019). Moreover, leaf fall still occurred after the EOS in the rainy season in 2018, and a SOS with new leaf flushing was detected at the beginning of the next rainy season in December of 2019. All species showed a negative correlation to daily VPD since all are deciduous and lost their leaves, halting transpiration precisely in the periods when VPD was highest.

Despite their approximate correspondence (Figure 6), as both variables exhibit a strong seasonality, the correlation between F_{dn} and GCC obtained from the proximal RS was poor (see Table 4, last column; the highest correlation (0.51) was found for *A. colubrina* and

C. nordestinum, species with high wood density). In fact, F_{dn} and F_{dnc} have a stronger relationship with all RS indices, especially with NDVI (Table 4). $NDVI_{modis}$ data better explained the variation in F_{dnc} than $NDVI_{S2}$, Table 4 shows a Pearson correlation coefficient of 0.92 ($p < 0.01$) versus a value of 0.88 ($p < 0.01$).

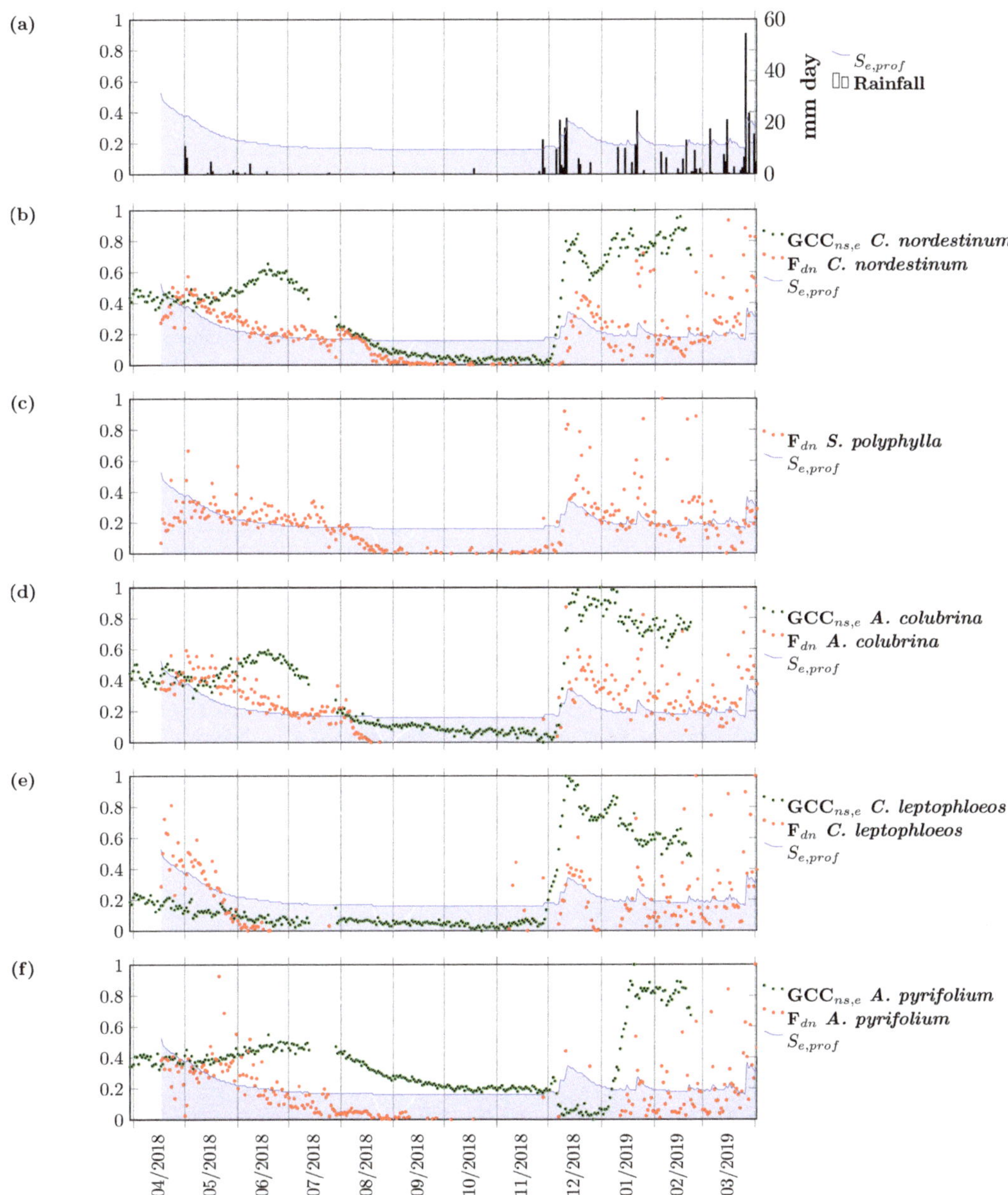

Figure 6. (**a**) Rainfall (mm day^{-1}) and soil water profile relative saturation ($S_{e,prof}$, dimensionless); and normalized near surface species GCC ($GCC_{ns,e}$, dimensionless) , and species averaged sap flow density (F_{dn}, dimensionless), together with $S_{e,prof}$ for (**b**) *C. nordestinum* and $S_{e,prof}$; (**c**) *S. polyphylla*; (**d**) *A. colubrina*; (**e**) *C. leptophloeos*; (**f**) *A. pyrifolium*.

Table 4. Pearson correlation coefficient between F_{dn} and $NDVI_{modis}$, $NDVI_{S2}$, GCC_{S2}, $SAVI_{S2}$ and GCC_{ns}, respectively.

F_{dn}	$NDVI_{modis}$	$NDVI_{S2}$	GCC_{S2}	$SAVI_{S2}$	GCC_{ns}
A. colubrina	0.83 *	0.80 *	0.71 *	0.74 *	0.36 *
A. pyrifolium	0.76 *	0.78 *	0.67 *	0.79 *	−0.14
C. leptophloeos	0.31	0.49	0.55 *	0.56 *	0.06
C. nordestinum	0.85 *	0.71 *	0.68 *	0.63 *	0.51 *
S. polyphylla	0.82 *	0.73 *	0.71 *	0.62 *	–
F_{dnc}	0.92 *	0.88 *	0.84 *	0.81 *	

obs.: * = $p < 0.5$.

The transition dates (in format DD/MM/YY) calculated for each species were the following: *A. pyrifolium*: EOS = 10/08/18, SOS = 10/01/19; *C. nordestinum*: EOS = 05/08/18, SOS = 08/12/18; *C. leptophloeos*: EOS = 12/06/18, SOS = 11/12/18; *A. colubrina*: EOS = 03/08/18, SOS = 07/12/18. Based on visual inspections of the images, we found that during the period between EOS and SOS, all individuals of the four species were leafless. The development dates for the tower region estimated by the sensors (Sentinel-2 and MODIS) can be seen in Table 5. We found that $NDVI_{modis}$ was the index with the highest correlation to species' water use, and its transition dates were the most similar to the GCC_{ns} results (Table 5).

Table 5. VI Transition dates (day of year/year) and bias in relation to near-surface GCC.

Transition Dates	GCC_{ns}	GCC_{S2}	$NDVI_{S2}$	$SAVI_{S2}$	$NDVI_{modis}$
EOS	238/18	220/18 (−18)	196/18 (−42)	191/18 (−47)	229/18 (−9)
SOS	339/18	306/18 (−33)	305/18 (−34)	321/18 (−18)	309/18 (−20)

Both the seasonal evolution of transpiration and the phenological response of the vegetation show an asymptotic limit in relation to $S_{e,prof}$ (Figure 7a,b), most likely because we are using $\theta_{s,prof}$ and $\theta_{r,prof}$ to normalise the data (please see the Supplementary Materials for an explanation of these parameters and of θ_{pwp} and θ_{fc} used below). During a considerable part of the year, the moisture content of the soil layers that contain most of the roots are below wilting point (θ_{pwp}), yet above the residual moisture content (θ_{fc}). Additionally, not all of the water stored in the soil pores is used for transpiration, as part of it is stored below the root zone or lost to soil evaporation.

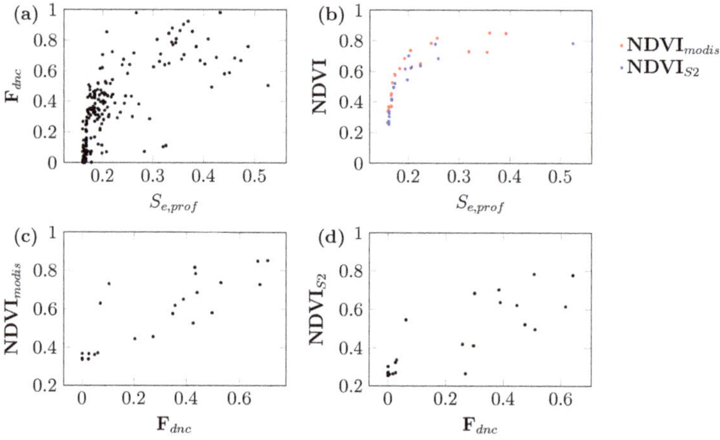

Figure 7. Scatter plots of (**a**) community level normalized sap flow (F_{dnc}) versus S_{prof} (dimensionless), (**b**) $NDVI_{modis}$ and $NDVI_{S2}$ versus S_{prof}, (**c**) $NDVI_{modis}$ versus F_{dnc}, (**d**) $NDVI_{S2}$ versus F_{dnc}.

4. Discussion

This study explored the sensitivity of remotely sensed indices to measure water use in the Caatinga. Our findings indicate that (i) the water use is highly dependent on soil moisture conditions; (ii) differences in wood characteristics seem to directly affect differences in when species sap flow ceases; (iii) with increasing resolution, from satellites with low resolutions to cameras that allow monitoring of individual tree species, new sources of uncertainties, previously masked by low spatial resolution, reduce correlations between RS indices and individual and community measurements of water use. Notwithstanding, the high resolution of the cameras has the benefit of allowing isolated events, such as the SOS and EOS, to be detected for different species.

It is remarkable how strongly F_{dnc} is related to $S_{e,prof}$, especially when considering daily values (Figure 6), which shows relative saturation together with rainfall, as well as greenness and sap flow per species. The strong seasonality, reflected in the highly variable sap flow values observed for all monitored Caatinga plant species, is evident and in agreement with previous studies [3,4,46]. Seasonality is particularly evident with regards to the timing of leaf shedding (EOS) and the drop in soil water storage (S_w) (Figure 5d). Moreover, the fact that seasonally varying values of F_{dnc} are only very weakly and negatively related to changes in VPD corroborates the idea that this vegetation is relatively independent of the atmospheric demand when considering intra-annual variation. This is not the case for the diurnal variation of sap flow (data not shown), which remains dependent on VPD as shown by Butz et al. (2018). These findings most likely reflect the fact that we considered a single value of F_{dnc} and VPD for each day, thus the diurnal dependence (where sap flow increases as VPD increases) between these variables was not considered. The same effect occurs between the intra-annual variation of R_n and F_{dn} (Figure 5).

The inverse relationship between VPD and ET is also partially caused by the fact that, during the dry season, all trees shed their leaves and water uptake reduces to near-zero, despite the increase in VPD. Furthermore, during leaf-on periods with low atmospheric humidity, it is the strong negative dependence of stomatal opening on VPD that counteracts the concurrent increase in driving force, as mentioned in the introduction. Although the VPD values tend to be lower during the rainy season, thereby reducing potential transpiration, ET is high due to an increased water availability and related presence of leaves. Moreover, there is a strong relationship between VPD and T_{air} (Figure 5c), given the temperature depends on the saturated vapour pressure as calculated by Teten's formula [47]; around 57% of the VPD variation is caused by the variation in T_{air}.

F_{dnc} appears to be independent of $S_{e,prof}$ when it reaches values higher than 0.25 (for reasons explained in the Results section, see Figure 7b). When $S_{e,prof}$ was below 0.25, there was a clear reduction in sap flow, which became even more pronounced with values below 0.18. When $S_{e,prof}$ values were around 0.16, F_{dnc} became negligible, and NDVI reached its minimum (0.34 for $NDVI_{modis}$ and 0.26 $NDVI_{S2}$), indicating a minimum NDVI for the monitored Caatinga species. Although F_{dnc} was poorly correlated to water availability when $S_{e,prof}$ values were >0.25 (indicating that the lack of radiation is a stronger driver), values of $S_{e,prof}$ remained below 0.25 for more than 80% of the monitored period, indicating that for at least 80% of the year, water availability was the dominant environmental driver of transpiration. This dependence of F_{dnc} on $S_{e,prof}$ indirectly explains the high correlation between F_{dnc} and NDVI, as shown in Table 4, since NDVI also strongly depends on $S_{e,prof}$ (Figure 7b).

In the dry season, when F_{dn} was approaching 0, for four of the monitored species (Figure 6) the timing difference seems to be due to differences in wood density, W_d. However, for the species with high F_{dn}, this difference is subtle. As can be surmised from Figure 6, the order in which F_{dn} approached 0 was inversely correlated with the order of W_d: *C. leptophloeos* (0.28 g·cm^{-3}) at the end of April; *A. pyrifolium* (0.54 g·cm^{-3}) at the end of June; *A. colubrina* (0.59 g·cm^{-3}) in mid July; *C. nordestinum* (0.66 g·cm^{-3}) in mid to late July. Although this study only compared the W_d of four species, our results indicate that low W_d is a limiting factor for these plants, particularly considering the duration in which they can maintain foliage, since the early senescence observed for low W_d species may be part of the plant's water conservation strategy as is stomatal control, which ultimately prevents the plant from cavitation [15–17]. Thus, the high W_d promoted by thick cell walls plays an important role in protecting plants from cavitation, allowing the plant to continue extracting water even from relatively dry soils.

Since W_d is directly related to the capacity of Caatinga plant tissues to store water [3], species with higher wood densities, which have larger stores of water, may be able to survive extremely dry periods. Moreover, the species with the lowest W_d, *C. leptophloeos* has the shortest sap flow period (Figure 6) and exhibits a peculiar greenish photosynthetic bark [48], which increases plant survival during the dry season [49]. Unfortunately, our data set did not allow us to gather evidence of a higher leaf water potential maintained over a longer period for the species with lower wood density.

C. leptophloeos was not dominant in the monitoring plot and behaved differently, from a phenological perspective, from the other species: leaf flushing began earlier and occured at a slightly faster rate. This species also has a very early onset of senescence (Figure 6f), as was previously reported by [3] and attributed to the species low wood density and high water storage characteristics. Despite the fact that the high storage water capacity increases plant survival during dry periods, its low wood density suggests it cannot extract water from soils with very negative water potentials. Therefore, to avoid cavitation, this species shows an early EOS. Furthermore, it is expected that species with considerably different phenologies and low dominance will have a low correlation with the overall phenological timing of the plot-level vegetation. This explains the low correlations of *C. leptophloeos* F_{dn} with the vegetation indices, which reflects the entire plot (Table 4).

The GCC_{ns} values of *A. pyrifolium* decreased considerably in the dry period but did not reach their minimum until the onset of the dry season, contrary to the drop in F_{dn} and its ranking with regards to W_d. However, low GCC values indicate a long leafless period, which was confirmed by the visual inspection of digital imagery and the negligible values of transpiration. The young green leaves of *A. pyrifolium*, with high photosynthetic and transpiration rates, were produced one month after the onset of the rainy season, unlike the other species in which leaf flushing began only a few days after the rains. Moreover, it is worth mentioning that the individuals monitored using GCC were not the same ones monitored with the sap flow sensors, despite the fact that they belong to the same species.

The index adjusted to the soil, $SAVI_{S2}$, showed higher EOS bias (-47 days), compared to $NDVI_{S2}$ (-42 days) (Table 5), but a considerable lower SOS bias (-18 days) than $NDVI_{S2}$ (-42 days). Moreover, GCC_{ns} presented a higher correlation with $SAVI_{S2}$ (0.74) than $NDVI_{S2}$ (0.72) (Supplementary Materials, Table S2). Although the differences may seem small, taking into account that the dates for EOS and SOS extracted from GCC_{ns} are the most realistic representation of phenology in this work (since they were confirmed with visual analysis), we can say that $SAVI_{S2}$ better represents phenology in terms of leaf presence. However, regarding plant water use, $SAVI_{S2}$ showed a lower correlation with F_{dnc}, indicating that the best vegetation index representing phenology is not necessarily the best VI representing plant water use.

F_{dnc} exhibited a stronger linear relationship with NDVI derived from MODIS and Sentinel-2 (R^2 0.92 and R^2 0.88, respectively) than the other indices (Table 4; Figure 7). A plausible explanation for the lower performance of the $NDVI_{S2}$ is that with an increase in the resolution, the pixel heterogeneity and number of pixels per image will increase, which means that the Sentinel-2 images will be noisier. Another explanation is that the MODIS algorithm has already undergone numerous revisions, while Sentinel-2 processing may not be mature enough. In the present work, the automatic cloud detection and cirrus correction filter offered by sen2core was not totally effective, requiring a subsequent visual analysis, suggesting that the algorithm still needs improvement. The last and most likely reason is that the higher temporal resolution of MODIS increases the chance of resulting in a better image composition; therefore, the lower temporal resolution of Sentinel-2 has a higher risk of yielding poor quality images, even when accounting for the filters and quality ratings.

Compared to the near-surface GCC_{ns}, the GCC_{S2} data show a stronger correlation with the sap flow data. Theoretically linked to leaf age, both GCC indices decrease when senescence occurs for all monitored species, as expected. However, GCC_{ns} displayed a slight increase before the end of the dry season (Figure 6). This unexpected rise follows the first sporadic rains and produced a temporary disconnect with F_{dn} and soil water storage, causing the relatively poor correlations (Table 4). Other likely explanations for these lower performances include: signal contamination by shrub vegetation, wet/dry terrain, difficulty in selecting pixels, and even the wind causing leaf movements. However, an increase in GCC_{ns} clearly signals the beginning of leaf flushing, thus it could be used in studies investigating when the growing season begins, the rate of leaf establishment, and how long the growing season lasts. Therefore, near-surface GCC data are important as it allows for individual monitoring of tree species.

Our results regarding the transition dates and water use indicate that the $NDVI_{modis}$ provided the best fit and most reliable proxy for the community phenology, but both datasets performed well. The choice of VI is of paramount importance to understand the relation of phenology and plant water use at both the species and community levels [21]. Therefore, detection of the near remote phenology through the use of digital cameras (phenocams) can fill the gap between satellite and on-the-ground observations, improving the detection of which VI better represents the vegetation changes observed on the ground (see [21]). On the other hand, phenocam GCC also enhanced our understanding of the sap flow dynamics in individual plants. The fact that the decrease in sap flow is not coupled with the decrease in GCC_{ns} at the end of the rainy season in 2018 (Figure 6) indicates that the leaf drop of most species is preceded by the cessation of sap flow. As expected, both sap flow and GCC_{ns} remained low during the dry season. Nevertheless, while GCC_{ns} remained high and at approximately constant values after leaf flushing, the sap flow was variable during the 2018 rainy season. Regardless, the strong correlation between F_{dnc} and both NDVIs shows that the recorded phenological behaviour of the tree vegetation in the plot, as reflected by fluctuations in vegetation indices such as NDVI, represents the overall profile soil moisture conditions well and its effects on the canopy exchange processes for this Caatinga vegetation.

Our findings compare remote sensing products related to vegetation status at different temporal and spatial scales, with the aim to analyse the water use of Caatinga vegetation. Sentinel-2 and MODIS satellite products also provide data from various other spectral bands that have not been used in our analyses. Bands that could be used to generate other indices, for example, which could be explored with advanced statistical methods, and combined with in-situ data that are complementary to ours or with data from other satellite platforms. Therefore, our study addresses the hypotheses set out at the beginning of the paper, but points to paths in which future work could be further explored.

5. Conclusions

Given the hypotheses raised, we can conclude the following:

(1) Soil water availability explained the Caatinga phenology at an ecosystem level, but less so at the species level;

(2) Seasonal signals of vegetation indices GCC and NDVI derived from optical RS data collected for Caatinga vegetation were indeed strongly related to water use, as represented by the late-morning community level sap flow data. However, the individual GCC determined from proximal RS (phenocams) was poorly correlated to individual sap flow.

(3) The sap flow of the monitored species was entirely limited by soil water availability when relative saturation of the soil profile fell below 0.25, a situation that occurred for more than 80% of the observed year.

Our results provide meaningful information about the physiological responses of dry forests in the Caatinga area to environmental variables. They also show the ability of different scales of SR products in detecting variations in plant phenology, which is essential information to interpret and analyse the soil water availability and plant flux density. Moreover, our findings identified when to use SR data as a proxy for phenological and physiologal processes in the Caatinga vegetation.

Supplementary Materials: The following are available online at https://www.mdpi.com/2072-429 2/13/1/75/s1, Table S1: Soil characteristics estimated: θ_s, θ_{fc}, θ_{wp}, θ_r; the θ_{max} and θ_{min} observed for each monitored soil layer depth (in $cm^3\ cm^{-3}$) and; the r^2 of the Durner model versus observed water retention curve, for each layer, Figure S1: Scatter plot of measured individuals sapwood depth (mm) and their respective DBH (mm), Table S2: Pearson correlation table—near-surface GCC versus orbital VI. Figure S2: Community sap flow and phenological response: (A) Normalized sap flow for the plot community (F_{dnc}, dimensionless), GCC_{S2} and $SAVI_{S2}$. (B) F_{dnc}, $NDVI_{modis}$ and $NDVI_{S2}$.

Author Contributions: Conceptualization, R.A.P. and L.D.S.B.; Formal analysis, R.A.P.; Funding acquisition, L.D.S.B.; Investigation, R.A.P., D.J.V., R.S. and L.D.S.B.; Methodology, R.A.P., D.M.R., L.P.C.M. and L.D.S.B.; Project administration, L.D.S.B.; Resources, E.S. and L.D.S.B.; Software, R.A.P., D.M.R. and R.S.; Supervision, L.D.S.B.; Validation, R.A.P., R.S., R.L.B.N. and A.V.; Writing—original draft, R.A.P. and L.D.S.B.; Writing—review & editing, R.A.P., R.L.B.N., Í.A.C.C., A.V., L.P.C.M., D.M.R. and T.S.K. All authors have read and agreed to the published version of the manuscript.

Funding: This study was financed by the São Paulo Research Foundation (FAPESP) grant #2015/50488-5, by Coordenação de Aperfeiçoamento de Pessoal de Nível Superior-Brasil (CAPES)-Finance Code 001, and Newton Fund/NERC/FAPESP NE/N012488/1. A Verhoef and R Nobrega were supported by the Newton/NERC/FAPESP Nordeste project: NE/N012488/1. L.P.C.M. receives a research fellowship from the National Council for Scientific and Technological Development (CNPq). DMR receives a FAPESP fellowship (grant FAPESP #2017/17380-1).

Acknowledgments: We thank the Government Company of Agricultural Research from the Semi-arid (Embrapa semi-árido); the Academic Unit of Serra Talhada (UAST); the Academic Unit of Garanhuns (UAG) from Rural Federal University of Pernambuco (UFRPE); the National Observatory of Water and Carbon Dynamics in the Caatinga Biome-NOWCDCB, for, among others, allowing the use of tools, academic installations, providing technical assistance and labour to carry out the installation of sensors and maintenance of equipment; Luciano Paganucci de Queiroz and his team, for leading the work of the Herbarium at State University of Feira de Santana in the species identification; Jonathan Lloyd for assisting in the management of financial and human resources and for his contributions in developing the methods.

Conflicts of Interest: The authors declare no conflict of interest.

References

1. Pugnaire, F.I.; Morillo, J.A.; Peñuelas, J.; Reich, P.B.; Bardgett, R.D.; Gaxiola, A.; Wardle, D.A.; van der Putten, W.H. Climate change effects on plant-soil feedbacks and consequences for biodiversity and functioning of terrestrial ecosystems. *Sci. Adv.* **2019**, *5*, eaaz1834. [CrossRef] [PubMed]
2. Francesconi, W.; Srinivasan, R.; Pérez-Miñana, E.; Willcock, S.P.; Quintero, M. Using the Soil and Water Assessment Tool (SWAT) to model ecosystem services: A systematic review. *J. Hydrol.* **2016**, *535*, 625–636. [CrossRef]
3. Lima, A.L.A.; de Sá Barretto Sampaio, E.V.; de Castro, C.C.; Rodal, M.J.N.; Antonino, A.C.D.; de Melo, A.L. Do the phenology and functional stem attributes of woody species allow for the identification of functional groups in the semiarid region of Brazil? *Trees* **2012**, *26*, 1605–1616. [CrossRef]
4. Alberton, B.; da Silva Torres, R.; Silva, T.S.F.; Rocha, H.; Moura, M.S.B.; Morellato, L. Leafing Patterns and Drivers across Seasonally Dry Tropical Communities. *Remote Sens.* **2019**, *11*, 2267. [CrossRef]
5. Murphy, P.G.; Lugo, A.E. Ecology of Tropical Dry Forest. *Annu. Rev. Ecol. Syst.* **1986**, *17*, 67–88. [CrossRef]
6. Särkinen, T.; Iganci, J.R.; Linares-Palomino, R.; Simon, M.F.; Prado, D.E. Forgotten forests-issues and prospects in biome mapping using Seasonally Dry Tropical Forests as a case study. *BMC Ecol.* **2011**, *11*. [CrossRef]
7. Pennington, R.T.; Lavin, M.; Oliveira-Filho, A. Woody Plant Diversity, Evolution, and Ecology in the Tropics: Perspectives from Seasonally Dry Tropical Forests. *Annu. Rev. Ecol. Evol. Syst.* **2009**, *40*, 437–457. [CrossRef]
8. Pennington, R.T.; Lehmann, C.E.; Rowland, L.M. Tropical savannas and dry forests. *Curr. Biol.* **2018**, *28*, R541–R545. [CrossRef]
9. Machado, I.C.S.; Barros, L.M.; Sampaio, E.V.S.B. Phenology of Caatinga Species at Serra Talhada, PE, Northeastern Brazil. *Biotropica* **1997**, *29*, 57–68. [CrossRef]

10. de Souza, M.J.N.; de Oliveira, J.G.B.; Lins, R.C.; Jatobá, L. Overview of the Brazilian caatinga. *Rev. CiêNcia TróPico* **1992**, *20*, 173–198.
11. Sampaio, E.V. Overview of the Brazilian caatinga. In *Seasonally Dry Tropical Forests*; Cambridge University Press: Cambridge, UK, 1995; pp. 35–63.
12. Lemos, J.R.; Rodal, M.J.N. Fitossociologia do componente lenhoso de um trecho da vegetação de caatinga no Parque Nacional Serra da Capivara, Piauí, Brasil. *Acta Bot. Bras.* **2002**, *16*, 23–42. [CrossRef]
13. Reddy, S. Climatic classification: the semi-arid tropics and its environment—A review. *Pesquisa Agro* **1983**, *18*, 823–847.
14. Lima, A.L.A.; Rodal, M.J.N. Phenology and wood density of plants growing in the semi-arid region of northeastern Brazil. *J. Arid. Environ.* **2010**, *74*, 1363–1373. [CrossRef]
15. Sobrado, M.A. Trade-off between water transport efficiency and leaf life-span in a tropical dry forest. *Oecologia* **1993**, *96*, 19–23. [CrossRef] [PubMed]
16. Stratton, L.; Goldstein, G.; Meinzer, F.C. Stem water storage capacity and efficiency of water transport: their functional significance in a Hawaiian dry forest. *Plant Cell Environ.* **2000**, *23*, 99–106. [CrossRef]
17. Hacke, U.G.; Sperry, J.S.; Pockman, W.T.; Davis, S.D.; McCulloh, K.A. Trends in wood density and structure are linked to prevention of xylem implosion by negative pressure. *Oecologia* **2001**, *126*, 457–461. [CrossRef]
18. Reich, P.B.; Wright, I.J.; Cavender-Bares, J.; Craine, J.M.; Oleksyn, J.; Westoby, M.; Walters, M.B. The Evolution of Plant Functional Variation: Traits, Spectra, and Strategies. *Int. J. Plant Sci.* **2003**, *164*, S143–S164. [CrossRef]
19. Swenson, N.G.; Enquist, B.J. Ecological and evolutionary determinants of a key plant functional trait: Wood density and its community-wide variation across latitude and elevation. *Am. J. Bot.* **2007**, *94*, 451–459. [CrossRef]
20. Chave, J.; Coomes, D.; Jansen, S.; Lewis, S.L.; Swenson, N.G.; Zanne, A.E. Towards a worldwide wood economics spectrum. *Ecol. Lett.* **2009**, *12*, 351–366. [CrossRef]
21. Richardson, A.D.; Hufkens, K.; Milliman, T.; Aubrecht, D.M.; Chen, M.; Gray, J.M.; Johnston, M.R.; Keenan, T.F.; Klosterman, S.T.; Kosmala, M.; et al. Tracking vegetation phenology across diverse North American biomes using PhenoCam imagery. *Sci. Data* **2018**, *5*. [CrossRef]
22. Ferreira, T.R.; Silva, B.B.D.; Moura, M.S.B.D.; Verhoef, A.; Nóbrega, R.L. The use of remote sensing for reliable estimation of net radiation and its components: A case study for contrasting land covers in an agricultural hotspot of the Brazilian semiarid region. *Agric. For. Meteorol.* **2020**, *291*, 108052. [CrossRef]
23. de Queiroga Miranda, R.; Nóbrega, R.L.B.; de Moura, M.S.B.; Raghavan, S.; Galvíncio, J.D. Realistic and simplified models of plant and leaf area indices for a seasonally dry tropical forest. *Int. J. Appl. Earth Obs. Geoinf.* **2020**, *85*, 101992. [CrossRef]
24. Coaguila, D.N.; Hernandez, F.B.T.; de Teixeira, C.A.H.; Franco, R.A.M.; Leivas, J.F. Water productivity using SAFER-Simple Algorithm for Evapotranspiration Retrieving in watershed. *Rev. Bras. Eng. Agric. Ambient.* **2017**, *21*, 524–529. [CrossRef]
25. de C. Teixeira, A.H. Determining Regional Actual Evapotranspiration of Irrigated Crops and Natural Vegetation in the São Francisco River Basin (Brazil) Using Remote Sensing and Penman-Monteith Equation. *Remote Sens.* **2010**, *2*, 1287–1319. [CrossRef]
26. Alberton, B.; Torres, R.; Cancian, L.F.; Borges, B.D.; Almeida, J.; Mariano, G.C.; dos Santos, J.; Morellato, L.P.C. Introducing digital cameras to monitor plant phenology in the tropics: Applications for conservation. *Perspect. Ecol. Conserv.* **2017**, *15*, 82–90. [CrossRef]
27. Kandasamy, S.; Baret, F.; Verger, A.; Neveux, P.; Weiss, M. A comparison of methods for smoothing and gap filling time series of remote sensing observations: Application to MODIS LAI products. *Biogeosci. Discuss.* **2012**, *9*, 17053–17097. [CrossRef]
28. Ren, R.; von der Crone, J.; Horton, R.; Liu, G.; Steppe, K. An improved single probe method for sap flow measurements using finite heating duration. *Agric. For. Meteorol.* **2020**, *280*, 107788. [CrossRef]
29. Goldstein, G.; Meinzer, F.C.; Bucci, S.J.; Scholz, F.G.; Franco, A.C.; Hoffmann, W.A. Water economy of Neotropical savanna trees: Six paradigms revisited. *Tree Physiol.* **2008**, *28*, 395–404. [CrossRef]
30. Butz, P.; Hölscher, D.; Cueva, E.; Graefe, S. Tree Water Use Patterns as Influenced by Phenology in a Dry Forest of Southern Ecuador. *Front. Plant Sci.* **2018**, *9*. [CrossRef]
31. Granier, A. Une nouvelle méthode pour la mesure du flux de sève brute dans le tronc des arbres. *Ann. Des Sci. For.* **1985**, *42*, 193–200. [CrossRef]
32. Granier, A. Evaluation of transpiration in a Douglas-fir stand by means of sap flow measurements. *Tree Physiol.* **1987**, *3*, 309–319. [CrossRef] [PubMed]
33. Ping, L.; Layrent, U.; Ping, Z. Granier's thermal dissipation probe (TDP) method for measuring sap flow in trees: Theory and practice. *Acta Bot. Sin.* **2004**, *46*, 631–646.
34. Pasqualotto, G.; Carraro, V.; Menardi, R.; Anfodillo, T. Calibration of Granier-Type (TDP) Sap Flow Probes by a High Precision Electronic Potometer. *Sensors* **2019**, *19*, 2419. [CrossRef]
35. Grossiord, C.; Christoffersen, B.; Alonso-Rodríguez, A.M.; Anderson-Teixeira, K.; Asbjornsen, H.; Aparecido, L.M.T.; Berry, Z.C.; Baraloto, C.; Bonal, D.; Borrego, I.; et al. Precipitation mediates sap flux sensitivity to evaporative demand in the neotropics. *Oecologia* **2019**, *191*, 519–530. [CrossRef]
36. Soil-Survey-Staff. *Soil Taxonomy: A Basic System of Soil Classification of Making and Interpreting Soil Surveys*; USDA. Agriculture Handbook, 436; Natural Resources Conservation Service: Washington, DC, USA, 1999; p. 869.
37. Jacomine, P.K.T.; Cavalcanti, A.C.; Burgos, N.; Pesso, S.C.P.; Silveira, C.O. *Levantamento Exploratório-Reconhecimento de Solos do Estado de Pernambuco*; Boletim Técnico, 26-Pedologia, 14 2V; EMBRAPA-Divisão de Pesquisa Pedológica: Recife, Brazil, 1973.

38. Kottek, M.; Grieser, J.; Beck, C.; Rudolf, B.; Rubel, F. World Map of the Köppen-Geiger climate classification updated. *Meteorol. Z.* **2006**, *15*, 259–263. [CrossRef]
39. Rubel, F.; Brugger, K.; Haslinger, K.; Auer, I. The climate of the European Alps: Shift of very high resolution Köppen-Geiger climate zones 1800–2100. *Meteorol. Z.* **2017**, *26*, 115–125. [CrossRef]
40. Climate-Data.org. Clima Serra Talhada. Available online: https://pt.climate-data.org/america-do-sul/brasil/pernambuco/serra-talhada-42488/ (accessed on 16 September 2019).
41. Moonlight, P.; Banda, R.K.; Phillips, O.; Dexter, K.; Pennington, R.; Baker, T.; de Lima, H.C.; Fajardo, L.; González-M, R.; Linares-Palomino, R.; et al. Expanding tropical forest monitoring into Dry Forests: The DRYFLOR protocol for permanent plots. *Plants People Planet* **2020**. [CrossRef]
42. Vajsová, B.; Fasbender, D.; Wirnhardt, C.; Lemajic, S.; Devos, W. Assessing Spatial Limits of Sentinel-2 Data on Arable Crops in the Context of Checks by Monitoring. *Remote Sens.* **2020**, *12*, 2195. [CrossRef]
43. Yan, L.; Roy, D.; Li, Z.; Zhang, H.; Huang, H. Sentinel-2A multi-temporal misregistration characterization and an orbit-based sub-pixel registration methodology. *Remote Sens. Environ.* **2018**, *215*, 495–506. [CrossRef]
44. da Silva, P.F.; Lima, J.R.D.S.; Antonino, A.C.D.; Souza, R.; de Souza, E.S.; Silva, J.R.I.; Alves, E.M. Seasonal patterns of carbon dioxide, water and energy fluxes over the Caatinga and grassland in the semi-arid region of Brazil. *J. Arid. Environ.* **2017**, *147*, 71–82. [CrossRef]
45. Costa, M.H.; Biajoli, M.C.; Sanches, L.; Malhado, A.C.M.; Hutyra, L.R.; da Rocha, H.R.; Aguiar, R.G.; de Araújo, A.C. Atmospheric versus vegetation controls of Amazonian tropical rain forest evapotranspiration: Are the wet and seasonally dry rain forests any different? *J. Geophys. Res.* **2010**, *115*. [CrossRef]
46. de Oliveira, C.C.; Zandavalli, R.B.; Lima, A.L.A.; Rodal, M.J.N. Functional groups of woody species in semi-arid regions at low latitudes. *Austral Ecol.* **2014**, *40*, 40–49. [CrossRef]
47. Tetens, O. Uber einige meteorologische Begriffe. *Z. Geophys.* **1930**, *6*, 297–309.
48. Killeen, T.J.; Jardim, A.; Mamani, F.; Rojas, N. Diversity, composition and structure of a tropical semideciduous forest in the Chiquitanía region of Santa Cruz, Bolivia. *J. Trop. Ecol.* **1998**, *14*, 803–827. [CrossRef]
49. Mahr, D. Commiphora: An Introduction to the Genus. *Cactus Succul. J.* **2012**, *84*, 140–154. [CrossRef]

Article

Alternative Vegetation States in Tropical Forests and Savannas: The Search for Consistent Signals in Diverse Remote Sensing Data

Sanath Sathyachandran Kumar [1,*], Niall P. Hanan [1], Lara Prihodko [2], Julius Anchang [1], C. Wade Ross [1], Wenjie Ji [1] and Brianna M Lind [1]

[1] Department of Plant and Environmental Sciences, New Mexico State University, Las Cruces, NM 88003, USA; nhanan@nmsu.edu (N.P.H.); anchang@nmsu.edu (J.A.); cwross@nmsu.edu (C.W.R.); wenjieji@nmsu.edu (W.J.); lind@nmsu.edu (B.M.L.)

[2] Department of Animal and Range Sciences, New Mexico State University, Las Cruces, NM 88003, USA; prihodko@nmsu.edu

* Correspondence: sanath@nmsu.edu

Received: 25 January 2019; Accepted: 1 April 2019; Published: 4 April 2019

Abstract: Globally, the spatial distribution of vegetation is governed primarily by climatological factors (rainfall and temperature, seasonality, and inter-annual variability). The local distribution of vegetation, however, depends on local edaphic conditions (soils and topography) and disturbances (fire, herbivory, and anthropogenic activities). Abrupt spatial or temporal changes in vegetation distribution can occur if there are positive (i.e., amplifying) feedbacks favoring certain vegetation states under otherwise similar climatic and edaphic conditions. Previous studies in the tropical savannas of Africa and other continents using the MODerate Resolution Imaging Spectroradiometer (MODIS) vegetation continuous fields (VCF) satellite data product have focused on discontinuities in the distribution of tree cover at different rainfall levels, with bimodal distributions (e.g., concentrations of high and low tree cover) interpreted as alternative vegetation states. Such observed bimodalities over large spatial extents may not be evidence for alternate states, as they may include regions that have different edaphic conditions and disturbance histories. In this study, we conduct a systematic multi-scale analysis of diverse MODIS data streams to quantify the presence and spatial consistency of alternative vegetation states in Sub-Saharan Africa. The analysis is based on the premise that major discontinuities in vegetation structure should also manifest as consistent spatial patterns in a range of remote sensing data streams, including, for example, albedo and land surface temperature (LST). Our results confirm previous observations of bimodal and multimodal distributions of estimated tree cover in the MODIS VCF. However, strong disagreements in the location of multimodality between VCF and other data streams were observed at 1 km scale. Results suggest that the observed distribution of VCF over vast spatial extents are multimodal, not because of local-scale feedbacks and emergent bifurcations (the definition of alternative states), but likely because of other factors including regional scale differences in woody dynamics associated with edaphic, disturbance, and/or anthropogenic processes. These results suggest the need for more in-depth consideration of bifurcation mechanisms and thus the likely spatial and temporal scales at which alternative states driven by different positive feedback processes should manifest.

Keywords: Savanna; alternative stable states; MODIS VCF; land surface temperature; albedo

---

1. Introduction

Global distributions of vegetation structure, species composition, and functional composition are governed primarily by climate and phylogenetic processes [1–5]. In the seasonal tropics, precipitation patterns and ecohydrological interactions are the principal drivers of large scale vegetation patterns [6–8], with increasing precipitation generally corresponding to increasing tree density, canopy cover, and tree height [8–10]. While continental-scale average tree density, cover, and height may increase with precipitation, at fine spatial scales, local factors including soil type and hydrological interactions, fire and herbivory, and agriculture and wood harvest lead to considerable spatial heterogeneity in vegetation structure [3,7,11–15]). Thus, landscapes with similar climate are not necessarily similar in vegetation structure (e.g., tree cover, density, and height), and patch mosaics emerge at different spatial scales, reflecting different edaphic conditions and disturbance histories.

More interesting in the context of this paper, however, is the degree to which spatial variability in a landscape is (a) proportional to the frequency, intensity, and time since disturbance (e.g., a harvest event that removes a fraction of tree cover in a particular location) and subsequent regrowth ("linear patch dynamics"), or is (b) amplified by positive feedbacks that lead to alternative state dynamics in response to disturbance events ("bifurcating patch dynamics"). The classic example of an amplifying feedback producing bifurcating patch dynamics (between forest and savannas in mesic systems, and savanna and grasslands in drier savannas) is the so-called fire-trap, where an initial loss (or gain) of tree cover promotes (or suppresses) grass growth, providing more (or less) fuel for fires and increased (or decreased) tree mortality in a continuing cycle of tree loss (or gain) [16–23].

Numerous studies have examined satellite observations in search of empirical support for the prevalence of alternative vegetation states [18,19,24–27] in spatial and temporal data [23,28]. The majority of studies that presented evidence for alternate states were based on interpretation of frequency distributions of tree cover over large spatial regions with similar mean annual rainfall (implicitly assuming similar local edaphic and ecological conditions). Regions following linear patch dynamics would be expected to result in unimodal histograms while, in sharp contrast, regions following bifurcating patch dynamics would result in bi-modal or multimodal histograms, with each mode defining an alternative state caused by the presence of a positive feedback. The earlier studies [18,19,24–27] used tree cover estimates from the MODIS vegetation continuous field product (VCF) [29,30] and found that bimodal and multimodal tree cover distributions were common, particularly in the semi-arid and mesic tropical savanna regions, suggesting the presence of alternative states. In these studies, potential spatial variability in ecological, edaphic, and anthropogenic factors within rainfall zones were mostly ignored. Further, the MODIS VCF retrieval was based on classification and regression tree (CART) algorithms and trained using pre-averaged binning techniques that have been linked to possible artifacts (artificial clumping of predictions) that could be wrongly interpreted as alternative states not present in reality [31–33]. Thus, the true extent and prevalence of bifurcating patch dynamics in the tropical savannas remains unknown.

Remote sensing data are invaluable for our understanding of savanna ecology and the importance of positive feedbacks at landscape, continental, and global scales. However, to advance our understanding of alternative vegetation states, it is critical that we identify not only the specific processes potentially leading to alternative states but also the spatial scale at which the emergent patterns should be detectable and thus the appropriate spatial resolution for data used in its detection. We suggest that the controversy and uncertainty in the extent to which VCF-based analyses have proven the existence of alternative states in tropical savannas can be resolved by systematic comparisons among diverse remotely sensed data streams which, in theory, should also respond to changing tree cover. We posit that bifurcations (i.e., discontinuities) in the frequency distributions of tree cover should also logically be expressed in other remotely-sensed data streams, including albedo and surface temperature. More particularly, not only should the bimodal distributions in the VCF tree cover be detected in other physically based remote sensing variables, but those bifurcations should exhibit spatial consistency (i.e., occur in the same geographic locations). Consistency among diverse data-streams would provide strong evidence supporting earlier conclusions from analysis of the VCF product and remove potential bias associated with any one product. Inconsistencies, on the other hand, would require reevaluation of the earlier results and more in-depth consideration of the mechanisms of bifurcation and thus the likely spatial and temporal scales at which alternative states driven by different positive feedback processes should manifest. In this work, we examine MODIS derived satellite VCF, albedo, and land surface temperature to determine the degree of similarity in the occurrence of multimodality and inference of alternative stable states across Sub-Saharan Africa (Figure 1). For this analysis, the Sub-Saharan study area includes the African continent from 22° North to 35° South, including Madagascar but excluding the Arabian Peninsula (Figure 1). We ask the following questions: (1) are the apparent forest-savanna and savanna-grassland bifurcations detected in earlier VCF analysis also present in albedo and surface temperature data?; and (2) to what extent are emergent bifurcations in VCF, albedo, and land surface temperature dependent on the spatial scale of analysis?

We report results of three distinct multi-scale analyses (at the continental scale, based on zones of similar rainfall, and at the landscape scale) designed to quantify the presence of bimodality as a diagnostic of forest-savanna and savanna-grassland bifurcations. Further, we undertake a comprehensive analysis to identify evidence of multiple vegetation states across different data streams [VCF, near-infrared (NIR) albedo, and LST] at the same geographic locations, noting that the independent (i.e., non-VCF) products are derived using fundamental physical principles with relatively little empirical model fitting, while VCF estimates rely on an empirical CART approach. For this analysis, the Sub-Saharan study area includes the African continent south of the Sahara Desert, including Madagascar (Figure 1). The dominant (~66%) ecosystems in this region are tropical and subtropical grassland, savanna, and shrublands.

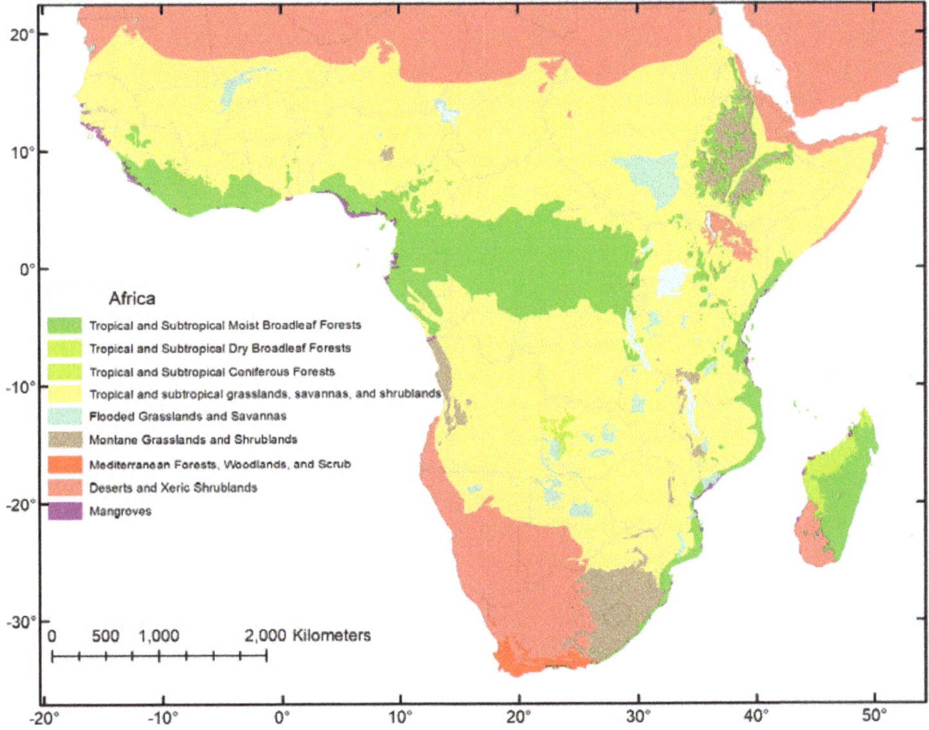

Figure 1. Terrestrial ecosystems [34] over the Sub-Saharan study area.

2. Data

2.1. Precipitation Data

We use the Climate Hazards Group InfraRed Precipitation with Station data (CHIRPS) [35]. CHIRPS is a 30+ year (1981 onwards to current) quasi-global (50°S–50°N) rainfall dataset defined at monthly time-steps for 0.05° × 0.05° spatial grids. CHIRPS combines satellite based precipitation estimates with in-situ station data to create high quality reliable rainfall time series with reduced systematic bias relative to other available global precipitation datasets [35–37], particularly over Sub-Saharan Africa. The mean annual precipitation (MAP) used in this study is computed using 30 years of CHIRPS monthly precipitation data (1981 to 2011; Figure 2).

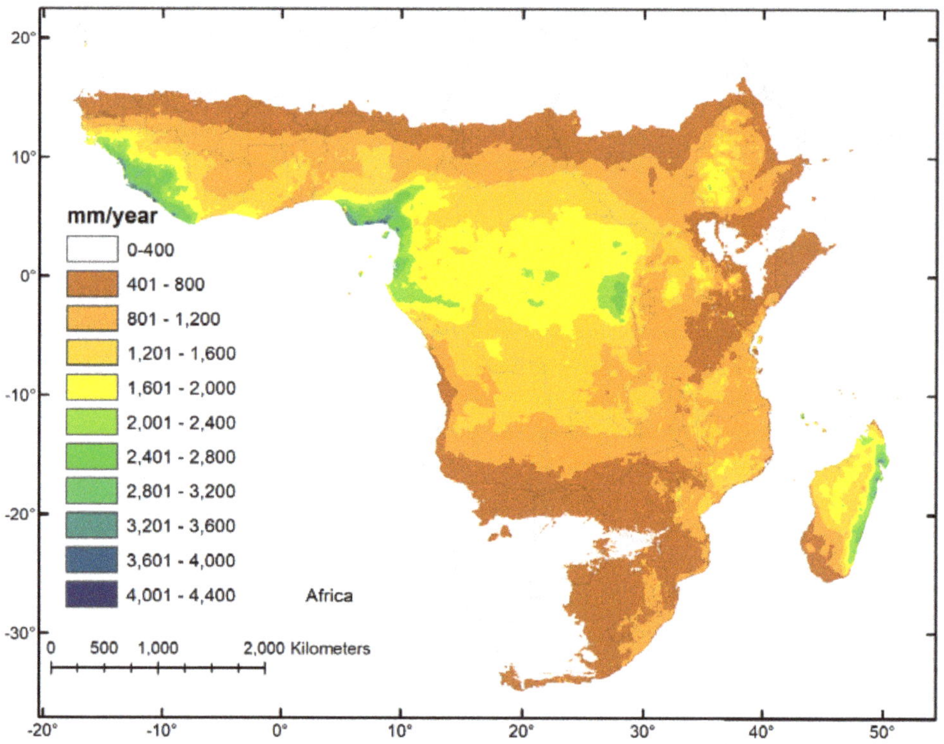

Figure 2. Long-term mean annual precipitation [MAP (mm yr^{-1})] in Sub-Saharan Africa using 30 years (1981–2011) of data from Climate Hazards Group InfraRed Precipitation with Station data (CHIRPS).

2.2. MODIS Vegetation Continuous Fields

The most recent MOD44B collection 6 VCF data available at a spatial scale of 250 m are used in this study [29,30]. VCF estimates are generated using a CART based algorithm and MODIS optical bands spanning blue to shortwave infrared wavelengths. A suite of metrics derived using annual multi-spectral reflectance observations and vegetation indices are used as predictor variables. The CART is trained using estimates of tree cover, non-tree vegetation, and non-vegetated area observations from field data that are up-scaled using Landsat data [29,30]. The VCF product also includes standard deviation estimates for retrieved tree cover and non-vegetated percent cover along with quality and cloud or water status information. In this study, a spatially explicit VCF map for the study region is created using only the highest quality non-cloudy 2005 data over Sub-Saharan Africa (Figure 3A). The 2005 MOD44B 250 m data are nearest neighbor resampled to a spatial resolution of

1 km to match the resolution of other MODIS data streams (NIR albedo and LST) used in this study using the nearest neighbor method. We use the nearest neighbor approach since averaging pixels tends to reduce spatial heterogeneity and thus suppress identification of alternative vegetation modes.

2.3. MODIS NIR Albedo

The albedo defines the ratio of radiant energy scattered away from a surface in all directions to the down-welling irradiance incident upon that surface. The albedo over vegetated regions is a function of the local illumination conditions that includes solar geometry and atmospheric conditions, the reflectance properties of the vegetation (leaf optical properties, leaf area index, and distribution of leaf angles), and the underlying soil [38,39]. The MODIS albedo retrieval algorithm first characterizes the surface anisotropy by combining multi-date multi-band atmospherically corrected surface reflectance data to derive the albedo (MCD43B3 V5, 1 km spatial and 16 day temporal resolutions) defined in three spectral regions: visible (VIS: 0.3–0.7 μm), near-infrared (NIR: 0.7–5.0 μm), and shortwave (SWIR: 0.3–5.0 μm) albedo. Changes in the VIS and NIR albedo have also been related to season and vegetation growth [40], while the SWIR albedo is more influenced by non-vegetative components of the land cover. The VIS and NIR albedo are highly correlated; we use the NIR albedo for this study [39], as it is less affected by the atmosphere. The MCD43B3 albedo data include a solar angle specific black sky albedo, solar angle independent diffuse white sky albedo, and associated quality flags. The median NIR albedo map is computed using 45 eight day composites over Sub Saharan Africa for the year 2005. The median (instead of the mean) is used because it is less sensitive to outliers in the data sets (Figure 3B).

2.4. MODIS Land Surface Temperature

Land surface temperature is a function of both net radiation and the fluxes of latent and sensible heat that are mediated by vegetation cover [41–47]. Observations [46,47] suggest that land surface temperatures depend on the abundance and type of vegetation, with regions with higher vegetation density—such as forests—showing lower surface temperatures than sparsely vegetated surfaces. Hence, surface temperature is expected to reflect underlying tree cover distributions. The observations in the thermal infrared (TIR: 10–13 μm) bands in conjunction with observations in the middle infrared (MIR: 3 μm) are used here to deduce both the land surface temperatures and emissivity using physically based models. The MYD11A2 [48] V6 daytime LST (1 km spatial resolution and eight day composite) includes LST observations and emissivity estimates. LST observations for Sub Saharan Africa in the year 2005 are used to compute the median daytime LST for analysis in this study. The median (instead of the mean) is preferred to reduce sensitivity to outliers (Figure 3C).

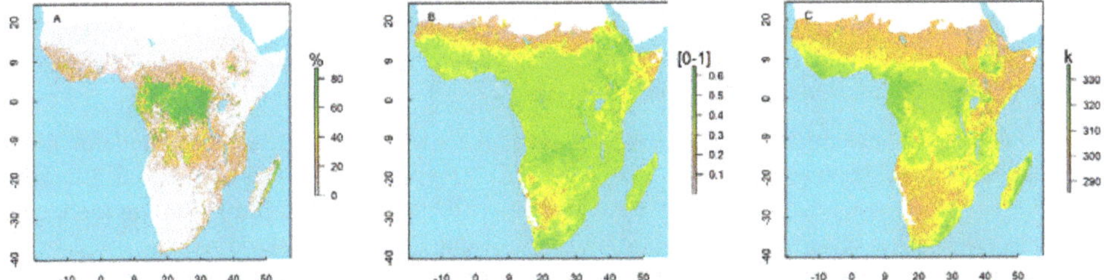

Figure 3. Illustration of all data streams used in this study. Spatially explicit maps of: (**A**) tree cover estimates from MODerate Resolution Imaging Spectroradiometer (MODIS) vegetation continuous fields (VCF), (**B**) median 2005 near-infrared (NIR) white sky albedo, and (**C**) median 2005 land surface temperature (LST, K).

3. Methods

Past studies that suggested empirical evidence of alternate vegetation states analyzed tree cover distributions in rainfall zones defined using MAP intervals over large geographic areas (see, for example, the rainfall intervals in Figure 2). Here, we report results of three distinct multi-scale analyses designed to quantify the presence of bimodality as a diagnostic of savanna bifurcations: (1) a non-spatially explicit continental-scale analysis, (2) a non-spatially explicit rainfall zone analysis, and (3) a spatially explicit landscape-scale analysis. Further, we undertake a comprehensive analysis to identify evidence of multiple states across different data streams (VCF, NIR albedo, and LST) at the same geographic locations, noting that the independent (i.e., non-VCF) products are derived using fundamental physical principles with relatively little empirical model fitting, while VCF estimates rely on an empirical CART approach. We also note that while MODIS VCF is known to be unreliable for low tree cover values, distinctions of high and low tree cover may be more reliable [31–33]. Quantitative evidence for the presence or absence of unimodality (i.e., multimodality) can be inferred statistically using Hartigan's dip test [48] on a population of observations [49,50]. This work uses the Hartigan's dip test for populations of observations at multiple scales as evidence for or against unimodality. Hartigan's dip test is sensitive to skew [24,51], thus the VCF data are arcsine transformed to reduce skew. Hartigan's dip test p values of less than 0.05 and 0.01 are considered to be moderately and highly significant indicators of multimodality, respectively. The following sections detail the multi-scale analysis approach followed in this work.

3.1. Non-Spatially Explicit Continental Scale Analysis

For this initial analysis, observations over all of Sub-Saharan Africa are examined both qualitatively and quantitatively. Histograms of each data set (VCF, albedo, and LST) are used to quantitatively infer the presence or absence of multiple vegetation states using the Hartigan's dip test. Scatter plots of each data set with long-term MAP as the dependent variable are also visually inspected for evidence of discontinuities indicative of multimodality.

3.2. Non-Spatially Explicit Rainfall Zone Analysis

Rainfall zones defined by ranges of MAP at continental scales are often spatially extended and, in some cases, spatially disconnected across the African continent (Figure 1). We examine the histograms of MODIS VCF, albedo, and LST in 200 mm rainfall zones (e.g., 0–200 mm/yr, 200–400 mm/yr, etc.) for all of Sub-Saharan Africa, replicating earlier analyses that used VCF tree cover stratified by rainfall to infer bifurcation dynamics [18,19,24–27]. We anticipate that bimodality detected in VCF data should also be present in the albedo and LST measurements. Hartigan's dip test p values are tabulated for each precipitation zone, and similarities among data sets are used to identify consistent/inconsistent signals of alternative vegetation states.

3.3. Spatially Explicit Landscape-Scale Analysis

We finally conduct a finer-scale, spatially explicit analysis to detect evidence for bimodality at the landscape level (local spatial extents of 15 km × 15 km). The premise is that bimodality in savanna structure (i.e., tree cover) detected at these local scales will provide more compelling evidence for positive feedbacks and bifurcations occurring within the same local climate and broad edaphic, biotic, and anthropogenic environments. This is in contrast to the continental scale rainfall zone analyses (above) where observed bimodality might reflect much coarser scale differences in soils, topography, and anthropogenic activity not reflected in coarse scale MAP stratifications. Strong evidence for alternative states is inferred if landscapes with multimodality are spatially consistent among independent data sets (i.e., occur in the same geographic locations). Conversely, the lack of spatial consistency might be cause for caution in interpreting apparent bimodality as evidence for the existence of alternative vegetation states in those locations. To test for spatial consistency at finer pixel level scales, spatially explicit signals of multimodality for each data product for each pixel are inferred using the dip test over its surrounding n × n pixel window (n = 15 × 15 = 225, 1 km^2 pixels), ignoring pixels labeled as cloudy and pixels labeled as water in the VCF data. The number of locations that shows evidence for multimodality for each product on its own and in conjunction with other products are tabulated. Spatial consistency between two variables is inferred by generating a confusion matrix and interpreting the Cohen's kappa [52] agreement/disagreement metric.

4. Results

4.1. Non-Spatially Explicit Continental Scale Analysis

Figure 3 shows maps of MODIS VCF, NIR albedo, and LST analyzed in this study. All datasets exhibit broadly similar spatial patterns at continental scales, with the drylands distinct from the mesic savannas and forest zones. Spearman's correlation between the MODIS VCF-NIR albedo is moderately high at −0.43, while it is higher for MODIS VCF-LST at −0.71 and LST-NIR albedo at 0.64. However, the histograms (Figure 4) are markedly different for each of the data streams. The MODIS VCF and LST show at least two distinct modes, while NIR albedo appears unimodal visually. The MODIS VCF shows distinct peaks in the frequency of tree cover at <20% and ~80%, while the MODIS LST shows three distinct peaks between 300 K and 315 K. Hartigan's dip test indicates that all histograms show highly significant (*p* value << 0.001) statistical evidence for multimodality.

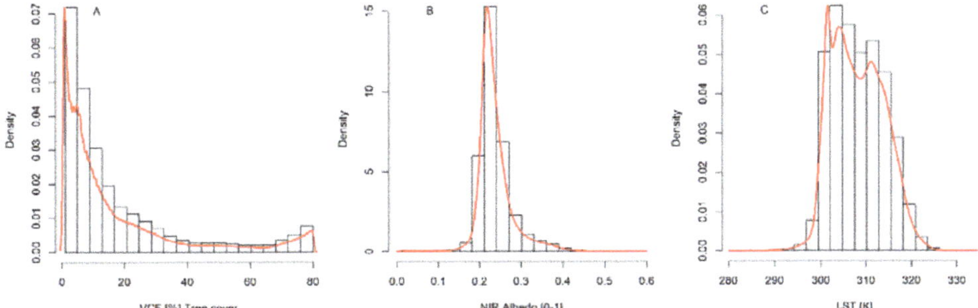

Figure 4. Histograms showing data distributions for (**A**) tree cover, (**B**) NIR albedo, and (**C**) LST. A total of 21,081,051 1 km^2 locations with MAP ≥ 100 mm are used in the histogram analysis, with 20 equal bins spanning the range of each dataset. The red line traces the density (>512 equal bins) for visual identification of multiple modes not easily seen in the histogram. Multimodal distributions are visually evident in the VCF and LST, but less so in the NIR albedo. Hartigan's dip test results indicate highly significant departure from unimodal distribution for all data streams.

Figure 5 shows scatter plots illustrating the relationships between the MODIS data streams and MAP. MODIS VCF shows a distinct bimodality (disconnected islands of red, orange, and yellow at a similar MAP in Figure 5A) between low (<20%) and high (~80%) tree cover, with bimodality for MAP regions >1200 mm/yr. However, such bimodality is not visually evident in the NIR albedo, which shows a broad exponential decline with MAP associated with increasing vegetation cover but without visually distinct discontinuities (Figure 5B). LST shows a decreasing trend with increased MAP (Figure 5C), as more trees tend to reduce the daytime land surface temperature, but the bimodality observed in the tree cover data is not visually evident in the LST data. These observations corroborate past researcher observations of bimodality in VCF, but they also suggest that the discontinuities seen in the VCF tree cover are not readily apparent in other related MODIS data at the 1 km scale.

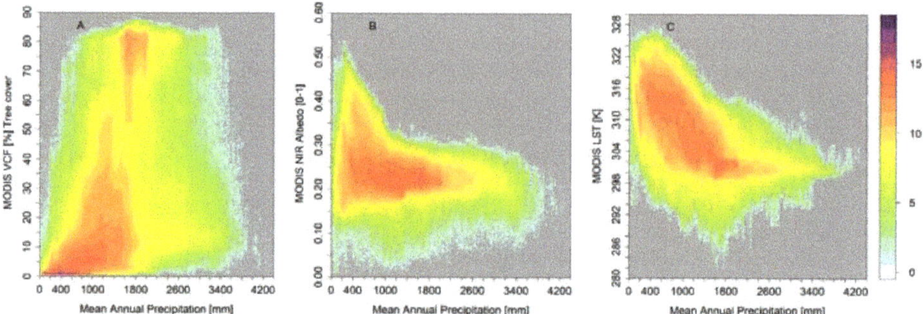

Figure 5. Scatter plots of each data stream with respect to long term MAP. (**A**) VCF tree cover, (**B**) NIR albedo, and (**C**) LST. The density of points is color-coded to show increasing point-density in warmer (red) colors.4.2. Non-Spatially Explicit Rainfall Zone Analysis

Similar to past work [26], we analyze tree cover distributions in 200 mm MAP precipitation intervals, providing 22 MAP zones between 0 and 4400 mm MAP. Hartigan's dip test results applied to all MODIS data over each zone are shown in Table 1. Individual histograms for each precipitation interval are shown in supplemental Figures S1–S3 for reference. It can be inferred from Table 1 that the VCF tree cover shows evidence for multimodality over the entire MAP range. Visual examination of the tree cover data (Figure 5) shows distinct bifurcation between 1200 and 2000. However, the dip test (Table 1) and Figure S1 suggest multimodality over almost the entire range of MAP. The dip test results for LST also indicate the presence of multimodality over most of the range of MAP, while multimodality in the albedo data is indicated in relatively fewer MAP ranges at low and intermediate rainfall.

Table 1. Hartigan's dip test for multimodality in three independent remote sensing datasets assessed within 200 mm MAP zones of Sub-Saharan Africa. Statistical significance is color coded, with red highly significant ($p < 0.01$), yellow moderately significant ($p < 0.05$), and gray not significant. These data organized by rainfall zones suggest multimodality is common, however, see text and Figure 4 for analysis of the spatial organization of bifurcation in these datasets.

MAP Range [mm]	0 200	200 400	400 600	600 800	800 1000	1000 1200	1200 1400	1400 1600	1600 1800	1800 2000	2000 2200	2200 2400	2400 2600	2600 2800	2800 3000	3000 3200	3200 3400	3400 3600	3600 3800	3800 4000	4000 4200	4200 4400
VCF	0.00	0.00	0.00	0.00	0.00	0.00	0.00	0.00	0.00	0.00	0.00	0.00	0.00	0.00	0.00	0.00	0.00	0.00	0.00	0.00	0.00	0.18
LST	0.00	0.00	0.00	0.00	0.00	0.00	0.00	0.00	0.00	0.00	0.00	0.00	0.00	0.00	0.01	0.02	0.15	0.49	0.51	1.00	0.33	0.11
Albedo	0.00	0.00	0.99	0.96	0.36	0.16	0.11	0.01	0.00	0.04	0.37	0.92	0.78	0.38	0.99	0.99	1.00	0.99	0.39	0.91	0.66	0.70

To further understand the results in Table 1, we show the tree cover, albedo, LST histograms, and associated spatial maps for the regions of the continent with mean annual rainfall between 1600–1800 mm MAP, where all three datasets show evidence for bifurcations (Figure 6). The histogram for the VCF clearly shows a marked bimodality between the highest (~80%) and lowest (<20%) VCF values. However, the potential problems in using such MAP intervals and histograms to analyze for savanna bifurcations is evident from the spatially explicit map in Figure 6. In particular, when we examine these data spatially, it becomes evident that the apparent bifurcations in the VCF histogram occur in distinctly different regions. This suggests that histograms based on rainfall zones are multimodal not because of local-scale feedbacks and emergent bifurcations (the definition of alternative states), but because of regional scale differences in woody dynamics associated with edaphic, disturbance, and anthropogenic regimes.

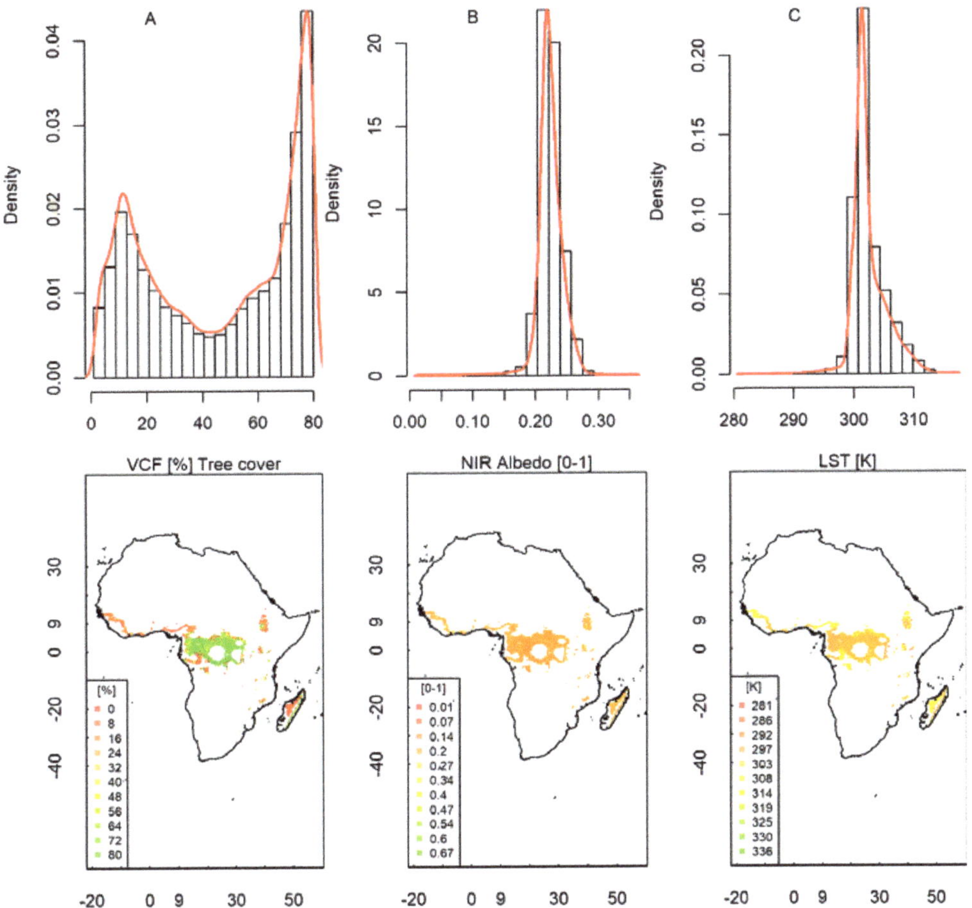

Figure 6. Histograms and maps of (**A**) tree cover (%), (**B**) NIR albedo, and (**C**) LST (K) in the 1600–1800 mm/yr mean annual precipitation zones of Africa, showing that bimodality seen in the tree cover datasets occurs in spatially disconnected regions. While the histograms in this MAP interval are visually and statistically bimodal, the apparent bifurcations occur in spatially distant locations on the fringes of the moist tropical forests of West Africa (e.g., the low tree cover mode in red) and the fringes of the Congo Basin (e.g., the high tree cover mode in green). Similar separation of low and high tree cover modes is apparent between eastern and western Madagascar. The LST and NIR albedo multimodality, while much less pronounced than in the tree cover data, is also geographically separated between West Africa and Congo Basin locations.

4.2. Spatially Explicit Landscape-Scale Analysis

Figure 7 shows results of our analysis of multimodality within 15 × 15 km (i.e., 225 pixels) moving windows across all of Africa, with red and yellow colors showing locations with statistical evidence for multimodality assessed using the Dip test. The VCF data have the highest number of locations across most of Sub-Saharan Africa, with evidence for bimodal or multimodal distributions in most areas at this scale. At this scale, the occurrence of multimodality in the albedo and LST data streams is much less common, with a general lack of spatial consistency between VCF, albedo, and LST datasets (Table 2). Quantitatively, the percent agreement of multimodality among all three data streams (VCF, LST, and albedo) conveys that only 0.52% of the locations show multimodality in the VCF data.

Analysis of Cohen's kappa (Table 2) quantifies this apparent lack of spatial consistency in the locations where multimodality is observed. The exclusion of VCF to examine the degree of spatial coherence between LST and albedo marginally improves the spatial consistency (Table 2), but there is still considerable disagreement between LST and albedo in the locations with multimodality.

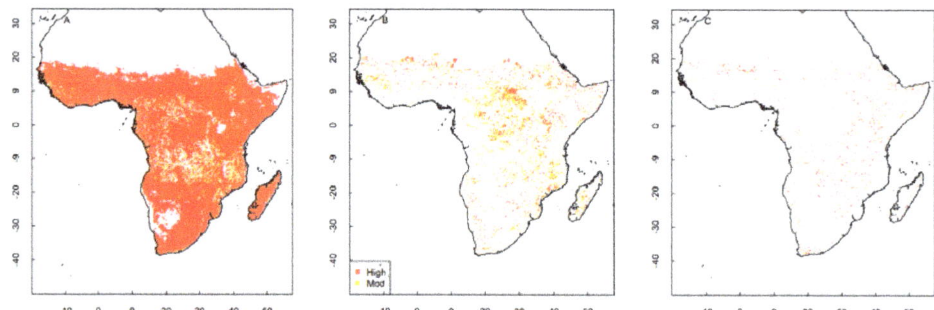

Figure 7. Spatially explicit signals of local bifurcations (non-unimodality) calculated within window sizes of 15 × 15 km for data shown in Figure 3, (**A**) tree cover, (**B**) albedo, (**C**) LST. Red and yellow colors show 15 × 15 km landscapes with high and moderate statistical evidence for local bifurcations (i.e., Hartigan's unimodality dip test is rejected in these locations).

Table 2. Tests for spatial consistency in detections of multimodality among three remote sensing datasets, showing the number of 15 km × 15 km locations across Sub-Saharan Africa with evidence for multimodality individually (columns 2–4) and in conjunction with other products (columns 5–8, where, for example, VCF and LST quantify locations scored as multimodal in both VCF and LST). Only locations that have a MAP $\geq$ 100 mm are considered. Statistical significance is based on Hartigan's dip test p value < 0.05. The Cohen's kappa (κ), a measurement of agreement between data sets, is given in parenthesis. Kappa values close to unity indicate strong agreement, while values close to zero imply low or poor agreement, and negative values imply strong disagreement. The percent agreement of multimodality among all data streams over the entire continent is also tabulated.

	VCF	LST	Albedo	VCF and LST	VCF and Albedo	LST and Albedo	VCF and LST and Albedo
Number of multimodal locations and (Cohen's Kappa κ)	15,913,877	657,792	2,360,289	475,702 (−0.005)	1,774,261 (−0.117)	118,256 (0.026)	84,095 0.52%

5. Discussion

In this study, we examine the distributions of MODIS VCF tree cover estimates in Sub-Saharan Africa alongside data on albedo and LST to determine the degree of similarity among datasets in the occurrence of multimodality and inference of alternative stable states. While we argue that albedo and LST should be closely correlated with tree cover, we also recognize that the darkening and cooling effect of trees in a landscape may also occur with dense herbaceous vegetation, potentially reducing the effectiveness of both NIR-albedo and LST as indices of tree cover.

The above caveat notwithstanding, our results (Figures 4–7) show clearly that bimodality is common in MODIS VCF when visualized at a continental scale, when sorted into MAP bands (consistent with past studies, [18,19,24–26]; Table 2), and when analyzed at landscape scales (15 × 15 km blocks; Table 2). However, analysis of NIR albedo and LST datasets suggest multimodality is much less common in these datasets. Examination of the spatial distribution of high and low tree cover within an example MAP interval (Figure 6) highlights the potential dangers in the approach. In particular, while Figures 4 and 5 suggest strong bimodality in tree cover, examination of the spatial distribution of the data reveals that the high and low modes are geographically separated, indicating regional differences in tree cover related to possible edaphic or anthropogenic impacts on mean tree cover rather than positive feedbacks giving rise to bifurcations at local scales under similar edaphic conditions.

Past studies presenting empirical evidence for alternative states at different levels of MAP did not examine independent data sets, nor did they consider the spatial organization of tree cover modes within each MAP band. Based on the conclusion that detection of multimodality in spatially extensive MAP intervals may not be appropriate for the diagnosis of bifurcations at regional scales, we carry out spatially explicit analyses designed to detect bifurcations at landscape scales (15 × 15 km) in the three independent data streams. Multimodality is suggested across most of Sub-Saharan Africa in the VCF dataset but exists in much more constrained regions in the NIR albedo and LST datasets. Metrics of spatial consistency in multimodality among data-streams based on Cohen's kappa (Table 2) indicate that VCF is distinct from LST and albedo (i.e., very little spatial consistency), with more agreement (but still considerable differences) in the locations of bimodality in the albedo and LST data streams. These results support concerns raised elsewhere [31–33] that the VCF product may not be appropriate for the diagnosis of alternative states (i.e., bifurcations) because of the calibration approach utilized and the discontinuities inherent in CART-based model-fitting approaches.

6. Conclusions

We posed the following two questions: (1) are the apparent forest-savanna and savanna-grassland bifurcations detected in earlier VCF analysis also present in albedo and surface temperature data?; and (2) to what extent are emergent bifurcations in VCF, albedo, and land surface temperature dependent on the spatial scale of analysis? Our results show that apparent bifurcation in the VCF data, diagnosed via statistical tests for multi-modality, far exceeds bifurcation detections in albedo and land surface temperature data. Further, our multi-scale analysis highlights that the "rainfall zone" analysis approach adopted by earlier authors (i.e., aggregating continental data using rainfall bins; Section 4.2) tends to identify multimodality associated with regional differences in tree cover rather than detecting landscape-scale bifurcations. Our explicit landscape scale analysis finds that, while apparent bifurcations are ubiquitous in the VCF data, they are far less common in both the albedo and the land surface temperature data.

The lack of consistent signatures of bimodality in satellite estimates of tree cover, land surface temperature, and albedo highlights the need for caution in analyses suggesting that alternative states are widespread in tropical savannas. A more considered review of the associated ecological theory points to several explanations for this difference of opinion: (1) while positive feedbacks relating to fire, tree-grass competition, and herbivory can—in theory—lead to the emergence of areas of high and low tree cover under similar climate [12,18,53], in biologically and edaphically complex landscapes, the processes potentially leading to bifurcations may be buffered by other interactions [54]; (2) in connected landscapes, spatial interactions may reduce the climate space where bifurcations emerge, making bifurcations less common than expected based on simple non-spatial models [18,55]; and (3) the positive feedbacks potentially driving bifurcations in tree cover are processes that operate at distinct spatial scales that may not be captured by the relatively coarse grain of many satellite tree cover datasets. For example, while facilitation and competition among plants in arid and semi-arid systems is a well-known cause of the patterned vegetation states known as tiger-bush (i.e., stripes of dense vegetation separated by stripes of bare soil [56]), the processes involved (i.e., redistribution of water and root competition) occur at relatively fine spatial scales (<~100 m). We should therefore not anticipate being able to detect this type of alternative vegetation state in data with a spatial grain >~50 m. While remote sensing data are invaluable for our understanding of savanna ecology at landscape, continental, and global scales, to advance our understanding of alternative vegetation states, it is critical that we identify not only the specific processes potentially leading to alternative states but also the spatial scale at which the emergent patterns should be detectable.

Supplementary Materials: The following are available online at http://www.mdpi.com/2072-4292/11/7/815/s1, Figure S1: Histograms of MODIS tree cover (VCF) over regions stratified by mean annual precipitation (MAP in mm/y). Each histogram was generated using 20 equally spaced bins spanning the range of the data in each region Red lines trace the kernel density estimates to help visualize the shape of the histograms., Figure S2: Histograms of MODIS NIR albedo over regions stratified by mean annual precipitation (MAP in mm/y). Each histogram was generated using 20 equally spaced bins spanning the range of the data in each region Red lines trace the kernel density estimates to help visualize the shape of the histograms, Figure S3: Histograms of MODIS Land Surface Temperature over regions stratified by mean annual precipitation (MAP in mm/y). Each histogram was generated using 20 equally spaced bins spanning the range of the data in each region Red lines trace the kernel density estimates to help visualize the shape of the histograms.

Author Contributions: S.S.K., N.P.H. and L.P. conceived research, S.S.K. conducted the analyses with inputs from J.A., C.W.R., W.J. and B.M.L. All authors contributed towards writing the manuscript.

Funding: Research was supported by the US National Aeronautic and Space Administration (NASA) as part of the NASA Carbon Cycle Science program (Grant # NNX17AI49G).

Acknowledgments: We thank the editor and the two anonymous reviewers for their insightful suggestions that greatly improved our manuscript.

Conflicts of Interest: The authors declare no conflict of interest.

References

1. Bond, W.J. What limits trees in C4 grasslands and savannas? *Annu. Rev. Ecol. Evol. Syst.* **2008**, *39*, 641–659. [CrossRef]

2. Hill, M.J.; Hanan, N.P. *Ecosystem Function in Savannas: Measurement and Modeling at Landscape to Global Scales*; CRC Press: Boca Raton, FL, USA, 2010; ISBN 1-4398-0471-0.

3. House, J.I.; Archer, S.; Breshears, D.D.; Scholes, R.J.; NCEAS Tree–Grass Interactions Participants. Conundrums in mixed woody–herbaceous plant systems. *J. Biogeogr.* **2003**, *30*, 1763–1777. [CrossRef]

4. Prentice, I.C.; Cramer, W.; Harrison, S.P.; Leemans, R.; Monserud, R.A.; Solomon, A.M. Special paper: A global biome model based on plant physiology and dominance, soil properties and climate. *J. Biogeogr.* **1992**, 117–134. [CrossRef]

5. Woodward, F.I.; Lomas, M.R.; Kelly, C.K. Global climate and the distribution of plant biomes. *Philos. Trans. R. Soc. B Biol. Sci.* **2004**, *359*, 1465–1476. [CrossRef]

6. Bond, W.J.; Midgley, G.F.; Woodward, F.I.; Hoffman, M.T.; Cowling, R.M. What controls South African vegetation—Climate or fire? *S. Afr. J. Bot.* **2003**, *69*, 79–91. [CrossRef]

7. Lehmann, C.E.; Archibald, S.A.; Hoffmann, W.A.; Bond, W.J. Deciphering the distribution of the savanna biome. *New Phytol.* **2011**, *191*, 197–209. [PubMed]

8. Sankaran, M.; Hanan, N.P.; Scholes, R.J.; Ratnam, J.; Augustine, D.J.; Cade, B.S.; Gignoux, J.; Higgins, S.I.; Le Roux, X.; Ludwig, F. Determinants of woody cover in African savannas. *Nature* **2005**, *438*, 846–849.

9. Axelsson, C.R.; Hanan, N.P. Patterns in woody vegetation structure across African savannas. *Biogeosciences* **2017**, *14*, 3239–3252. [CrossRef]

10. Yin, Z.; Dekker, S.C.; van den Hurk, B.J.J.M.; Dijkstra, H.A. The climatic imprint of bimodal distributions in vegetation cover for western Africa. *Biogeosciences* **2016**, *13*, 3343–3357. [CrossRef]

11. Mayer, A.L.; Khalyani, A.H. Grass trumps trees with fire. *Science* **2011**, *334*, 188–189. [CrossRef]

12. Scheffer, M.; Carpenter, S.; Foley, J.A.; Folke, C.; Walker, B. Catastrophic shifts in ecosystems. *Nature* **2001**, *413*, 591. [CrossRef] [PubMed]

13. Veenendaal, E.M.; Torello-Raventos, M.; Miranda, H.S.; Sato, N.M.; Oliveras, I.; van Langevelde, F.; Asner, G.P.; Lloyd, J. On the relationship between fire regime and vegetation structure in the tropics. *New Phytol.* **2018**, *218*, 153–166. [CrossRef] [PubMed]

14. Hanan, N.; Lehmann, C. *Tree-Grass Interactions in Savannas: Paradigms, Contradictions, and Conceptual Models*; Taylor and Francis Group: Boca Raton, FL, USA, 2010.

15. Ji, W.; Hanan, N.; Browning, D.; Monger, C.; Peters, D.; Bestelmeyer, B.; Archer, S.; Ross, C.; Lind, B.; Anchang, J.; et al. Constraints on shrub cover and shrub-shrub competition in a U.S. Southwest desert. *Ecosphere* **2019**, *10*, e02590. [CrossRef]

16. De Michele, C.; Accatino, F. Tree Cover Bimodality in Savannas and Forests Emerging from the Switching between Two Fire Dynamics. *PLoS ONE* **2014**, *9*, e91195. [CrossRef] [PubMed]

17. Van de Koppel, J.; Herman, P.M.J.; Thoolen, P.; Heip, C.H.R. Do Alternate Stable States Occur In Natural Ecosystems? Evidence from a Tidal Flat. *Ecology* **2001**, *82*, 3449–3461. [CrossRef]

18. Staal, A.; Dekker, S.C.; Xu, C.; van Nes, E.H. Bistability, spatial interaction, and the distribution of tropical forests and savannas. *Ecosystems* **2016**, *19*, 1080–1091. [CrossRef]

19. Staver, A.C.; Archibald, S.; Levin, S.A. The global extent and determinants of savanna and forest as alternative biome states. *Science* **2011**, *334*, 230–232. [CrossRef]

20. Stromayer, K.A.; Warren, R.J. Are overabundant deer herds in the eastern United States creating alternate stable states in forest plant communities? *Wildl. Soc. Bull.* **1997**, 227–234.

21. Touboul, J.D.; Staver, A.C.; Levin, S.A. On the complex dynamics of savanna landscapes. *Proc. Natl. Acad. Sci. USA* **2018**, *115*, E1336–E1345. [CrossRef]

22. Tredennick, A.T.; Hanan, N.P. Effects of Tree Harvest on the Stable-State Dynamics of Savanna and Forest. *Am. Nat.* **2015**, *185*, E153–E165. [CrossRef]

23. Weissmann, H.; Kent, R.; Michael, Y.; Shnerb, N.M. Empirical analysis of vegetation dynamics and the possibility of a catastrophic desertification transition. *PLoS ONE* **2017**, *12*, e0189058. [CrossRef] [PubMed]

24. Hirota, M.; Holmgren, M.; Van Nes, E.H.; Scheffer, M. Global Resilience of Tropical Forest and Savanna to Critical Transitions. *Science* **2011**, *334*, 232. [CrossRef] [PubMed]

25. Murphy, B.P.; Bowman, D.M. What controls the distribution of tropical forest and savanna? *Ecol. Lett.* **2012**, *15*, 748–758. [CrossRef] [PubMed]

26. Staver, A.C.; Archibald, S.; Levin, S. Tree cover in sub-Saharan Africa: Rainfall and fire constrain forest and savanna as alternative stable states. *Ecology* **2011**, *92*, 1063–1072. [CrossRef] [PubMed]

27. Wuyts, B.; Champneys, A.R.; House, J.I. Amazonian forest-savanna bistability and human impact. *Nat. Commun.* **2017**, *8*, 15519. [CrossRef] [PubMed]

28. Ratajczak, Z.; Nippert, J.B. Comment on "Global Resilience of Tropical Forest and Savanna to Critical Transitions". *Science* **2012**, *336*, 541. [CrossRef]

29. Hansen, M.C.; DeFries, R.S.; Townshend, J.R.G.; Carroll, M.; Dimiceli, C.; Sohlberg, R.A. Global percent tree cover at a spatial resolution of 500 meters: First results of the MODIS vegetation continuous fields algorithm. *Earth Interact.* **2003**, *7*, 1–15. [CrossRef]

30. Dimiceli, C.; Carroll, M.; Sohlberg, R.; Kim, D.H.; Kelly, M.; Townshend, J.R.G. *MOD44B MODIS/Terra Vegetation Continuous Fields Yearly L3 Global 250 m SIN Grid V006*; NASA EOSDIS Land Processes DAAC: Sioux Falls, SD, USA, 2015.

31. Gerard, F.; Hooftman, D.; van Langevelde, F.; Veenendaal, E.; White, S.M.; Lloyd, J. MODIS VCF should not be used to detect discontinuities in tree cover due to binning bias. A comment on Hanan et al. (2014) and Staver and Hansen (2015). *Glob. Ecol. Biogeogr.* **2017**, *26*, 854–859. [CrossRef]

32. Hanan, N.P.; Tredennick, A.T.; Prihodko, L.; Bucini, G.; Dohn, J. Analysis of stable states in global savannas: Is the CART pulling the horse? *Glob. Ecol. Biogeogr.* **2014**, *23*, 259–263. [CrossRef]

33. Staver, A.C.; Hansen, M.C. Analysis of stable states in global savannas: Is the CART pulling the horse?—A comment. *Glob. Ecol. Biogeogr.* **2015**, *24*, 985–987. [CrossRef]

34. Olson, D.M.; Dinerstein, E.; Wikramanayake, E.D.; Burgess, N.D.; Powell, G.V.; Underwood, E.C.; D'amico, J.A.; Itoua, I.; Strand, H.E.; Morrison, J.C.; et al. Terrestrial Ecoregions of the World: A New Map of Life on EarthA new global map of terrestrial ecoregions provides an innovative tool for conserving biodiversity. *BioScience* **2001**, *51*, 933–938. [CrossRef]

35. Funk, C.; Peterson, P.; Landsfeld, M.; Pedreros, D.; Verdin, J.; Shukla, S.; Husak, G.; Rowland, J.; Harrison, L.; Hoell, A. The climate hazards infrared precipitation with stations—A new environmental record for monitoring extremes. *Sci. Data* **2015**, *2*, 150066. [CrossRef] [PubMed]

36. McNally, A.; Shukla, S.; Arsenault, K.R.; Wang, S.; Peters-Lidard, C.D.; Verdin, J.P. Evaluating ESA CCI soil moisture in East Africa. *Int. J. Appl. Earth Obs. Geoinf.* **2016**, *48*, 96–109. [CrossRef] [PubMed]

37. Toté, C.; Patricio, D.; Boogaard, H.; van der Wijngaart, R.; Tarnavsky, E.; Funk, C. Evaluation of satellite rainfall estimates for drought and flood monitoring in Mozambique. *Remote Sens.* **2015**, *7*, 1758–1776. [CrossRef]

38. Schaaf, C.B.; Wang, Z. *MCD43A3 MODIS/Terra+ Aqua BRDF/Albedo Daily L3 Global 500 m V006*; LP DAAC: Sioux Falls, SD, USA, 2015.

39. Strahler, A.H.; Muller, J.P.; Lucht, W.; Schaaf, C.; Tsang, T.; Gao, F.; Li, X.; Lewis, P.; Barnsley, M.J. MODIS BRDF/albedo product: Algorithm theoretical basis document version 5.0. *MODIS Doc.* **1999**, *23*, 42–47.

40. Houldcroft, C.J.; Grey, W.M.; Barnsley, M.; Taylor, C.M.; Los, S.O.; North, P.R. New vegetation albedo parameters and global fields of soil background albedo derived from MODIS for use in a climate model. *J. Hydrometeorol.* **2009**, *10*, 183–198. [CrossRef]
41. Lyons, E.A.; Jin, Y.; Randerson, J.T. Changes in surface albedo after fire in boreal forest ecosystems of interior Alaska assessed using MODIS satellite observations. *J. Geophys. Res. Biogeosci.* **2008**, *113*. [CrossRef]
42. Carlson, T.N.; Gillies, R.R.; Perry, E.M. A method to make use of thermal infrared temperature and NDVI measurements to infer surface soil water content and fractional vegetation cover. *Remote Sens. Rev.* **1994**, *9*, 161–173. [CrossRef]
43. Prihodko, L.; Goward, S.N. Estimation of air temperature from remotely sensed surface observations. *Remote Sens. Environ.* **1997**, *60*, 335–346. [CrossRef]
44. Still, C.J.; Pau, S.; Edwards, E.J. Land surface skin temperature captures thermal environments of C3 and C4 grasses. *Glob. Ecol. Biogeogr.* **2014**, *23*, 286–296. [CrossRef]
45. Weng, Q.; Lu, D.; Schubring, J. Estimation of land surface temperature–vegetation abundance relationship for urban heat island studies. *Remote Sens. Environ.* **2004**, *89*, 467–483. [CrossRef]
46. Mildrexler, D.J.; Zhao, M.; Running, S.W. A global comparison between station air temperatures and MODIS land surface temperatures reveals the cooling role of forests. *J. Geophys. Res. Biogeosci.* **2011**, *116*. [CrossRef]
47. Roy, D.P.; Kumar, S.S. Multi-year MODIS active fire type classification over the Brazilian Tropical Moist Forest Biome. *Int. J. Digit. Earth* **2017**, *10*, 54–84. [CrossRef]
48. Wan, Z.; Hook, S.; Hulley, G. *MOD11A2 MODIS/Terra Land Surface Temperature/Emissivity 8-Day L3 Global 1km SIN Grid V006*; NASA EOSDIS Land Processes DAAC: Sioux Falls, SD, USA, 2015.
49. Hartigan, P.M. Computation of the Dip Statistic to Test for Unimodality. *J. R. Stat. Soc. Ser. C Appl. Stat.* **1985**, *34*, 320–325.
50. Livina, V.N.; Kwasniok, F.; Lenton, T.M. Potential analysis reveals changing number of climate states during the last 60 kyr. *Clim. Past* **2010**, *6*, 77–82. [CrossRef]
51. Freeman, J.B.; Dale, R. Assessing bimodality to detect the presence of a dual cognitive process. *Behav. Res. Methods* **2013**, *45*, 83–97. [CrossRef] [PubMed]
52. Cohen, J. A coefficient of agreement for nominal scales. *Educ. Psychol. Meas.* **1960**, *20*, 37–46. [CrossRef]
53. Scheffer, M.; Hirota, M.; Holmgren, M.; Van Nes, E.H.; Chapin, F.S. Thresholds for boreal biome transitions. *Proc. Natl. Acad. Sci. USA* **2012**, *109*, 21384–21389. [CrossRef] [PubMed]
54. Petraitis, P. *Multiple Stable States in Natural Ecosystems*; OUP: Oxford, UK, 2013; ISBN 0-19-166833-8.
55. Bel, G.; Hagberg, A.; Meron, E. Gradual regime shifts in spatially extended ecosystems. *Theor. Ecol.* **2012**, *5*, 591–604. [CrossRef]
56. Rietkerk, M.; Van de Koppel, J. Regular pattern formation in real ecosystems. *Trends Ecol. Evol.* **2008**, *23*, 169–175. [CrossRef]

 *remote sensing*

Article

A Healthy Park Needs Healthy Vegetation: The Story of Gorongosa National Park in the 21st Century

Hannah Herrero [1,*], Peter Waylen [2], Jane Southworth [2], Reza Khatami [2], Di Yang [3] and Brian Child [2]

[1] Department of Geography, University of Tennessee, 1000 Philip Fulmer Way, Room 315, Knoxville, TN 37996-0925, USA

[2] Department of Geography, University of Florida, 3141 Turlington Hall, Gainesville, FL 32611, USA; prwaylen@ufl.edu (P.W.); jsouthwo@ufl.edu (J.S.); seyedghkhatami@ufl.edu (R.K.); bchild@ufl.edu (B.C.)

[3] Mt. Natural Heritage Program, University of Montana, 1515 E. 6th Ave, Helena, MT 59620, USA; di.yang@mso.umt.edu

* Correspondence: hherrero@utk.edu; Tel.: +1-865-974-6043

Received: 6 December 2019; Accepted: 29 January 2020; Published: 3 February 2020

Abstract: Understanding trends or changes in biomass and biodiversity around conservation areas in Africa is important and has economic and societal impacts on the surrounding communities. Gorongosa National Park, Mozambique was established under unique conditions due to its complex history. In this study, we used a time-series of Normalized Difference Vegetation Index (NDVI) to explore seasonal trends in biomass between 2000 and 2016. In addition, vegetation directional persistence was created. This product is derived from the seasonal NDVI time series-based analysis and represents the accumulation of directional change in NDVI relative to a fixed benchmark (2000–2004). Trends in precipitation from Climate Hazards Group InfraRed Precipitation with Station data (CHIRPS) was explored from 2000–2016. Different vegetation covers are also considered across various landscapes, including a comparison between the Lower Gorongosa (savanna), Mount Gorongosa (rainforest), and surrounding buffer zones. Important findings include a decline in precipitation over the time of study, which most likely drives the observed decrease in NDVI. In terms of vegetation persistence, Lower Gorongosa had stronger positive trends than the buffer zone, and Mount Gorongosa had higher negative persistence overall. Directional persistence also varied by vegetation type. These are valuable findings for park managers and conservationists across the world.

Keywords: remote sensing; vegetation dynamics; vegetation persistence; conservation; savannas; Africa

---

1. Introduction

Savannas are commonly defined as grassland with scattered trees, but in practice are a mix of grass, shrub, and trees [1–3] which comprise over 55% of the southern African landscape, and approximately 20% of Earth's landcover [1]. Savannas are controlled by key factors: precipitation, herbivory, fire, and humans [4–7]. Savannas are water-limited systems, therefore precipitation is the main driving factor for the ecosystem [5,8]. Up to a precipitation threshold of 750 mm/year, grasses dominate, between 750-950 mm/year, more mixed systems occur, and above 950 mm/year, woodlands dominate [5,7–9]. Decreases in precipitation alone can lead to declines in vegetation, and when coupled with rising temperatures the resulting lower availability of soil moisture causes even further declines in vegetation [10–12]. Savannas are an important conservation landscape in southern Africa. Understanding and conserving these systems is vital as they support high faunal and floral biodiversity. Savannas are also an important biome affecting the carbon cycle, supporting high human populations, and contributing 14% of global net primary productivity [8,13–17]. Conservation areas are a key resource in southern Africa both in terms of their ecological importance (biodiversity, biomass, carbon, etc.) and as socioeconomic drivers for the surrounding communities [18].

Remote sensing of these savanna systems is a challenge due to the heterogeneity of savanna landscapes, but given their ecological and socioeconomic importance, they are a target for development of innovative image analysis techniques. One such approach to evaluating savannas with imagery includes continuous time series analysis [19,20]. The normalized difference vegetation index (NDVI) can be used as a proxy for vegetation biomass, health, and abundance [20–26]. The value of this vegetation index may be enhanced with new time series measures to provide better understanding of trends in vegetation health over time and space [19]. One such measure, derived from NDVI, is that of vegetation persistence, which has demonstrated its usefulness in drylands globally [20]. This is an NDVI-based time series approach to compare vegetation greenness in a given season and year to a baseline period. This methodology allows the user to detect inter-annual changes of NDVI in a spatially explicit manner and identify key trends on the landscape, with a statistical significance attached to each pixel [19,20].

Gorongosa National Park (GNP) in central Mozambique contains two important ecosystem types—savanna and montane rainforest. This paper examines the complex savanna system of the lower part of GNP (herein referred to as "Lower Gorongosa") and compares it to the montane rainforest of Mount Gorongosa [27]. Like many mountainous areas [28], the latter provides a haven for biodiversity much of which remains to be discovered, as evidenced by the extraordinary discoveries from the rapid biodiversity collection event held by E.O. Wilson in 2011 [29]. GNP is chosen for analysis because of the interesting conditions under which it was established and also due to the fact that it has been restored in the last decade. GNP was originally established in 1920 under Portuguese colonial rule, and the park served as headquarters for both sides at some point during both the War for Independence and the Civil War, which lasted from 1964 to 1994. As a result, between 90% and 99% of all mammals in the park were extirpated by crossfire, lack of sustenance, and poaching, and this lack of mammals over time may affect vegetation [30]. Following peace, a variety of illegal hunting activities continued for years in the park, pressuring already drastically decreased populations. In 2004, a relationship developed between the Mozambican government and an American NGO initiating the Gorongosa Restoration Project, the first conservation collaboration of its kind in the world. This project has three main goals: science and conservation, sustainable tourism, and human development. These goals include the restoration of the ecosystem to a pre-war baseline of wildlife [31]. A significant management presence began in the park in 2008 with the aim to restock animals, which might then affect vegetation. The park is a natural experiment for the impact of herbivores on vegetation. Several studies note that the absence of large herbivores enables trees to attain maturity here in GNP, unlike other places in southern Africa where bush densification has been the trend [32,33]. The primary focus of restoration has been wildlife, however, a healthy park also needs healthy vegetation to sustain its wildlife populations. Therefore, the purpose of this paper is to evaluate trends in vegetation biomass during pre- (2000–2008) and post- (2009–2016) management periods in GNP.

The objective of this study is to understand the trends in vegetation in Lower Gorongosa and to compare them to the buffer zone and Mount Gorongosa. The questions we addressed are: 1. Is there a difference in the precipitation between the period since restoration management has been imposed and the prior period? 2. Has there been a trend in the seasonal precipitation time period between 2000 and 2016? 3. Can trends in vegetation density and biomass be detected in the period 2000 to 2016? 4. Does the vegetation directional persistence methodology detect significant changes in NDVI as an indicator of vegetation change, in the period 2000 to 2016?

2. Materials and Methods

2.1. Study Area

The GNP occupies approximately 4000 km^2 in central Mozambique (18.2°S and 34.0°E) (Figure 1). Annual minimum and maximum temperatures on average range between 15 °C and 30 °C, in the dry and wet seasons respectively. The ecosystem of the original area of the park, Lower Gorongosa (LG), is markedly different to that of the more newly incorporated Mount Gorongosa (MG), which lies to the northwest at 700 m elevation. Based on the park management report, LG contains various savannas ranging from "open" savanna (floodplain grassland, 21% of LG) in the central part of the park, to "mixed" savanna (grass-shrub-open tree, 44% of LG), and "closed" savanna (thick woodlands dominated by several different tree species, 35% of LG). Montane rain forest vegetation dominates the flanks of MG, while short grassland/rock covers the summit. The protected area on MG is critical to the health of both LG and people living in the surrounding area, as it constitutes the dominant source of stream flow. The areas surrounding parks, known as buffer zones, provide a comparison to vegetation trends occurring inside the park because they are not as protected as parks, but have similar environmental conditions affecting them. They therefore help better understand how the park is functioning in the larger context of the landscape. Based on literature indicating that a buffer zone analysis should encompass approximately 3–4 times the size of the park [34], a 20 km buffer zone around the park was considered in this work. There is some limited human activity in this zone where communities live. Six study areas are addressed: (1) LG (first as a whole), and then further divided into land cover types- (2) open savanna, (3) mixed savanna, and (4) closed savanna, (5) a 20 km buffer zone around LG and (6) MG.

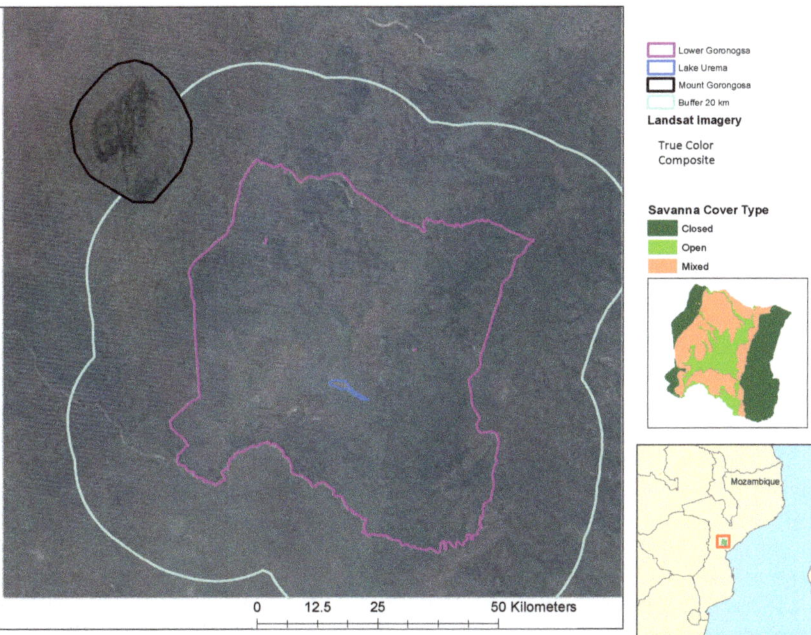

Figure 1. Study area map of low Lower Gorongosa, Mount Gorongosa and buffer area shown on a true color composite Landsat image. In the top inset map, open savanna, mixed savanna, and closed savanna are shown.

2.2. Vegetation Change Characteristics

This section addresses factors used to measure vegetation change and related drivers.

2.2.1. Mean NDVI

The Normalized Difference Vegetation Index (NDVI) is used to establish trends in vegetation health and density over time. The metric is a band ratio between red (R) (highly absorptive in healthy vegetation) and near infrared (NIR) (highly reflective in healthy vegetation) energy [35]:

$$NDVI = \frac{NIR - R}{NIR + R} \tag{1}$$

NDVI ranges from −1 to +1. Higher values indicate the presence of higher vegetation density, and values below 0 indicate no vegetation [35]. NDVI composite products from the Moderate Resolution Imaging Spectroradiometer (MODIS) sensor, i.e., MOD13Q1.006 Terra Vegetation Index, were used in this research due to their fine temporal resolution. The data span March 2000 to November 2016, have a spatial resolution of 250 m, and are a 16-day composite of the highest NDVI value per pixel. The data are atmospherically corrected and masked for clouds, cloud shadows, water, and heavy aerosols. Furthermore, as water produces negative NDVI values, the Lake Urema system in LG is also masked out and removed from further analysis. Data are aggregated into seasonal composites based on per-pixel seasonal mean values for the following seasons; December, January, February (wet), March, April, May (wet), June, July, August (dry), and September, October, November (dry). Data are further aggregated by mean pixel value for the seasons into different periods corresponding to different on-the-ground management practices in GNP: 2000–2004 (no management), 2005–2008 (management contracts negotiated), 2009–2012 (beginning of on the ground management), and 2013–2016 (fully implemented on the ground management). These can be grouped broadly into 2000–2008, pre-management and 2009–2016, post-management [29]. The analysis is separated into each of the six study areas. Spatial extents of the three savanna types in LG (open, mixed, and closed) were provided by the park management team.

2.2.2. Seasonal Precipitation Totals

As savannas are water limited systems, monitoring trends in precipitation is critical in understanding changes in vegetation. Estimates of monthly precipitation totals from 2000 to 2016 were extracted from the Climate Hazards Group InfraRed Precipitation with Station data (CHIRPS) gridded dataset (0.05° resolution). Given the coarse spatial resolution of the data and the relatively small study area, a mean value for each time period was extracted over the whole study area [36]. Total seasonal precipitation for the park is created for each year from the monthly estimates observations. The annual migration of the Intertropical Convergence Zone (ITCZ) is considered to be the principal driver of intra-annual precipitation [37,38]. Given the delay between the arrival of precipitation and green up of vegetation, seasonal estimates of vegetation are lagged by one month [39]. Therefore, the seasonal precipitation-vegetation pairs are respectively as follows: (NDJ-DJF; FMA-MAM; MJJ-JJA; ASO-SON). Mean annual precipitation (2000–2016) is 1075 mm, and individual annual totals range from 665 mm in 2015 to 1415 mm in 2001.

Analysis of seasonal precipitation totals proceeds in several ways. Totals are expressed as being above or below the long-term (2000–2016) mean and the subsequent counts of the number of above/below mean years in each time period recorded. To investigate if total seasonal precipitation has increased/decreased between the two management periods, a hypergeometric test was used. The hypergeometric distribution determines the probability of experiencing a given number of years with values above (or below) mean characteristic at random [40]. It therefore highlights whether the observed number of years (above/below mean) is significantly greater than/less than, would be expected at random. Years during which a seasonal total fall above the mean are considered "successes" and during which it falls below the mean are considered "failures." The probability distribution of the number of successes (failures) is given as:

$$p(x) = \frac{\binom{k}{x}\binom{N-k}{n-x}}{\binom{N}{n}} \tag{2}$$

where N is the number of years in the record (i.e., $N = 17$ during 2000–2016), n is the sample size (the number of years in a particular period during which a set of management practices is followed), and k is the total number of observations in the entire record with the desired property (success or failure) [40]. The sample time periods for the hypergeometric distribution was chosen based on the main management time frame of 2000–2008, where there was not a management presence on the ground, and 2009–2016, where there was a heavy management presence on the ground.

Precipitation values were also graphed as a time series, normalized for seasonality, anomalies were determined, and linear trends were added. Precipitation is also quantitatively compared to mean NDVI, as savanna vegetation-precipitation association is well established [5,7–9,12,20–26].

2.2.3. Vegetation Directional Persistence

Newly developed directional persistence (DP), a continuous time series metric, is calculated on a per-pixel basis and is the cumulative direction of change over time in NDVI calculated in comparison to a benchmark value [19,20]. This approach illuminates trends in vegetation over time and space [19,20]. We implemented the DP analysis using the time-series data from the MOD13Q1.006 Terra Vegetation Index products outlined above in Section 2.2.1. To calculate DP, per-pixel seasonal mean NDVI over the period 2000–2004 was calculated as the benchmark. Then, all subsequent seasonal values 2005–2016 ($N = 12$) were compared to it at a pixel level, generating the equivalent of a random walk in each pixel. Under the null hypothesis of persistent vegetation, i.e., no change in vegetation, there is an equal chance of an observation above the benchmark (success) or below it (failure) [19]. For a "success" in a particular year the pixel receives a value of +1, if a "failure", it receives a −1. These values are accumulated through time, resulting in a net value of DP in each pixel. In the case of a N of 12, a pixel value of +12 would indicate that every year had a NDVI value higher than the benchmark. Likewise, a pixel value of −12 would indicate that every year had a NDVI value lower than the benchmark. The DP values can be assigned to statistical significance levels based on the hypergeometric distribution. Given the preliminary nature of this exploration and the comparatively short (2005–2016) period of evaluation, a 0.10 level of significance, or critical DP of +/−6, is chosen for testing. The respective mean directional persistence, and the percentage of pixels returning significant (positive or negative) persistence are calculated for each of the six study areas.

3. Results

3.1. Precipitation Anomolies and Trends

Figure 2 highlights trends in seasonal precipitation, with the blue and red bars indicating the years with precipitation above and below the overall mean, respectively. Each season except ASO, returns approximately equal numbers of years above and below the long-term mean. However, it appears that there are more below-mean years later in the time series. Additionally, the magnitude of the negative anomalies are high compared to those of the positive anomalies. There were some periods of consecutive seasonal rainfall below the mean, such as in ASO from 2004 to 2010, however application of the hypergeometric test revealed no statistically significant differences at the 0.1 level in the number of positive and negative anomalies in any season or time period. This means that the number of years with rainfall above or below the mean were not statistically different from what we might expect at random when we compare the pre and post management periods.

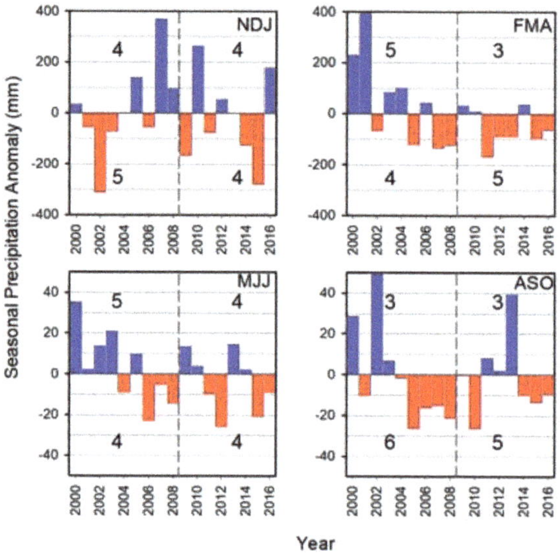

Figure 2. Seasonal precipitation anomalies 2000–2016 for November, December, January (NDJ), mean 550 mm; February, March, April (FMA), mean 384 mm; May, June, July (MJJ), mean 80 mm; August, September, October (ASO), mean 61 mm. The numbers in the graphs are count data for above and below mean precipitation years in the time period.

The time series in Figure 3 suggests a decreasing trend in precipitation over time, particularly in the latter part of the time series. In Figure 3a, when seasonal anomalies are represented by their standard deviates, there is a statistically significant decline over this time period at the 0.1 level of significance. In Figure 3b, where seasonal totals were separated out, there was a statistically significant decline over time in February, March, April (FMA) and May, June, July (MJJ) at the 0.05 significance level.

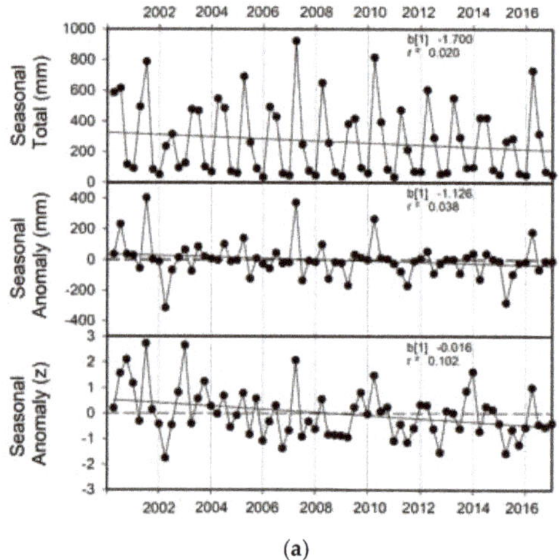

(a)

Figure 3. *Cont.*

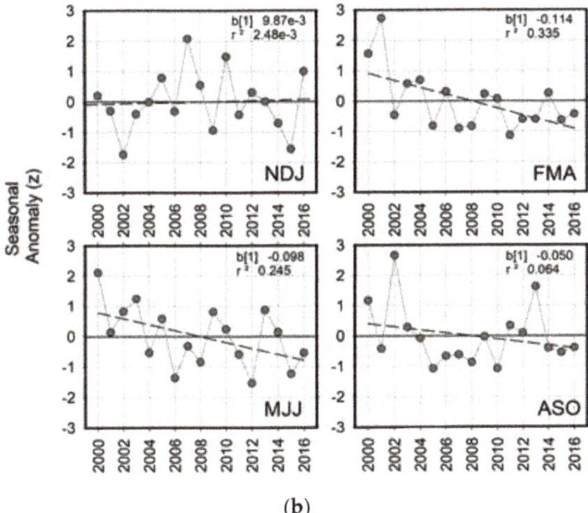

(b)

Figure 3. Precipitation time series 2000-2016 for (**a**) continuous raw seasonal totals, seasonal anomalies (mm), and seasonal anomalies expressed as standard deviates (z) and (**b**) separated seasonal anomalies for November, December, January (NDJ), February, March, April (FMA), May, June, July (MJJ), and August, September, October (ASO).

3.2. Changes of NDVI

Mean NDVI values computed by season (MAM, JJA, SON, DJF) and time period (2000–2004, 2005–2008, 2009–2012, and 2013–2016) are mapped across the LG and MG (Figure 4). This clearly shows the role of seasonality, with higher NDVI values in the wet season than in the dry season. Even within the wet/dry seasons the impact of soil moisture accumulation (MAM is greener than DJF), and depletion (JJA is greener than SON) is apparent. Each season also appears to have some variability of NDVI during the time periods. In general, the first period (2000–2004) had the highest NDVI across the study areas, the most recent period (2013–2016) had the lowest NDVI across the study areas, and the two periods between (2005–2008 and 2009–2012) were variable.

Daily NDVI values, extracted by pixel from MODIS data from 2000-2016, are examined to give further insight into the vegetation dynamics of LG, revealing several important findings, including notable shifts in both the magnitude and timing of anomalies in each period (Figure 5a). There was an overall decline in dry season NDVI after the first time period (2000–2004), which may have even greater negative effects on flora and fauna as this is already the most difficult part of the year for them to survive. In Figure 5a, during the initial period (2000–2004), the data supports that the landscape was greener than the long-term mean during the dry season. During the second period (2005–2008), which was a time of less precipitation, the data shows that it was much less green during the dry season than the long-term mean. During the third time period (2009–2012), NDVI was greener than the long-term mean during the wet season. In the fourth time period (2013–2016), it was less green in the first half of the wet season, and greener during the second half of the wet season than the long-term mean. Figure 5b highlights the magnitude of the positive and negative anomalies (seen in Figure 5a) in both absolute number and percentage.

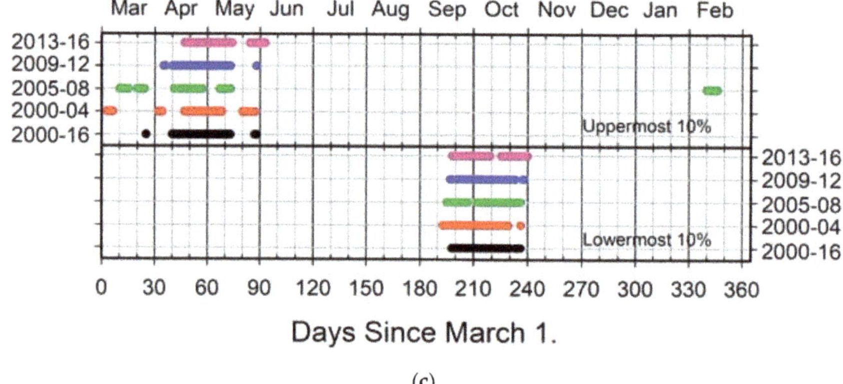

(c)

Figure 5. Daily NDVI (2000-04, 2005-08, 2009-12, 2013-16) of Lower Gorongosa (**a**) positive and negative anomalies compared to the long term mean (**b**). NDVI anomalies displaying the frequency of anomaly both as an absolute count and as a percentage of days exhibiting negative and positive anomalies from the long term mean in each of the four periods (left to right and top to bottom 2000–2004, 2005–2008, 2009–2012, 2013–2016), and histograms of the distribution of the magnitudes of those anomalies. (**c**) The dates on which the uppermost (0.732) and lowermost (0.343) 10% of NDVI values occurred during the entire period of study and period of interest. The day count on the *x*-axis starts on 1 March, given our seasons for the year start with March–April–May, then June–July–August, then September–October–November, then December into January–February of the following year.

Time series of mean NDVI across the six study areas highlight several important findings. All data reveal decreasing trends, though not all are statistically significant (Figure 6). There are also some similarities between the LG and the 20 km buffer zone. However, the rate of decline within the buffer zone is steeper than that of LG, suggesting that the conservation and management practices within the park have been effective in mitigating some potential decline [34].

The overall higher NDVI exhibited by MG than LG is attributable to the differing land cover. LG contains varying types of drier savanna, and MG supports montane rainforest. There is a more pronounced and significant decline in NDVI on MG (Figure 6a). This may be due in part to the rainforest of MG supporting higher NDVI and the impacts of human-driven processes such as deforestation being particularly pronounced. Given the land cover types, there is also much less interannual variability in MG when compared to LG. Regardless, the same trends of declining NDVI and precipitation are seen across this landscape.

NDVI was also graphed by land cover type within LG and MG. All land cover types were clearly separable by mean NDVI values. The lowest NDVI values occur in the open savanna vegetation class (0.36–0.69). These increase to 0.41-0.76 in mixed savanna and 0.51–0.81 in closed savanna. The highest NDVI occurs in the heavily forested landscape on MG (0.62–0.82) even though there is a significant decline over time (Figure 6a). Given that these three savanna land cover classes within LG are respectively dominated by grassland, shrubland, and woodland, this separation of vegetation abundance (NDVI) makes sense biologically. There is a significant decline in the NDVI of open grassland savanna in LG (Figure 6b) and greater interannual variability than in closed woodland because grasslands are more dependent, in the short term, on precipitation than woodlands are. The smallest and non-significant decline is found in the closed savanna land cover class in LG (Figure 6c) as woody plants tend to be hardier/less vulnerable, which is also supported by the literature [5,8,20,41].

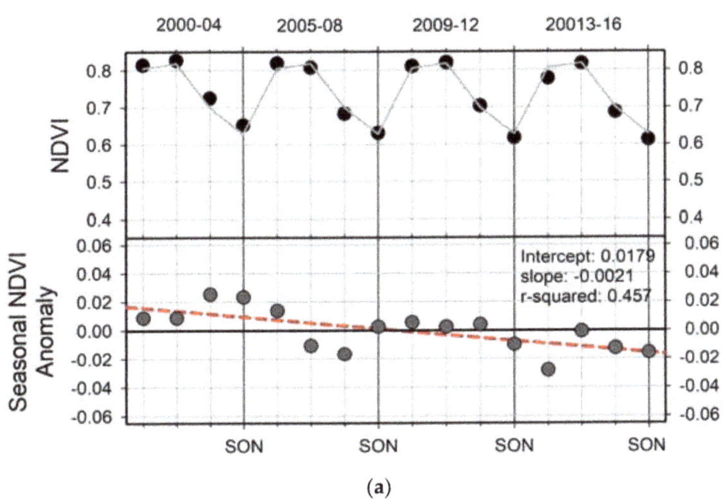

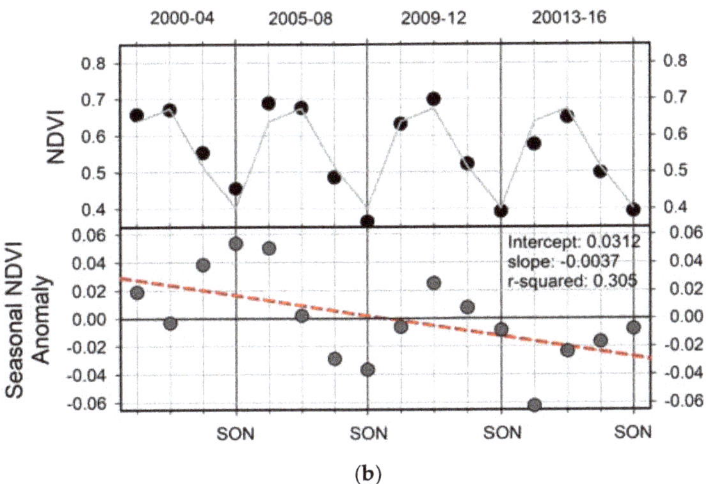

Figure 6. *Cont.*

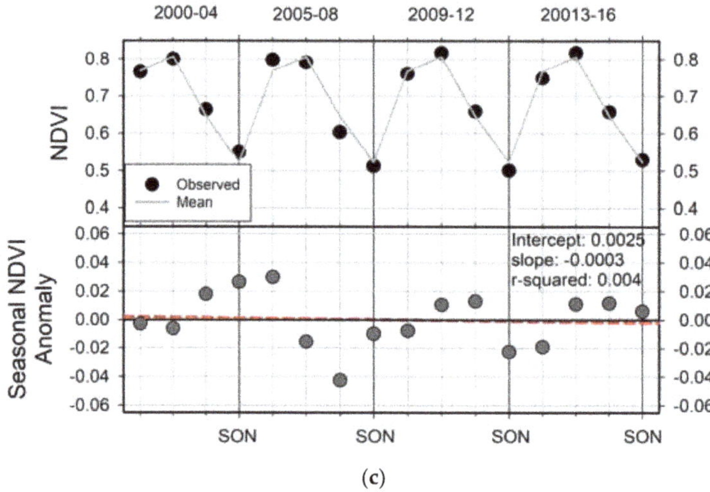

(c)

Figure 6. Time series of mean NDVI seasonally (December–January–February (DJF); March–April–May (MAM); June–July–August (JJA); September–October–November (SON)) from 2000–2016 for (**a**) Mount Gorongosa (significant trend), (**b**) open savanna (significant trend), (**c**) and closed savanna (no significant trend) other non-significant trends (in mixed savanna, LG, and the buffer) are not shown.

In Figure 7a, seasonal precipitation and lagged NDVI are shown. Dry season NDVI is much more sensitive (steeper slope) to small increments in precipitation than the rainy season NDVI. In Figure 7b, there is a significant relationship between precipitation and vegetation in terms of interannual variability (e.g., if there was a wetter than normal wet season versus an unusually high NDVI). This relationship was significant across all seasons of comparison (precipitation to NDVI respectively): FMA to MAM (r-squared = 0.466, slope = 0.000416, and is significant at the 0.01 level); MJJ to JJA (r-squared = 0.146, slope = 0.002981, and is significant at the 0.10 level); ASO to SON (r-squared = 0.570, slope = 0.005185, and is significant at the 0.01 level); and NDJ to DJF (r-squared = 0.247, slope = 0.000441, and is significant at the 0.05 level), thus supporting the notion of savanna as a water limited system and precipitation as the main driver of vegetation growth [5].

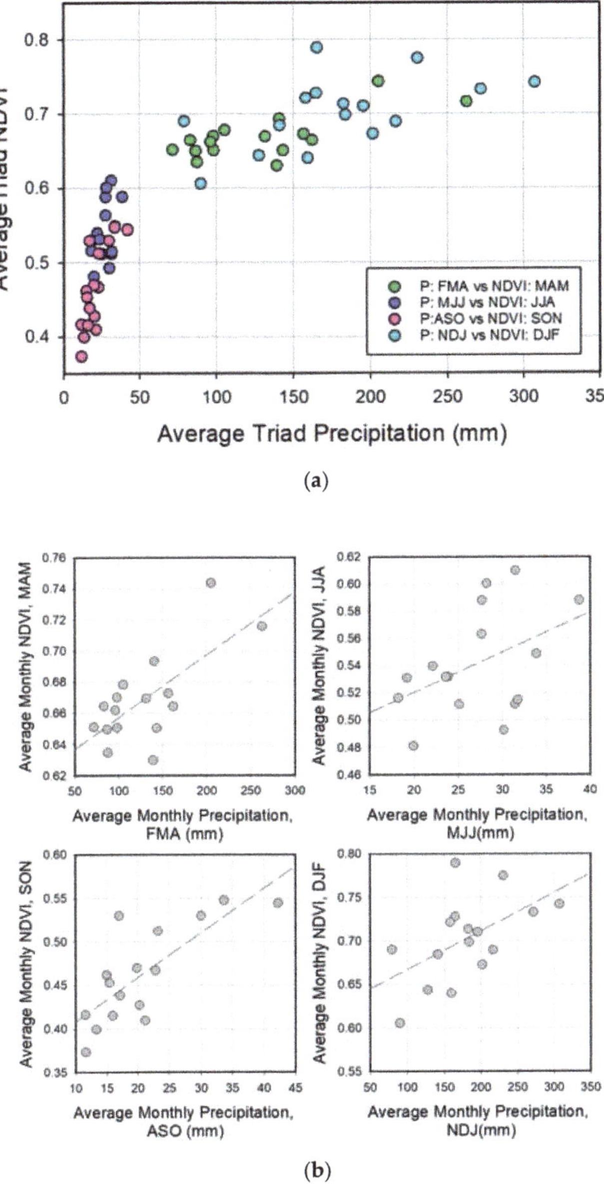

Figure 7. Precipitation and lagged NDVI (respectively) (**a**) plotted by season together with different colors for: February, March, April (FMA) to March, April, May (MAM); May, June, July (MJJ) to June, July, August (JJA); August, September, October (ASO) to September, October, November (SON); and November, December, January (NDJ) to December, January, February (DJF) (**b**) by season individually with fitted regression lines.

3.3. Tendencies in Vegetation Directional Persistence

Maps of DP for these study areas (Figure 8) show declines in vegetation, particularly during the dry season and in central LG (which is the open savanna cover, Figure 1) relative to the initial baseline period. This area corresponds with open savanna (grassland/floodplain) land cover. There is a dominating negative trend in vegetation persistence during the dry season, particularly in the central part of LG and on MG. During the wet season fewer significant values (positive or negative) of DP prevail. Again similar tendencies are found both inside LG and in the buffer area, though the buffer area has greater negative values of DP.

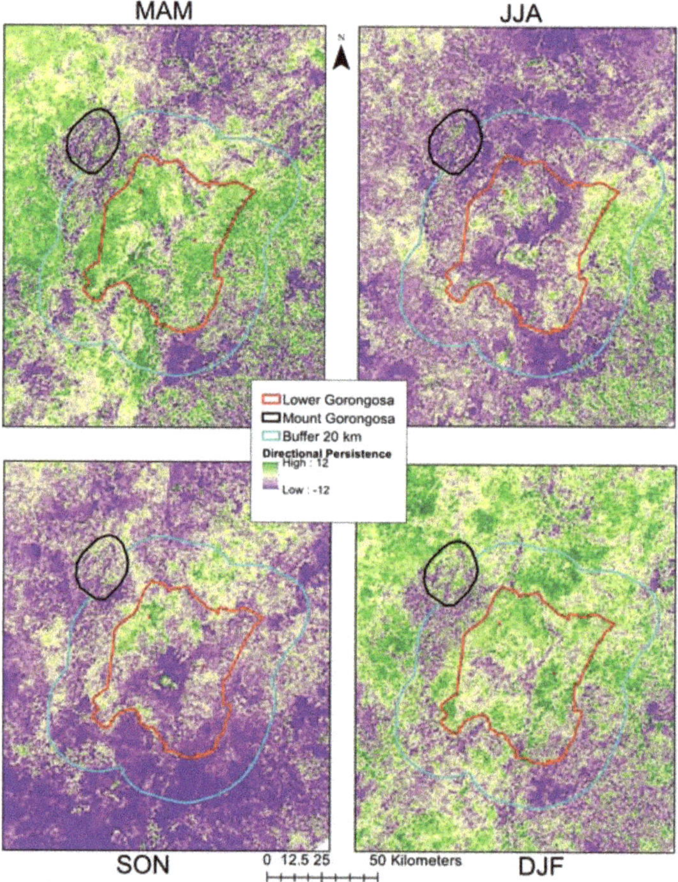

Figure 8. Directional persistence (2005–2016 compared to initial baseline period 2000–2004 at a pixel level) in Lower Gorongosa (including the three savanna types, open, mixed, and closed), Mount Gorongosa, and the buffer zone highlighted by season (MAM = March, April, May, JJA = June, July, August, SON = September, October, November, DJF = December, January, February).

The seasonal properties of mean directional persistence of all six study areas are presented in Figure 9. Across all study areas, the absolute value of mean persistence is larger during the dry season (more negatively deviated from zero), than the wet season (less positively deviated from zero). This suggests the tendency for NDVI in the dry season to be more negative through time than the baseline value to a greater magnitude than in the wet season. When comparing LG, the buffer zone, and MG, LG had the most positive/least negative mean DP values across seasons. This indicates that, of the six study areas, LG had higher percentages of positive vegetation persistence over time. MG with negative mean DP values across all seasons, shows the most marked decline in vegetation persistence over time. Within the three land cover types in LG, closed savanna had the most positive/least negative mean DP values, whereas open savanna had the most negative DP values (Figure 9). OS appears to suffer most in terms of vegetation declines, as found elsewhere in the literature.

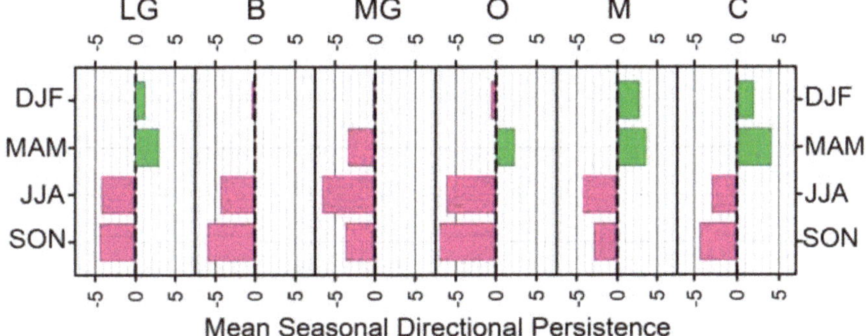

Figure 9. Mean directional persistence values for Lower Gorongosa (LG), the buffer zone (B), Mount Gorongosa (MG), and the three Lower Gorongosa land cover types: open savanna (O), mixed savanna (M), and closed savanna (C) by season (December-January-February (DJF); March-April-May (MAM); June-July-August (JJA); September-October-November (SON)).

Percentage significant positive and negative DP (+/−6) is also calculated for each study area (Figure 10). LG has higher percentages of significant positive DP and lower percentages of significant negative DP than the buffer zone. MG returns higher percentages of significant negative DP and lower percentages of significant positive DP than any other study area. In the dry season, as much as 70% of the MG landscape has a significant negative DP value. Of the three LG land cover types, open savanna exhibits the highest percentages of significant negative DP. In the dry season, as much as 75% of the OS landscape has a significant negative DP value. OS also has the lowest percentages of significant positive DP across all seasons.

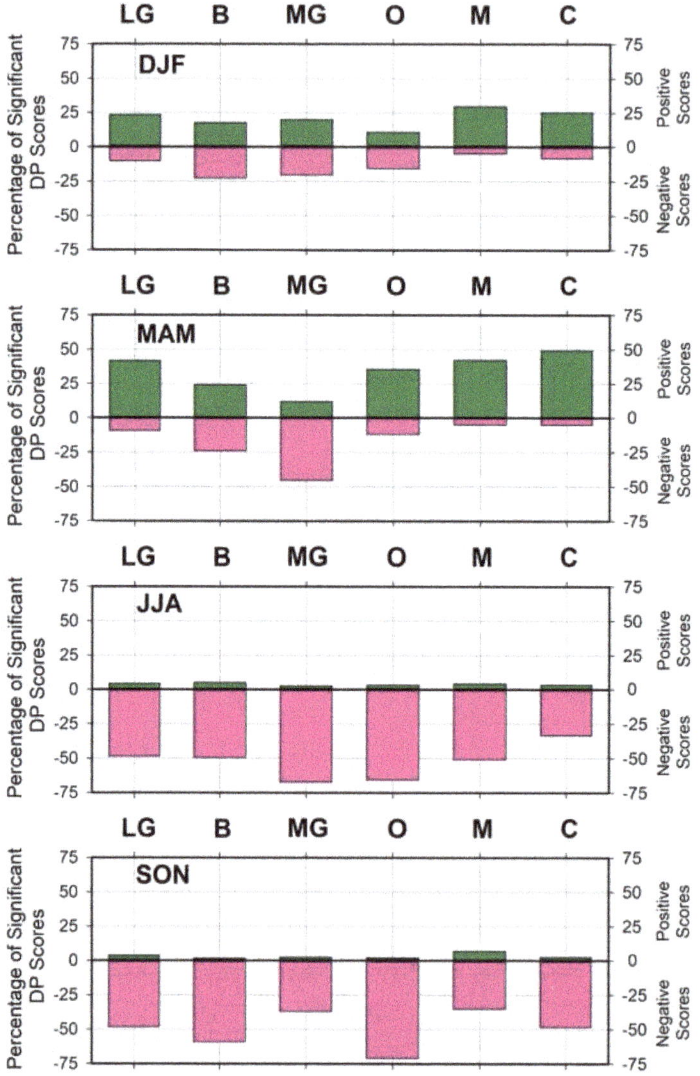

Figure 10. Percentages of significant directional persistence (positive and negative) across the landscapes for: Lower Gorongosa (LG), the buffer zone (B), Mount Gorongosa (MG), and the three Lower Gorongosa land cover types: open savanna (O), mixed savanna (M), and closed savanna (C) by season (December–January–February (DJF); March–April–May (MAM); June–July–August (JJA); September–October–November (SON)).

4. Discussion

This study evaluates the relationship between savanna vegetation and precipitation, the role of protected areas in vegetation conservation, and the importance of distinguishing vegetation type for evaluation of its health/persistence. We found that there was a decline in precipitation in Gorongosa National Park over the 2000 to 2016 period, which could be linked to a decline in NDVI. However, we found that there were differences in decline of NDVI amongst savanna land cover classes within LG, with OS having the steepest decline. Compared to the buffer zone, inside the protected area of

LG had less vegetation decline. MG had the greatest decline of all six study areas in terms of NDVI, as well as the highest negative DP. This may be due to more localized drivers outlined below, but with further conservation efforts, we hope to see more positive trends in the future.

Degradation has been of concern for several decades in savannas [42,43]. Degradation can include several phenomena, including a decrease in biomass of vegetation (seen in this study in and around GNP) and bush encroachment. Bush encroachment is a type of degradation that results in an increase in lower quality short woody plants [12,44,45]. Important drivers of savanna vegetation and therefore potential degradation include, decline in precipitation (already seen here), changes in herbivory (increases to a certain population level may mitigate degradation, but beyond that level, may actually exacerbate it), decreased fire frequency, and humans [4–7,42–47]. Under continued climate change and rewilding, these systems will need to be carefully monitored for further degradation [15,48]. One type of bush encroachment reported in the park, particularly in the vulnerable floodplain grasslands, is that of *Mimosa pigra*. This is due to the previous absence of herbivory (almost no herbivores survived the civil war, allowing encroachers to increase). Now with the heavy reintroduction of herbivores, the park is seeing a decrease in these woody encroachers [84]. However, given the relatively recent rewilding, our longer-term study is seeing the trend of degradation in OS. Though a constant fire regime (the fire regime remains consistent pre- and post-war) has enabled large trees to be able to survive and thrive here [29,32,33]. A common way to mitigate the effects of some drivers on the savanna protected area landscape is to implement buffer zones. Buffer zones are areas surrounding parks that act as a more neutral space for activity than the broader landscape. In buffer zones there may be some human activity, but generally it is limited [49]. Creation of these areas increases the distance between protected zones and potential disturbances, limits edge effects, and thereby increases the conservation value of the protected area [50,51].

Jung et al. (2010) [11] point to a global decline in precipitation leading to reduced soil moisture and evapotranspiration. This study found some decrease in precipitation in this region of southeastern Africa during the 21st century [10,44], linked to a decline in vegetation health [10–12], [52,53]. The observed decline in vegetation health (using NDVI as a proxy) further strengthens the link between vegetation biomass in savannas and its main driver, precipitation [5,7–9,12,20–26]. Declines in precipitation and vegetation health suggest that management has not yet had a significant impact on vegetation [10,53] and is still subordinate to the effects of precipitation.

Global greening has been detected over recent decades by remote sensing [53–60]. The most important factors influencing greening appears to be an increase in carbon dioxide and nitrogen levels, land use, temperature, and weather [52,61–63]. Some limited studies have found an opposite trend, specifically that of forest browning [53,64]. This research matches more with the browning literature, with decreases in vegetation amount over time. Negative directional persistence in the forests of MG is consistent with literature on browning hotspots globally, especially in subequatorial Africa [10]. de Jong et al., (2013) stress the importance of the scale of the study, and, at a localized scale, direct human interventions, such as deforestation in this landscape, may have more effect than climate. Since rainforests are less climatically driven than savannas [9,54], the significant decline in NDVI on MG may reflect its recent incorporation into the national park and on-going struggles with deforestation for conversion to agriculture and timber harvesting [29]. Therefore, this study provides a local exception to the literature on global greening [53–60]. de Jong et al., (2013) report much less greening in more open savanna cover types, and Southworth et al., (2015) [20] note significant negative trends in dryland vegetation of Mozambique from 1982 to 2010.

These results also highlight the importance of understanding specific vegetation types that make up the heterogeneous savannas [44], whose impacts may otherwise have been masked by contrasting trends. The various rates of decline of NDVI across differing vegetation types underline the importance of understanding the vegetation structural groups within LG and MG [31,65–68]. While all vegetation types examined exhibit declining NDVI, particularly in the dry season, the decline is not equal. In LG, open grassland shows the greatest and most significant decline, whereas the declines of woody vegetation cover types (mixed savanna and closed savanna) are not statistically significant. This finding further supports the proposal that open grasslands are the most vulnerable type of savanna because of their strong water-dependence and vulnerability to encroachment [5,8,10,20,41].

A principal purpose of protected areas is to conserve vegetation. Placement of the park on the landscape and proximity to humans can be as important in achieving this goal as biophysical variables, such as climate and soil [34,69–73]. The NDVI time series and directional persistence demonstrate that despite evidence of a strong link between precipitation and vegetation [11], protected areas can be effective in conserving vegetation. Vegetation inside protected areas tends to fare better than in surrounding areas [34]. Management practices of the LG protected area seem to be conserving vegetation better than in the buffer zones, as suggested by the smaller declines in NDVI compared to the surrounding buffer areas. MG is unique in the sense that management practices are not yet fully implemented and some deforestation persists, with concomitant detrimental impacts on NDVI as reported globally [20,24,25,74–84].

More locally, this study will help the park management team identify areas for conservation and those most vulnerable to vegetation decline like the grasslands and MG, both of which are key to tourism—the socioeconomic driver of this region. While precipitation data correspond very well with NDVI, a limitation of this study is that the data are resolved at a coarse spatial scale (0.05°) relative to the study area [36], and that gridded datasets are dependent on the availability of local observations and satellite data, which may not be robust in this system. Another important note concerning DP values is that this is an NDVI comparison to the baseline in number of years above or below the baseline, but this measure does not account for magnitude of difference.

In the context of the broader landscape, this park seems to be maintaining vegetation health, and proving its effectiveness as a protected area in this respect. There were similar, but more negative trends detected in the buffer zone. Going forward, the park management team's work with community projects should also have a positive effect in the buffer zone. Though, socioeconomically speaking, this new model of conservation is already mutually beneficial for conservation [29]. One effort to combat deforestation on MG and stimulate the local economy is the production of shade-grown coffee [29]. This year, the shade-grown coffee being grown on MG is being sold globally. The park is currently expanding this coffee by 100 hectares per year, and by the year 2025, they will have developed 1000 hectares of coffee plantation. This translates to over 5000 hectares of restored forest cover on and around MG [29].

Significant mammal restocking, both herbivore (i.e., antelope) and carnivore (i.e., painted wolves), is also underway in the park, and more is planned in the future. Given the importance of herbivory as a driver in savanna landscapes and possible bush encroachment, the monitoring and measurement of vegetation cover change in this region is essential. The park can serve as an interesting comparison study to many other parks in southern Africa that are already suffering from bush encroachment and degradation due to overstocking [42–46]. The role of GNP, both as a crucial conservation area in this landscape, and as an important area for monitoring and research given its unique historical trajectory, cannot be overstated. Future monitoring and research will be of key importance to the management of African parks.

5. Conclusions

This study finds a declining trend in precipitation in GNP from 2000–2016, though the pre- (2000–2008) and post- (2009–2016) management precipitation regimes are not found to be statistically different. There is a corresponding decrease in NDVI over time across the entire study area. Different vegetation types display varying rates of decline in NDVI with grasslands and rainforest exhibiting the steepest (and significant) decline. Woody savanna vegetation experienced the least decline in NDVI over time.

The directional persistence metric evidences significant declines in vegetation biomass 2005–2016 compared to the 2000–2004 baseline. Negative persistence is strongest on MG, possibly reflecting continuing deforestation issues and the area's recent incorporation under protection. Open grassland savanna in LG also experienced significant negative persistence. LG experiences smaller declines in vegetation biomass and smaller values of directional persistence compared to the buffer zone, suggesting that the park is functioning effectively in conserving vegetation. Overall, there have been declines in vegetation in and around GNP during the 21st century. These declines may be prompted by a decline in precipitation, a shift in the seasonal timing of this precipitation, and direct human action. However, LG fares better than the surrounding area, further strengthening the justification for protected areas. The "laboratory" provided by this particular park and its unique history will be useful not only to scientists, but also to managers of protected areas, particularly savannas, as they can separate out the differing impacts of natural changes (such as precipitation which are outside manager control) and manmade changes (such as linked to deforestation, management rules etc.).

Author Contributions: Conceptualization, H.H., J.S., and B.C.; methodology, J.S. and P.W.; software, R.K.; formal analysis, H.H., R.K., P.W., and D.Y.; data curation, R.K., H.H., and D.Y.; writing—original draft preparation, H.H. and J.S.; writing—review and editing, J.S., P.W., R.K., H.H., D.Y. and B.C. All authors have read and agreed to the published version of the manuscript.

Funding: This research received no external funding.

Acknowledgments: These authors would like to thank the Gorongosa National Park Management/Science Team for their input on vegetation covers in the park and background information.

Conflicts of Interest: The authors declare no conflict of interest.

References

1. Chapin, F.S., III; Matson, P.A.; Vitousek, P. *Principles of Terrestrial Ecosystem Ecology*; Springer: New York, NY, USA, 2011.
2. Hanan, N.; Lehmann, C. Tree–Grass Interactions in Savannas: Paradigms, Contradictions, and Conceptual Models. Available online: https://www.taylorfrancis.com/books/e/9780429132469/chapters/10.1201%2Fb10275-10 (accessed on 23 September 2019).
3. Scholes, R.J.; Walker, B.H. An. *African savanna: Synthesis of the Nylsvley Study*; Cambridge University Press: Cambridge, UK, 1993.
4. Andela, N.; Liu, Y.Y.; Van Dijk, A.I.J.M.; De Jeu, R.A.M.; McVicar, T.R. Global changes in dryland vegetation dynamics (1988–2008) assessed by satellite remote sensing: Comparing a new passive microwave vegetation density record with reflective greenness data. *Biogeosciences* **2013**, *10*, 6657–6676. [CrossRef]
5. Campo-Bescos, M.A.; Munoz-Carpena, R.; Kaplan, D.A.; Southworth, J.; Zhu, L.; Waylen, P.R. Beyond Precipitation: Physiographic Gradients Dictate the Relative Importance of Environmental Drivers on Savanna Vegetation. *PLoS ONE* **2013**, *8*, e72348. [CrossRef]
6. Lehmann, C.E.; Anderson, T.M.; Sankaran, M.; Higgins, S.I.; Archibald, S.; Hoffmann, W.A.; Hanan, N.P.; Williams, R.J.; Fensham, R.J.; Felfili, J.; et al. Savanna Vegetation-Fire-Climate Relationships Differ Among Continents. *Science* **2014**, *343*, 548–552. [CrossRef] [PubMed]
7. Sankaran, M.; Hanan, N.P.; Scholes, R.J.; Ratnam, J.; Augustine, D.J.; Cade, B.S.; Gignoux, J.; Higgins, S.I.; Le Roux, X.; Ludwig, F.; et al. Determinants of woody cover in African savannas. *Nature* **2005**, *438*, 846–849. [CrossRef] [PubMed]
8. Scholes, R.J.; Archer, S.R. Tree-Grass Interactions in Savannas. *Annu. Rev. Ecol. Syst.* **1997**, *28*, 517–544. [CrossRef]
9. Staver, A.C.; Archibald, S.; Levin, S. Tree cover in sub-Saharan Africa: Rainfall and fire constrain forest and savanna as alternative stable states. *Ecology* **2011**, *92*, 1063–1072. [CrossRef]
10. De Jong, R.; Verbesselt, J.; Zeileis, A.; Schaepman, M.E. Shifts in Global Vegetation Activity Trends. *Remote Sens.* **2013**, *5*, 1117–1133. [CrossRef]
11. Jung, M.; Reichstein, M.; Ciais, P.; Seneviratne, S.I.; Sheffield, J.; Goulden, M.L.; Bonan, G.; Cescatti, A.; Chen, J.; De Jeu, R.; et al. Recent decline in the global land evapotranspiration trend due to limited moisture supply. *Nature* **2010**, *467*, 951–954. [CrossRef]

12. Bunting, E.L.; Fullman, T.; Kiker, G.; Southworth, J. Utilization of the SAVANNA model to analyze future patterns of vegetation cover in Kruger National Park under changing climate. *Ecol. Model.* **2016**, *342*, 147–160. [CrossRef]

13. Houghton, E. *Climate Change 1995: The Science of Climate Change: Contribution of Working Group I to the Second Assessment Report of the Intergovernmental Panel on Climate Change*; Cambridge University Press: Cambridge, UK, 1996.

14. Houghton, R.A.; Hackler, J.L.; Lawrence, K.T. The U.S. Carbon Budget: Contributions from Land-Use Change. *Science* **1999**, *285*, 574–578. [CrossRef]

15. Monserud, R.A.; Tchebakova, N.M.; Leemans, R. Global vegetation change predicted by the modified Budyko model. *Clim. Chang.* **1993**, *25*, 59–83. [CrossRef]

16. Ojima, D.S.; Valentine, D.W.; Mosier, A.R.; Parton, W.J.; Schimel, D.S. Effect of land use change on methane oxidation in temperate forest and grassland soils. *Chemosphere* **1993**, *26*, 675–685. [CrossRef]

17. Scholes, R.J.; Hall, D.O. The carbon budget of tropical savannas, woodlands, and grasslands. *SCOPE-Sci. Comm. Probl. Environ. Int. Counc. Sci. Unions* **1996**, *56*, 69–100.

18. Child, B. Zimbabwe's CAMPFIRE programme: Using the high value of wildlife recreation to revolutionize natural resource management in communal areas. *Commonw. For. Rev.* **1993**, *72*, 284–296.

19. Waylen, P.; Southworth, J.; Gibbes, C.; Tsai, H. Time Series Analysis of Land Cover Change: Developing Statistical Tools to Determine Significance of Land Cover Changes in Persistence Analyses. *Remote Sens.* **2014**, *6*, 4473–4497. [CrossRef]

20. Southworth, J.; Zhu, L.; Bunting, E.; Ryan, S.J.; Herrero, H.; Waylen, P.R.; Hill, M.J. Changes in vegetation persistence across global savanna landscapes, 1982–2010. *J. Land Use Sci.* **2016**, *11*, 7–32. [CrossRef]

21. Carlson, T.N.; Ripley, D.A. On the relation between NDVI, fractional vegetation cover, and leaf area index. *Remote Sens. Environ.* **1997**, *62*, 241–252. [CrossRef]

22. Bégué, A.; Vintrou, E.; Ruelland, D.; Claden, M.; Dessay, N. Can a 25-year trend in Soudano-Sahelian vegetation dynamics be interpreted in terms of land use change? A remote sensing approach. *Glob. Environ. Chang.* **2011**, *21*, 413–420. [CrossRef]

23. Jiang, Z.; Huete, A.R.; Chen, J.; Chen, Y.; Li, J.; Yan, G.; Zhang, X. Analysis of NDVI and scaled difference vegetation index retrievals of vegetation fraction. *Remote Sens. Environ.* **2006**, *101*, 366–378. [CrossRef]

24. Lambin, E.F.; Ehrlich, D. Land-cover changes in sub-saharan Africa (1982–1991): Application of a change index based on remotely sensed surface temperature and vegetation indices at a continental scale. *Remote Sens. Environ.* **1997**, *61*, 181–200. [CrossRef]

25. Tucker, C.J. Red and photographic infrared linear combinations for monitoring vegetation. *Remote Sens. Environ.* **1979**, *8*, 127–150. [CrossRef]

26. Wang, J.; Rich, P.M.; Price, K.P.; Kettle, W.D. Relations between NDVI, Grassland Production, and Crop Yield in the Central Great Plains. *Geocarto Int.* **2005**, *20*, 5–11. [CrossRef]

27. Müller, T.; Mapaura, A.; Wursten, B.; Chapano, C.; Ballings, P.; Wild, R. Vegetation Survey of Mount Gorongosa. Available online: http://dev.gorongosa.forumone.com/sites/default/files/research/041-bfa_no.23_gorongosa_vegetation_survey.pdf (accessed on 30 January 2020).

28. Rahbek, C.; Borregaard, M.K.; Colwell, R.K.; Dalsgaard, B.; Holt, B.G.; Morueta-Holme, N.; Nogues-Bravo, D.; Whittaker, R.J.; Fjeldså, J. Humboldt's enigma: What causes global patterns of mountain biodiversity? *Science* **2019**, *365*, 1108–1113. [CrossRef] [PubMed]

29. Gorongosa National Park Website. Available online: www.gorongosa.org (accessed on 30 January 2020).

30. Adams, P. Opinion | A Comeback for African National Parks. Available online: https://www.nytimes.com/2019/02/20/opinion/africa-national-parks.html (accessed on 30 January 2020).

31. Tinley, K.L. Framework of the Gorongosa Ecosystem. Available online: https://www.gorongosa.org/our-story/science/reports/framework-gorongosa-ecosystem (accessed on 30 January 2020).

32. Daskin, J.H.; Stalmans, M.; Pringle, R.M. Ecological legacies of civil war: 35-year increase in savanna tree cover following wholesale large-mammal declines. *J. Anim. Ecol.* **2016**, *104*, 79–89. [CrossRef]

33. Herrero, H.V.; Southworth, J.; Bunting, E.; Child, B. Using Repeat Photography to Observe Vegetation Change Over Time in Gorongosa National Park. Available online: https://geog.ufl.edu/2017/06/26/using-repeat-photography-to-observe-vegetation-change-over-time-in-gorongosa-national-park/ (accessed on 30 January 2020).

34. DeFries, R.; Hansen, A.; Turner, B.L.; Reid, R.; Liu, J. Land Use Change Around Protected Areas: Management to Balance Human Needs and Ecological Function. *Ecol. Appl.* **2007**, *17*, 1031–1038. [CrossRef] [PubMed]

35. Measuring Vegetation (NDVI & EVI). Available online: https://earthobservatory.nasa.gov/features/MeasuringVegetation (accessed on 24 September 2019).

36. Funk, C.; Peterson, P.; Landsfeld, M.; Pedreros, D.; Verdin, J.; Shukla, S.; Husak, G.; Rowland, J.; Harrison, L.; Hoell, A.; et al. The climate hazards infrared precipitation with stations—a new environmental record for monitoring extremes. *Sci. Data* **2015**, *2*, 1–21. [CrossRef] [PubMed]

37. Nicholson, S.E. The nature of rainfall variability over Africa on time scales of decades to millenia. *Glob. Planet. Chang.* **2000**, *26*, 137–158. [CrossRef]

38. Dezfuli, A.K.; Zaitchik, B.F.; Gnanadesikan, A. Regional Atmospheric Circulation and Rainfall Variability in South Equatorial Africa. *J. Clim.* **2014**, *28*, 809–818. [CrossRef]

39. Zhu, L.; Southworth, J. Disentangling the Relationships between Net Primary Production and Precipitation in Southern Africa Savannas Using Satellite Observations from 1982 to 2010. *Remote Sens.* **2013**, *5*, 3803–3825. [CrossRef]

40. 3.2 Hypergeometric Distribution, 3.5, 3.9 Mean and Variance. Available online: http://www.math.ucsd.edu/~{}gptesler/186/slides/186_hypergeom_17-handout.pdf (accessed on 30 January 2020).

41. Jeltsch, F.; Weber, G.E.; Grimm, V. Ecological buffering mechanisms in savannas: A unifying theory of long-term tree-grass coexistence. *Plant. Ecol.* **2000**, *150*, 161–171. [CrossRef]

42. Campbell, A.; Child, G. The Impact of Man on the Environment of Botswana. *Botsw. Notes Rec.* **1971**, *3*, 91–110.

43. Child, G. *An Ecological Survey of Northeastern Botswana*; Food and Agriculture Organization of the United Nations: Rome, Italy, 1968.

44. Herrero, H.V.; Southworth, J.; Bunting, E. Utilizing Multiple Lines of Evidence to Determine Landscape Degradation within Protected Area Landscapes: A Case Study of Chobe National Park, Botswana from 1982 to 2011. *Remote Sens.* **2016**, *8*, 623. [CrossRef]

45. Moleele, N.M.; Ringrose, S.; Matheson, W.; Vanderpost, C. More woody plants? the status of bush encroachment in Botswana's grazing areas. *J. Environ. Manag.* **2002**, *64*, 3–11. [CrossRef] [PubMed]

46. Ringrose, S.; Matheson, W.; Wolski, P.; Huntsman-Mapila, P. Vegetation cover trends along the Botswana Kalahari transect. *J. Arid Environ.* **2003**, *54*, 297–317. [CrossRef]

47. Van Wilgen, B.W.; Govender, N.; Biggs, H.C.; Ntsala, D.; Funda, X.N. Response of Savanna Fire Regimes to Changing Fire-Management Policies in a Large African National Park. *Conserv. Biol.* **2004**, *18*, 1533–1540. [CrossRef]

48. Vitousek, P.M.; Mooney, H.A.; Lubchenco, J.; Melillo, J.M. Human Domination of Earth's Ecosystems. *Science* **1997**, *277*, 494–499. [CrossRef]

49. Wells, M.; Bradon, K. *People and Parks: Linking Protected Area Management with Local Communities*; World Bank: Washington, DC, USA, 1992.

50. Sayer, J. *Rainforest Buffer Zones: Guidelines for Protected Area Managers*; International Union for Conservation of Nature and Natural Resources: Gland, Switzerland, 1991.

51. Martino, D. Buffer Zones Around Protected Areas: A Brief Literature Review. Available online: https://escholarship.org/content/qt02n4v17n/qt02n4v17n.pdf (accessed on 30 January 2020).

52. De Jong, R.; de Bruin, S.; de Wit, A.; Schaepman, M.E.; Dent, D.L. Analysis of monotonic greening and browning trends from global NDVI time-series. *Remote Sens. Environ.* **2011**, *115*, 692–702. [CrossRef]

53. De Jong, R.; Verbesselt, J.; Schaepman, M.E.; De Bruin, S. Trend changes in global greening and browning: Contribution of short-term trends to longer-term change. *Glob. Chang. Biol.* **2012**, *18*, 642–655. [CrossRef]

54. Mitchard, E.T.; Saatchi, S.S.; Gerard, F.F.; Lewis, S.L.; Meir, P. Measuring Woody Encroachment along a Forest–Savanna Boundary in Central Africa. *Earth Interact.* **2009**, *13*, 1–29. [CrossRef]

55. Piao, S.; Wang, X.; Ciais, P.; Zhu, B.; Wang, T.A.O.; Liu, J.I.E. Changes in satellite-derived vegetation growth trend in temperate and boreal Eurasia from 1982 to 2006. *Glob. Chang. Biol.* **2011**, *17*, 3228–3239. [CrossRef]

56. Guay, K.C.; Beck, P.S.; Berner, L.T.; Goetz, S.J.; Baccini, A.; Buermann, W. Vegetation productivity patterns at high northern latitudes: A multi-sensor satellite data assessment. *Glob. Chang. Biol.* **2014**, *20*, 3147–3158. [CrossRef] [PubMed]

57. Brandt, M.; Mbow, C.; Diouf, A.A.; Verger, A.; Samimi, C.; Fensholt, R. Ground- and satellite-based evidence of the biophysical mechanisms behind the greening Sahel. *Glob. Chang. Biol.* **2015**, *21*, 1610–1620. [CrossRef] [PubMed]

58. Schimel, D.; Pavlick, R.; Fisher, J.B.; Asner, G.P.; Saatchi, S.; Townsend, P.; Miller, C.; Frankenberg, C.; Hibbard, K.; Cox, P. Observing terrestrial ecosystems and the carbon cycle from space. *Glob. Chang. Biol.* **2015**, *21*, 1762–1776. [CrossRef] [PubMed]

59. Garonna, I.; de Jong, R.; Schaepman, M.E. Variability and evolution of global land surface phenology over the past three decades (1982–2012). *Glob. Chang. Biol.* **2016**, *22*, 1456–1468. [CrossRef] [PubMed]

60. Bastin, J.F.; Berrahmouni, N.; Grainger, A.; Maniatis, D.; Mollicone, D.; Moore, R.; Patriarca, C.; Picard, N.; Sparrow, B.; Abraham, E.M.; et al. The extent of forest in dryland biomes. *Science* **2017**, *356*, 635–638. [CrossRef] [PubMed]

61. Olsson, L.; Eklundh, L.; Ardö, J. A recent greening of the Sahel—trends, patterns and potential causes. *J. Arid Environ.* **2005**, *63*, 556–566. [CrossRef]

62. Zhang, K.; Kimball, J.S.; Nemani, R.R.; Running, S.W.; Hong, Y.; Gourley, J.J.; Yu, Z. Vegetation Greening and Climate Change Promote Multidecadal Rises of Global Land Evapotranspiration. *Sci. Rep.* **2015**, *5*, 15956. [CrossRef]

63. Zhu, Z.; Piao, S.; Myneni, R.B.; Huang, M.; Zeng, Z.; Canadell, J.G.; Ciais, P.; Sitch, S.; Friedlingstein, P.; Arneth, A.; et al. Greening of the Earth and its drivers. *Nat. Clim. Chang.* **2016**, *6*, 791–795. [CrossRef]

64. Alcaraz-Segura, D.O.; Chuvieco, E.; Epstein, H.E.; Kasischke, E.S.; Trishchenko, A. Debating the greening vs. browning of the North American boreal forest: Differences between satellite datasets. *Glob. Chang. Biol.* **2010**, *16*, 760–770. [CrossRef]

65. Cunliffe, R.; Lynam, T. Preliminary Vegetation Classification and Mapping of Gorongosa National Park, Sofala Province, Mozambique. Available online: https://www.gorongosa.org/sites/default/files/research/018-cunliffe_lynam_veg._classification_and_mapping_1-05.pdf (accessed on 30 January 2020).

66. Stalmans, M. Tinley's Plant Species List for the Greater Gorongosa Ecosystem. Available online: http://dev.gorongosa.forumone.com/sites/default/files/research/052-stalmans_tinley_plant_species_list_gorongosa.pdf (accessed on 30 January 2020).

67. Stalmans, M.; Beilfuss, R. Long-Term Plan for Gorongosa National Park Vegetation Monitoring at Multiple Scales. Available online: https://www.researchgate.net/profile/Richard_Beilfuss/publication/313728082_Long-term_plan_for_Gorongosa_National_Park_vegetation_monitoring_at_multiple_scales/links/58a43c59a6fdcc0e0751a8a0/Long-term-plan-for-Gorongosa-National-Park-vegetation-monitoring-at-multiple-scales.pdf (accessed on 30 January 2020).

68. Stalmans, M.; Beilfuss, R. Landscapes of the Gorongosa National Park. Available online: https://www.researchgate.net/publication/314878798_Landscapes_of_the_Gorongosa_National_Park (accessed on 30 January 2020).

69. Barnes, M.D.; Glew, L.; Wyborn, C.; Craigie, I.D. Prevent perverse outcomes from global protected area policy. *Nat. Ecol. Evol.* **2018**, *2*, 759–762. [CrossRef]

70. Fuller, R.A.; McDonald-Madden, E.; Wilson, K.A.; Carwardine, J.; Grantham, H.S.; Watson, J.E.; Klein, C.J.; Green, D.C.; Possingham, H.P. Replacing underperforming protected areas achieves better conservation outcomes. *Nature* **2010**, *466*, 365–367. [CrossRef]

71. Newmark, W.D. Insularization of Tanzanian Parks and the Local Extinction of Large Mammals. *Conserv. Biol.* **1996**, *10*, 1549–1556. [CrossRef]

72. Rodrigues, A.S.; Andelman, S.J.; Bakarr, M.I.; Boitani, L.; Brooks, T.M.; Cowling, R.M.; Fishpool, L.D.; Da Fonseca, G.A.; Gaston, K.J.; Hoffmann, M.; et al. Effectiveness of the global protected area network in representing species diversity. *Nature* **2004**, *428*, 640–643. [CrossRef]

73. Saura, S.; Bertzky, B.; Bastin, L.; Battistella, L.; Mandrici, A.; Dubois, G. Protected area connectivity: Shortfalls in global targets and country-level priorities. *Biol. Conserv.* **2018**, *219*, 53–67. [CrossRef] [PubMed]

74. De Almeida, A.S.; Stone, T.A.; Vieira, I.C.G.; Davidson, E.A. Nonfrontier Deforestation in the Eastern Amazon. *Earth Interact.* **2010**, *14*, 1–15. [CrossRef]

75. Ingram, J.C.; Dawson, T.P. Technical Note: Inter-annual analysis of deforestation hotspots in Madagascar from high temporal resolution satellite observations. *Int. J. Remote Sens.* **2005**, *26*, 1447–1461. [CrossRef]

76. Jin, Y.; Sung, S.; Lee, D.K.; Biging, G.S.; Jeong, S. Mapping Deforestation in North Korea Using Phenology-Based Multi-Index and Random Forest. *Remote Sens.* **2016**, *8*, 997. [CrossRef]

77. Kumar, P.; Rani, M.; Pandey, P.C.; Majumdar, A.; Nathawat, M.S. Monitoring of Deforestation and Forest Degradation Using Remote Sensing and GIS: A Case Study of Ranchi in Jharkhand (India). Available online: https://s3.amazonaws.com/academia.edu.documents/27495816/03_2578_report0204_14_20.pdf?response-content-disposition=inline%3B%20filename%3DMonitoring_of_deforestation_and_forest_d.pdf&X-Amz-Algorithm=AWS4-HMAC-SHA256&X-Amz-Credential=AKIAIWOWYYGZ2Y53UL3A%2F20200130%2Fus-east-1%2Fs3%2Faws4_request&X-Amz-Date=20200130T055501Z&X-Amz-Expires=3600&X-Amz-SignedHeaders=host&X-Amz-Signature=77313dbc5b0b67d184a938083b116a87b73bdd2929f6bcdf96b205283ea987b6 (accessed on 30 January 2020).

78. Lunetta, R.S.; Knight, J.F.; Ediriwickrema, J.; Lyon, J.G.; Worthy, L.D. Land-cover change detection using multi-temporal MODIS NDVI data. *Remote Sens. Environ.* **2006**, *105*, 142–154. [CrossRef]

79. Manoharan, V.S.; Welch, R.M.; Lawton, R.O. Impact of deforestation on regional surface temperatures and moisture in the Maya lowlands of Guatemala. *Geophys. Res. Lett.* **2009**, *36*. [CrossRef]

80. Senay, G.B.; Elliott, R.L. Combining AVHRR-NDVI and landuse data to describe temporal and spatial dynamics of vegetation. *For. Ecol. Manag.* **2000**, *128*, 83–91. [CrossRef]

81. Sobrino, J.A.; Julien, Y. Global trends in NDVI-derived parameters obtained from GIMMS data. *Int. J. Remote Sens.* **2011**, *32*, 4267–4279. [CrossRef]

82. Van der Werf, G.R.; Morton, D.C.; DeFries, R.S.; Giglio, L.; Randerson, J.T.; Collatz, G.J.; Kasibhatla, P.S. Estimates of fire emissions from an active deforestation region in the southern Amazon based on satellite data and biogeochemical modelling. *Biogeosciences* **2009**, *6*, 235–249. [CrossRef]

83. Walsh, S.J.; Crawford, T.W.; Welsh, W.F.; Crews-Meyer, K.A. A multiscale analysis of LULC and NDVI variation in Nang Rong district, northeast Thailand. *Agric. Ecosyst. Environ.* **2001**, *85*, 47–64. [CrossRef]

84. Guyton, J.A.; Pansu, J.; Hutchinson, M.C.; Kartzinel, T.R.; Potter, A.B.; Coverdale, T.C.; Daskin, J.H.; da Conceição, A.G.; Peel, M.J.; Stalmans, M.E.; et al. Trophic rewilding revives biotic resistance to shrub invasion. *Nat. Ecol. Evol.* **2020**, 1–13. [CrossRef] [PubMed]

 *remote sensing*

Article

Functional Phenology of a Texas Post Oak Savanna from a CHRIS PROBA Time Series

Michael J. Hill [1,2,*], Andrew Millington [2], Rebecca Lemons [1] and Cherie New [1]

[1] Department of Earth System Science and Policy, University of North Dakota, Grand Forks, ND 58202, USA; rebecca.lemons@und.edu (R.L.); cherie.new@und.edu (C.N.)
[2] College of Science and Engineering, Flinders University, Sturt Road, Bedford Park, South Australia 5042, Australia; andrew.millington@flinders.edu.au
* Correspondence: michael.hill4@und.edu; Tel.: +61-413161853

Received: 19 September 2019; Accepted: 13 October 2019; Published: 15 October 2019

Abstract: Remnant midwestern oak savannas in the USA have been altered by fire suppression and the encroachment of woody evergreen trees and shrubs. The Gus Engeling Wildlife Management Area (GEWMA) near Palestine, Texas represents a relatively intact southern example of thickening and evergreen encroachment in oak savannas. In this study, 18 images from the CHRIS/PROBA (Compact High-Resolution Imaging Spectrometer/Project for On-Board Autonomy) sensor were acquired between June 2009 and October 2010 and used to explore variation in canopy dynamics among deciduous and evergreen trees and shrubs, and savanna grassland in seasonal leaf-on and leaf-off conditions. Nadir CHRIS images from the 11 useable dates were processed to surface reflectance and a selection of vegetation indices (VIs) sensitive to pigments, photosynthetic efficiency, and canopy water content were calculated. An analysis of temporal VI phenology was undertaken using a fishnet polygon at 90 m resolution incorporating tree densities from a classified aerial photo and soil type polygons. The results showed that the major differences in spectral phenology were associated with deciduous tree density, the density of evergreen trees and shrubs—especially during deciduous leaf-off periods—broad vegetation types, and soil type interactions with elevation. The VIs were sensitive to high densities of evergreens during the leaf-off period and indicative of a photosynthetic advantage over deciduous trees. The largest differences in VI profiles were associated with high and low tree density, and soil types with the lowest and highest available soil water. The study showed how time series of hyperspectral data could be used to monitor the relative abundance and vigor of desirable and less desirable species in conservation lands.

Keywords: savanna; post oak; vegetation index; ecosystem function; phenology; encroachment; evergreen; deciduous

---

1. Introduction

Midwestern oak savannas originally occupied the transition zone between the true prairie grasslands and the eastern deciduous forests in North America. They stretched from eastern Texas in the south to northern Minnesota and occupied as much as 20 M ha in a continuous band [1]. Despite their association with rolling landforms on light sandy soils, most of these savannas were cleared for grazing and arable agriculture in the 19th century, leaving only about 12,000 ha of highly fragmented remnants [2]. Much of the remaining southern remnants are post oak woodlands and savanna grasslands in varying states of intactness, conservation status, or extirpation. Post oak savanna and savanna grasslands occupy approximately 2.7 M ha of eastern Texas [3]. Other studies place the total area in Texas at around 4 M ha [4,5]. In eastern Texas, Post Oak Motte and Woodland occupies about 38% and Savanna Grassland occupies about 58% of this vegetation type, with other minor types making up the remainder [3].

The Gus Engeling Wildlife Management Area (GEWMA) represents a relatively intact example of post oak savanna and savanna grassland and includes hardwood bottomland forest and riparian areas [5]. As a result of fire suppression, the upland oak savanna dominated by post oak (*Quercus stellata*), red oak (*Q. falcata*), blackjack oak (*Q. marilandica*), and bluejack oak (*Q. incana*) has thickened to exclude understory grasses such as little bluestem (*Shizachyrium scoparium*) in many areas, and open areas have suffered from woody encroachment by native evergreen species such as eastern red cedar (*Juniperus virginiana*; [6]) and yaupon (*Ilex vomitoria*). The GEWMA is under an active program of land cover management aimed at the regeneration of open savannas by clearing sections of oak and replanting little bluestem grassland.

For the past 15 years, the spaceborne hyperspectral remote sensing of vegetation has depended upon two experimental sensors, EO-1 Hyperion (Earth Observer 1 Hyperion) and CHRIS/PROBA (Compact High-Resolution Imaging Spectrometer/Project for On-board Autonomy) for data at a spatial resolution (30–34 m pixels) for field and landscape-scale studies [7–11]. Although the CHRIS sensor only covers the 400–1000 nm wavelength range, it also provides nadir images and two forward and backward-looking off-nadir angles. It has been applied to a wide range of studies on classification [12], vegetation properties [13,14], vegetation mapping [15,16], water [17–19], canopy function [20], canopy radiative transfer [21,22], and vegetation indices [23].

Time series of hyperspectral data have proven valuable in vegetation classification and the detection of changes in functional processes. However, dense time series over a single site are relatively rare. The GEWMA site, with its thickening of post oak woodland and woody encroachment by evergreens, provides an opportunity to use CHRIS data to explore the extent to which suites of functionally based vegetation indices (VI) can distinguish differences in vegetation canopy properties and behaviors over a full seasonal growing cycle. A very large number of VIs have been constructed to target various aspects of photosynthetic function, canopy structure, fractional cover of photosynthetic and non-photosynthetic components, and canopy water content [24]. In general, narrow-band VIs acquired from visible and near infrared wavelengths can provide quantitative information about photosynthetic pigments such as chlorophyll, carotenoids and anthocyanins ([25–27], photosynthetic radiation use efficiency (RUE; [28,29]), canopy senescence using the chlorophyll to carotenoid ratio [30], and subtle differences in photosynthetic capacity based on many variants and more complex surrogates of the Normalized Difference Vegetation Index (NDVI; see [24]). Imagery from sensors such as Hyperion with short-wave infrared bands can be used to derive VIs with sensitivity to canopy water content and dynamics [31] and the fractional cover of photosynthetic and non-photosynthetic canopies and bare soil [32].

Previous work has explored the application of selected VIs derived from EO-1 Hyperion imagery to the identification of vegetation states in tropical savannas [33] and North American savannas and grasslands [34]. However, the image acquisition frequency in these studies was insufficient for the characterization of phenology. In this study, a sequence of 17 CHRIS images was acquired between June 2009 and October 2010, providing an opportunity to explore variation in canopy dynamics among deciduous and evergreen trees and shrubs and savanna grassland in seasonal leaf-on and leaf-off conditions for a valuable remnant of the threatened midwestern oak savannas. The objectives of the study were as follows:

(1) To characterize the seasonal phenology of deciduous post oak savanna and encroaching evergreen shrubs with functional VIs derived from CHRIS PROBA imagery;
(2) To explore the sensitivity of VIs for distinguishing differences in photosynthetic capacity and water status between deciduous savanna vegetation and evergreen shrubs;
(3) To illustrate the utility and sensitivity of combinations of VIs for describing differences in vegetation function across soil-vegetation associations in a post oak savanna.

2. Materials and Methods

2.1. Study Area

The Gus Engeling Wildlife Management Area (GEWMA) is located about 29 km northwest of Palestine, near Bethel in Anderson County, Texas (Figure 1a). The GEWMA was acquired by the State of Texas between 1950 and 1960, and comprises 4434 ha of prairie, post oak savanna, hardwood bottomlands, and riparian areas including pitcher plant bogs ([3,5,35]. The area is regarded as being less disturbed than much of Texas, since soils are too poor for arable agriculture. However, it was grazed as open range by cattle and hogs from the mid-1800s, becoming severely overgrazed prior to acquisition by the State [5]. The landscape has a rolling terrain with elevation ranging from 75 to 150 m and vegetation cover dominated by a mix of deciduous and evergreen trees and shrubs (Figure 1a,b). The climate is characterized by hot summers and mild winters, with average monthly maximum temperatures approaching 35 °C in August and average monthly minima approaching 0 °C in January (Figure 2). Average monthly rainfall reaches a peak in May and October; however, rainfall can be highly variable. Very heavy rainfall in May 2010 resulted in the flooding of the lower parts of GEWMA and the surrounding villages and towns (Figure 2).

The soils range from loamy sands and sandy loams on the terraces and lowlands to fine sands on the upland post, black and bluejack oak woodlands (Table 1). The Ecological Site Description (ESD) provided with soil maps by the Natural Resources Conservation Service (NRCS) of the United States Department of Agriculture (USDA) provides descriptions of original reference vegetation and state and transition frameworks for vegetation succession in response to disturbance, fire exclusion and management of herbivory (Table 1; Soil Survey Staff, NRCS, 3 October 2017). However, the detailed vegetation mapping in [5] and in Phase 2 of the Texas Vegetation Classification Project [3] (Figure 3b; Table 2) provides a better indication of the vegetation on GEWMA at the time of this study.

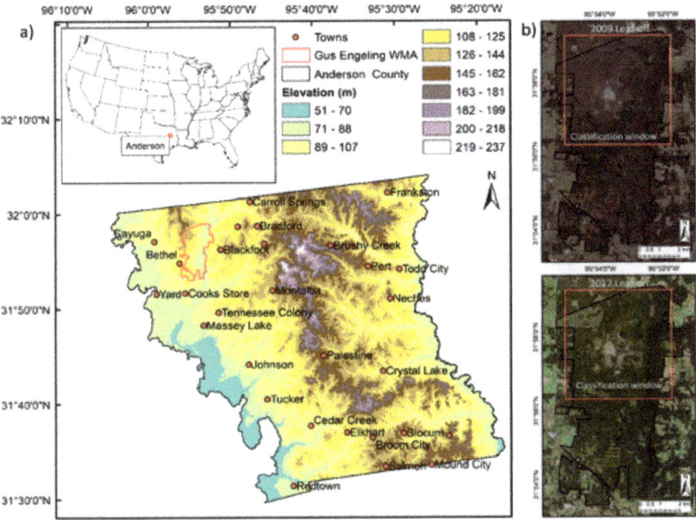

Figure 1. (**a**) Location of Gus Engeling Wildlife Management Area (GEWMA) in Anderson County, near Bethel, Texas showing the elevation of the county landscape. (https://tpwd.texas.gov/huntwild/hunt/wma). (**b**) National Agricultural Imagery Program (NAIP) images from 2009 and 2012 show the canopy cover in leaf-off and leaf-on states, the classification window (red box) used later in the study, and the structured clearing of thickened woodland to re-establish savanna in the north-west of the area (2012 image).

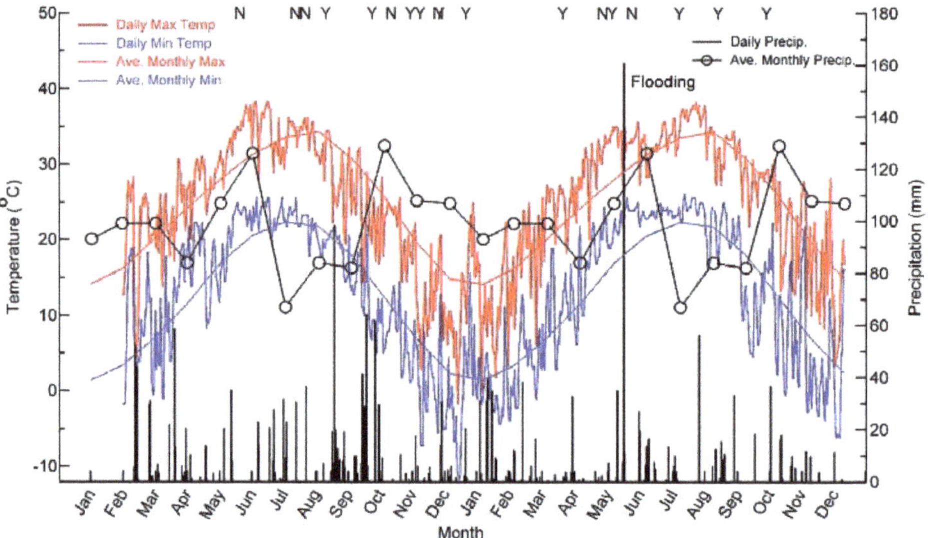

Figure 2. Average monthly and actual daily maximum and minimum temperatures and precipitation for Palestine, Texas for 2009 and 2019. Note the flooding rains in late May 2010. Letters indicate the timing of Compact High-Resolution Imaging Spectrometer (CHRIS) acquisitions with Y/N indicating whether the image was used in the analysis or not.

The vegetation canopy structure on GEWMA is shown in Figure 1b contrasting leaf-off and leaf-on images from 2009 and 2012. Note that site managers started to clear part of the thickened woodland in 2010 in order to restore some of the original oak savanna parkland environment. The result of this is visible in the 2012 image. A simplified version of the Phase 2 land cover map identifies four major and three minor vegetation associations on GEWMA: savanna grassland, post oak motte and woodland, seasonally flooded hardwood forest, floodplain hardwood evergreen forest; and juniper woodland, mesquite woodland, and post oak/red cedar motte and woodland (Figure 3b). The species composition of these main types is shown in Table 2, indicating the diversity of the genera and species involved. The proportions of these vegetation sites associated with each major ESD and NRCS soil types are shown in Table 1 (the minor land cover types with evergreen shrubs are all aggregated into one for this purpose). The savanna grassland, post oak motte and woodland are strongly concentrated on Darco, Kirvin, Teneha and Tonkawa fine sands, whilst floodplain hardwoods and seasonally flooded hardwoods are primarily concentrated on Thenas fine sandy loam, Naconiche loamy fine sand, and Nahatche and Pluck soils (Table 1).

Observations from field work carried out in November 2009 indicated that the upland oak savanna was thickened and dense with little or no understory (Figure 4a), the savanna grassland (Figure 4b,c) had areas of bare soil, thin areas of little bluestem and scattered evergreen shrubs, and the lowlands and terraces contained dense stands of hardwoods with evergreen shrub understories (Figure 4c background, Figure 4d roadside). The map developed in [5] identifies natural vegetation types with more spatially explicit detail but does not identify invasive evergreens aside from long-established eastern red cedar and mesquite stands.

Table 1. Dominant ecological site descriptions (provisional status), soil types and major Texas Land Cover Project classes for GEWMA.

[1] Dominant ESD Description	Soil Description (Code)	% Area	ASW 100 cm (%)	[2] Land Cover (%) POW; SG; HF; SFHF; JM
Clayey Upland	Kirvin–Sacul association, sloping; (KnE)	4.8	12.3	94; 0; 0; 0; 0; 0
	Sacul fine sandy loam, 1 to 5 percent slopes (SaC)	0.4	9.8	
Loamy Upland	Elrose fine sandy loam, 1 to 3 percent slopes (EiB);	0.2	13.1	71; 22; 0; 0; 4
	Kullit fine sandy loam, 1 to 3 percent slopes (KuB)	0.6	13.0	
Northern Sandy Loamy Upland	Rentzel fine sand, 0 to 5 percent slopes (ChC);	2.7	8.6	85; 8; 0; 0; 3
	Lilbert loamy fine sand, 0 to 3 percent slopes (FuB);	7.3	9.6	
	Lilbert loamy fine sand, 3 to 8 percent slopes (FuD);	2.6	9.2	
	Larue loamy fine sand, 1 to 3 percent slopes (LaB);	0.3	10.0	
	Rentzel loamy fine sand, 0 to 5 percent slopes (LeC);	0.3	8.6	
	Trep loamy fine sand, 1 to 5 percent slopes (TpC);	4.8	10.0	
Northern Deep Sandy Upland	Darco fine sand, 1 to 8 percent slopes (DaD);	28.5	7.8	89; 7; 0; 0; 2
	Darco, Kirvin, and Tenaha soils, sloping (DkF)	8.6	9.8	
Very Deep Sandy Upland	Tonkawa fine sand, 1 to 8 percent slopes (ArD)	7.3	6.0	66; 26; 0; 0; 3
Terrace	Annona fine sandy loam, 1 to 5 percent slopes (SsC);	0.3	14.6	95; 0; 3; 0; 0
	Annona soils, 3 to 10 percent slopes, eroded (SuD2)	0.4	14.6	
Stream Bottomland	Thenas fine sandy loam (Th)	4.4	13.8	9; 0; 76; 15; 0
Sandy Bottomland	Naconiche loamy fine sand, 0 to 5 percent slopes (PeC)	1.1	9.7	99; 0; 0; 0; 0
Loamy Bottomland	Nahatche and Pluck soils (Na)	21.9	16.1	3; 0; 43; 53; 0

[1] certain soil polygons do not have an Ecological Site Description (ESD) association: LuA (Lufkin fine sandy loam), W (Water); [2] POW – post oak woodland; SG – savanna grassland; HF – hardwood forest; SFH – seasonally flooded hardwood forest; JM – juniper or mesquite woodland.

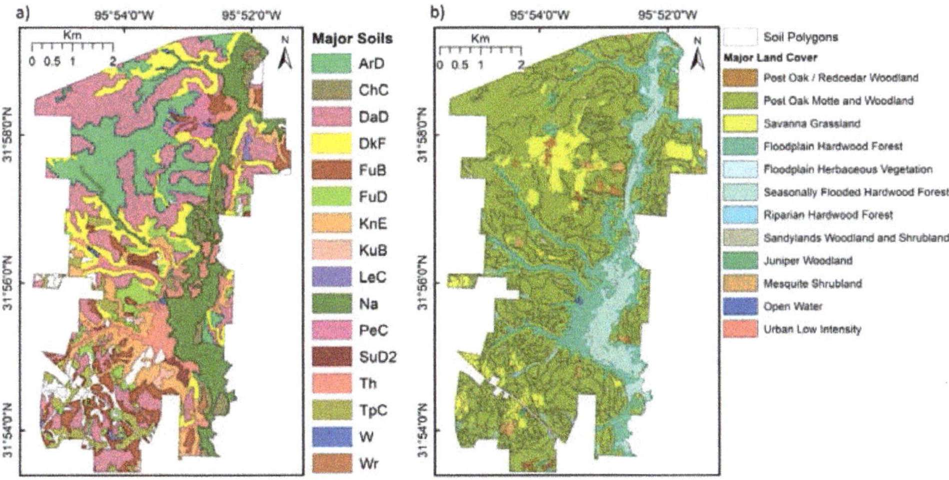

Figure 3. (**a**) Major soil types by soil polygon from the United States Department of Agriculture (USDA) Natural Resources Conservation Service (NRCS) for GEWMA; see Table 1 for descriptions (minor soil types in the southern area are shown as blank for simplicity). (**b**) Major land cover classes from the Texas Vegetation Classification Project ([2] for GEWMA (minor classes covering <200 pixels in the dissected southern area have been excluded for simplicity).

Table 2. Major species associated with the most important land cover types from the TNRIS classification and major encroaching evergreen species mapped in this study [3,5]. Species are listed from major to minor; however, woodlands and forests are very heterogeneous.

Land Cover Type	Species
Post oak motte and woodland (POW)	Overstory: *Quercus stellata, Q. marilandica, Q. nigra, Q. falcata, Q. incana, Q. fusiformis, Ulmus crassifolia, Celtis laevigata, Carya texana,* Mid-story: *Prosopis* spp., *Diospyros virginiana, Ilex vomitoria, I. decidua, U. alata, Sideroxylon lanuginosum, Callicarpa americana, Juniperus virginiana*
Savanna grassland (SG)	Herbaceous: *Shizachyrium scoparium, Sorghastrum nutans, Bothriochloa saccharoides, Stipa leucotricha, Sporobolus asper* var. *asper, Paspalum plicatulum* Scattered: *Q. stellata, Q. marilandica, Q. falcata, I. vomitoria, J. virginiana*
Floodplain hardwood forest (HF)	Overstory: *Fraxinus americana, Q. stellata, F. pennsylvanica, U. crassifolia, U. americana* Mid-story: *C. laevigata, Salix* spp.
Seasonally flooded hardwood forest (SHHF)	Overstory: *Q. marilandica, U. americana, Carya illinoinensis, F. pennsylvanica, Q. phellos, Q. lyrate, Liquidambar styraciflua*
Juniper or mesquite shrubland/woodland (JM)	*J. virginiana, Prosopis* spp., *Q. marilandica, Q. virginiana, U. crassifolia, U. alata, C. laevigata, D. texana*
Successional shrublands (Evergreens)	*J. virginiana, I. vomitoria, C. laevigata, U. crassifolia*

Figure 4. (**a**) Thickened post oak woodland on the sandy upland ArD or DaD soil type with virtually no understory. (**b**) Savanna grassland with sparse grasses, yaupon and eastern red cedar. (**c**) Degraded savanna grassland with bottomland hardwood forest in the background. (**d**) Dense mixed oak woodland with shrubby understory by the road. (Photos: M.J.H and A.M.).

2.2. Remote Sensing Data and Processing

2.2.1. Aerial Photography

To aid with the geometric correction of CHRIS data, characterization of the vegetation cover, and identification of evergreen vegetation, a National Agricultural Imagery Program (NAIP) 1 m resolution airborne image with red, green, blue and near infrared bands was acquired for January 2009 coinciding with the winter leaf-off period for deciduous trees. The four-band image was subjected to an unsupervised isodata classification to create 25 classes. These classes were then aggregated to five vegetation classes (water or shadow, deciduous trees, leaf/grass understory; bare or sparse herbaceous vegetation, and evergreen trees and shrubs) based on a visual interpretation of the NAIP image (Figure 5a). Potential for error was greatest in distinguishing deciduous tree cover from understory since the separation of trees from understory under leaf-off conditions was dependent upon trunk/shadow mixtures. Varying the aggregated class allocation for the most marginal class resulted in unacceptably high "omission" differences of -86% for deciduous tree cover and a "commission" difference of 140% for understory, which did not match our knowledge of the site based on measurements of canopy trees along two transects in central GEWMA, as well as qualitative surveys throughout the area. Using the most conservative aggregation for evergreens changed the evergreen percentage from around 20% to around 10%. However, this excluded classes with clear red edge signatures indicative of photosynthetic vegetation. Since this classification was undertaken at the NAIP pixel resolution of 1 m, but subsequent analysis was undertaken at a resolution of multiple CHRIS pixels, minor accuracy errors in this classification had very little impact on the results of the study.

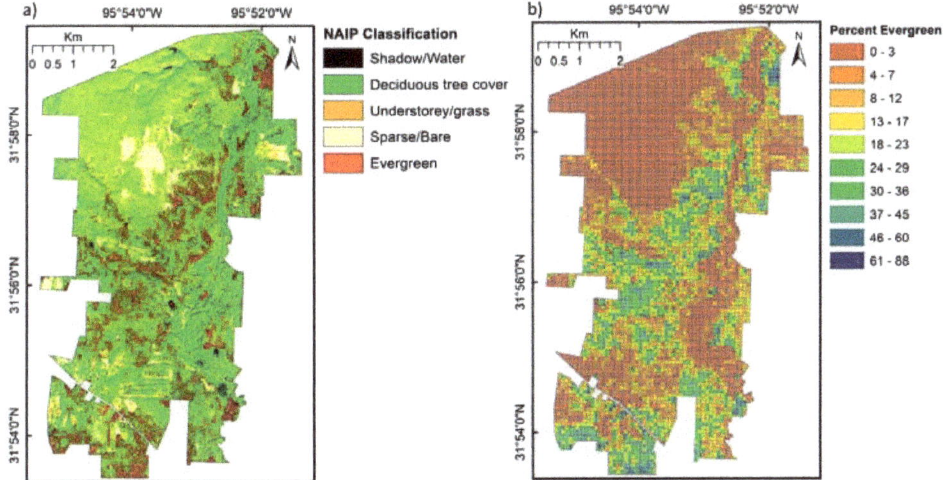

Figure 5. (**a**) Classification of the winter 2009 NAIP image into five classes specifically identifying evergreen trees; and (**b**) Fishnet polygons corresponding to a 5 × 5 grid of CHRIS Project for On-Board Autonomy (PROBA) pixels showing the proportion of evergreen trees.

2.2.2. CHRIS PROBA Data

The ESA sensor CHRIS/PROBA was launched on 22 October 2001 [8]. The CHRIS sensor on the PROBA-1 platform was designed to capture high-resolution images at multiple view angles in 62 narrow reflectance channels for wavelengths between 400 and 1000 nm [8] (Table 3). The sensor could be tasked in several modes which varied the swath width, pixel resolution and number of spectral channels. For this study, CHRIS data were acquired in Mode 1 which provided 62 spectral channels and five view angles at a nominal pixel resolution of 34 m (Table 3). The CHRIS data were acquired on 17 dates between June 2009 and October 2010 over the GEWMA study area (Table 4).

The CHRIS images were processed using the CHRIS Toolbox [36] in version 4.7 of the BEAM software package to process data from a variety of European Space Agency sources [37] and in ENVI®(ENvironment for Visualizing Images). Following the procedure in [36], the standard noise reduction, cloud identification and masking, and atmospheric correction with spectral calibration, aerosol optical thickness retrieval and column water vapor retrieval were carried out to obtain surface reflectance images. The atmospheric correction module of the CHRIS Toolbox [38] is based on MODTRAN4 (MODerate resolution atmospheric TRANsmission) [39]. Although the full set of multi-angular images was acquired and processed for almost all dates, only nadir images were used in this study, as the focus was on the temporal domain rather than the angular domain.

Nadir surface reflectance images were converted to tiff format, exported and imported into ENVI®. Only those images without cloud over the GEWMA site were subjected to further processing, and the 20 December image was excluded due to a targeting error. This left 11 useable images for further processing (Table 4). Images for each date were individually geometrically corrected using ground control points from the NAIP image. Due to the nature of the GEWMA environment with limited manmade features, CHRIS images were warped to an 18 m pixel resolution with a registration accuracy of approximately 1 pixel. Although this leads to the oversampling and smearing of the detected signal, this was not important in the context of this study, in which patch-based temporal profiles were the major target. Finally, the individually registered images were co-registered using Texas Road centerlines with an accuracy of approximately 2 pixels (36 m).

Table 3. CHRIS PROBA Specifications for Acquisition Mode 1.

Spatial Resolution	Image Size	Nominal View Angles	Spectral Bands	Spectral Range
34 m pixels at 556 km altitude	372 × 374 pixels (12.65 × 12.72 km)	+55°, +36°, 0° (nadir), −36°, −55°	62 bands 9–12 nm width except 20 nm at 930 and 950 nm	Min 406–992 nm Max 415–1003 nm Mid 411–997 nm

Table 4. CHRIS PROBA image acquisition dates, viewing geometry for nadir image, solar geometry and cloud contamination.

Date	Day (2009–2010)	Observation Zenith (°), Azimuth (°)	Solar Zenith (°), Azimuth (°)	Local Fly-by Time GMT -5/-6 (hr)	Cloud (%)	Image Used	[1] Leaf Stage
18 June 2009	169	5.73, 222.82	28.0, 100.55	11.25	32.0	N (site cloud)	On
9 August 2009	221	19.60, 136.99	35.0, 109.43	11.11	13.1	N (site cloud)	On
18 August 2009	230	15.84, 137.97	36.0, 113.95	11.12	17.4	N (site cloud)	On
6 September 2009	249	21.07, 315.08	38.0, 127.26	11.25	0	Y	On
19 October 2009	292	8.83, 139.26	51.0, 140.89	11.09	0	Y	Sen
6 November 2009	310	5.00, 140.00	56.0, 145.64	10.11	6.6	N	Sen
24 November 2009	328	5.00, 140.00	60.0, 147.91	10.12	0	Y	Off
3 December 2009	337	5.00, 140.00	61.0, 148.18	10.13	0	Y	Off
20 December 2009	358	5.00, 140.00	65.0, 145.06	10.04	0	N (off site)	Off
21 December 2009	359	5.00, 140.00	63.0, 147.30	10.15	0	Y	Off
25 January 2010	390	9.35, 224.82	62.0, 139.70	10.07	0	Y	Off
14 April 2010	469	12.29, 138.53	42.0, 113.72	10.50	0	Y	On
20 May 2010	505	2.99, 198.03	22.0, 100.50	10.50	0	N (missing data)	On
29 May 2010	514	3.57, 23.84	35.0, 97.84	10.51	0	Y	On
16 June 2010	532	9.21, 224.65	35.0, 94.62	10.51	19.1	N (site cloud)	On
30 July 2010	576	8.24, 141.47	40.0, 99.11	10.40	10.3	Y	On
4 September 2010	642	3.24, 205.52	45.0, 114.4	10.39	0	Y	On
19 October 2010	657	10.45, 224.75	56.0, 132.99	10.37	3.2	Y (site clear)	Sen

[1] Sen - senescing.

2.2.3. Vegetation Indices

A total of 51 narrow band VIs was calculated from the 62 channel surface reflectance images based on the formulae documented in [24] with wavelengths adjusted for CHRIS band centers. Given the nature of the study site, with very dense tree cover except for certain soils where savanna grassland occurred, the identification of the key VIs focused first on known pigment-sensitive wavelength combinations found to be effective in previous studies with Hyperion imagery [33,34], and then on whether the many different narrow-band indices creating variants on the NDVI [38] and targeting the red edge provided any additional sensitivity to augment NDVI. The VIs that utilized blue wavelengths below 480 nm often failed due to low-quality blue bands. Based on [33,34], visual assessment of images and pixel histograms, eight VIs were selected as indicators of vegetation canopy properties (Table 5). The NDVI was ultimately selected as the indicator of greenness/photosynthetic capacity since no advantage was gained from the other greenness VI variants. Although band centers were well aligned with narrow band reference values, CHRIS band widths at 6–12 nm can only broadly approximate the narrow bands used in the original derivation of some of these indices. For example, [28] used an average band width of 2.8 nm in deriving the PRI2. Nevertheless, the image value ranges for highly precise VIs such as PRI and PRI2 matched the ranges observed in the original studies [28,29].

Table 5. Vegetation indices selected as indicators. Formulae depict band reflectance (R) wavelengths in nanometers.

Narrow Band Vegetation Index [1]	Narrow Band Formula [1]	CHRIS PROBA Equivalent [2]	Context
ARI1 (Anthocyanin Reflectance Index)	$\left(\frac{1}{R_{550}}\right)-\left(\frac{1}{R_{700}}\right)$	$\left(\frac{1}{R_{551}}\right)-\left(\frac{1}{R_{703}}\right)$	Stress, senescence [26]
ARI2 (Anthocyanin Reflectance Index 2)	$R_{800}\left(\frac{1}{R_{550}}\right)-\left(\frac{1}{R_{700}}\right)$	$R_{800}\left(\frac{1}{R_{551}}\right)-\left(\frac{1}{R_{703}}\right)$	Stress, senescence [26]
CRI (Carotenoid Reflectance Index)	$\left(\frac{1}{R_{510}}\right)-\left(\frac{1}{R_{550}}\right)$	$\left(\frac{1}{R_{510}}\right)-\left(\frac{1}{R_{551}}\right)$	Senescence [27]
NDVI (Normalized Difference Vegetation Index)	$\frac{(R_{803}-R_{681})}{(R_{803}+R_{681})}$	$\frac{(R_{800}-R_{680})}{(R_{800}+R_{680})}$	Chlorophyll, photosynthetic capacity [40]
PRI (Photochemical Reflectance Index)	$\frac{(R_{531}-R_{570})}{(R_{531}+R_{570})}$	$\frac{(R_{530}-R_{572})}{(R_{530}+R_{572})}$	Photosynthetic efficiency [29]
PRI2 (Photochemical Reflectance Index 2)	$\frac{(R_{570}-R_{539})}{(R_{570}+R_{539})}$	$\frac{(R_{572}-R_{540})}{(R_{572}+R_{540})}$	Photosynthetic efficiency [28]
PSRI (Plant Senescence Reflectance Index)	$\frac{(R_{680}-R_{500})}{(R_{750})}$	$\frac{(R_{680}-R_{500})}{(R_{748})}$	Senescence, brownness [30]
WI (plant Water Index)	$\frac{R_{900}}{R_{970}}$	$\frac{R_{905}}{R_{965}}$	Canopy water content [31]

[1] From [24]; [2] band centers based on [41].

2.3. Analysis

2.3.1. Fishnet Polygon Extraction of Tree Cover Density and VI Profiles

In order to capture the broad patterns of vegetation canopy dynamics with such heterogeneous species mixtures (Table 2), the analysis was based on averaging across pixels at a patch scale. A fishnet polygon layer was constructed with a polygon cell size of 5 × 5 CHRIS pixels making a 90 × 90 m cell. Each fishnet polygon was attributed a unique number. Pixel counts of tree and other cover classes were then extracted from the classified NAIP image (with each fishnet polygon cell containing 8100 NAIP pixels), converted to cover class proportions and assigned to each fishnet polygon cell (Figure 5b; percentage of evergreen class shown). The fishnet was then merged with the soil polygon layer for GEWMA to create a combined layer with vegetation cover fractions and soil type for each fishnet polygon. The fishnet was then used to extract VI values for each fishnet polygon from each VI time series creating a database with cover fraction, soil type, and VI value for each 90 × 90 m polygon across GEWMA. To minimize the effects of uncertainty in the attribution of aggregated classes in the NAIP image classification, the time series profiles of each VI were then examined for average deciduous and evergreen tree cover at 10% intervals. Associations with the Texas Land Cover classes were also examined.

2.3.2. Combined Indicators of Vegetation Function

Based on image coverage and quality, two dates (September 2009—leaf-on—and 25 January 2010—leaf-off) were selected to explore information provided by combining VIs. The data were windowed to focus on the northern half of GEWMA, where a consistent gradient from the upland sandy savannas to the lowland hardwood floodplain and riparian areas could be sampled. The six best VIs from the phenological analysis were stacked (Anthocyanin Reflectance Index 2 (ARI2), Carotenoid Reflectance Index (CRI), NDVI, Photochemical Reflectance Index 2 (PRI2), Plant Senescence Reflectance Index (PSRI) and Water Index (WI)) for each date. An isodata unsupervised classification was carried out on the VI stacks for each date to create a 20-class image. The classifications were then aggregated to eight classes based on the dendrograms and class distances. The mean and standard deviation of VI index values, NAIP classification cover percentages and dominant soil type percentage were extracted for each classified image.

3. Results

The extracted time series profiles showed major differences in relation to the major land cover types and deciduous tree density levels and in relation to the proportion of evergreen trees and shrubs present.

3.1. Overall VI Profiles

The comparison of profiles for the selected VIs showed a clear separation between areas with trees and those without during leaf-on. For most leaf-on dates, savanna grassland VI profiles were separated from the 40% and 80% tree cover profiles by more than one standard deviation (Figure 6). The differences among the deciduous tree canopies largely disappeared during leaf-off.

Although increments in VI response to tree density were evident, there was a substantial overlap in standard deviations, with CRI and ARI1 and ARI2 exhibiting a higher variation in values within a tree density level and between dates. In general, magnitudes of pigment-sensitive VIs during leaf on could be ranked 80% > 40% hardwood > 40% post oak >> savanna/grassland (for PRI2, low values represent a high response). The largest range of response was exhibited by CRI during leaf-on, but the rankings among tree densities were the same. The PSRI exhibited little sensitivity to hardwood tree density, but a consistently higher response for post oak at 40% tree cover and savanna/grassland (Figure 6g). The compressed and low values for CRI, ARI and ARI2 for day 576 (July 30) are likely anomalous as they are not supported by parallel behaviors in NDVI, WI, PRI or PRI2. The WI showed large seasonal variation in canopy water content, but the main difference in profiles was between the savanna grassland and the areas with trees (Figure 6h).

3.2. Response of VIs to Evergreen Density

The effect of evergreens on the VI profiles was examined for the DaD soil type dominated by post oak woodland with total tree cover of >80% (see Table 1) and evergreen tree cover in increments of 10% up to the maximum observed of 40% (Figure 7). The pigment-sensitive VIs exhibited small increases in values in the leaf-off period between 350 and 430 days (mid-December 2009–mid-March) in response to increasing evergreen cover, except for the ARI (Figure 7). Although the increments due to evergreen density were evident, standard deviations within date and density levels were relatively large and exhibited a significant overlap. The PRI2 and NDVI exhibited lower values at 40% tree cover than at 80% tree cover in the leaf-on period of 2010 (Figure 7a,b). The PSRI and WI provided the clearest response to increasing evergreen percentage, with PSRI declining and WI increasing incrementally with each 10% increase in evergreens (Figure 7c,d). The responses to increasing the evergreen percentage were similar for the other soil types, but the magnitude of the difference between zero and 40% evergreens was generally smaller. Evergreen percentages were very low on the ArD soil type except when adjacent to bottomlands.

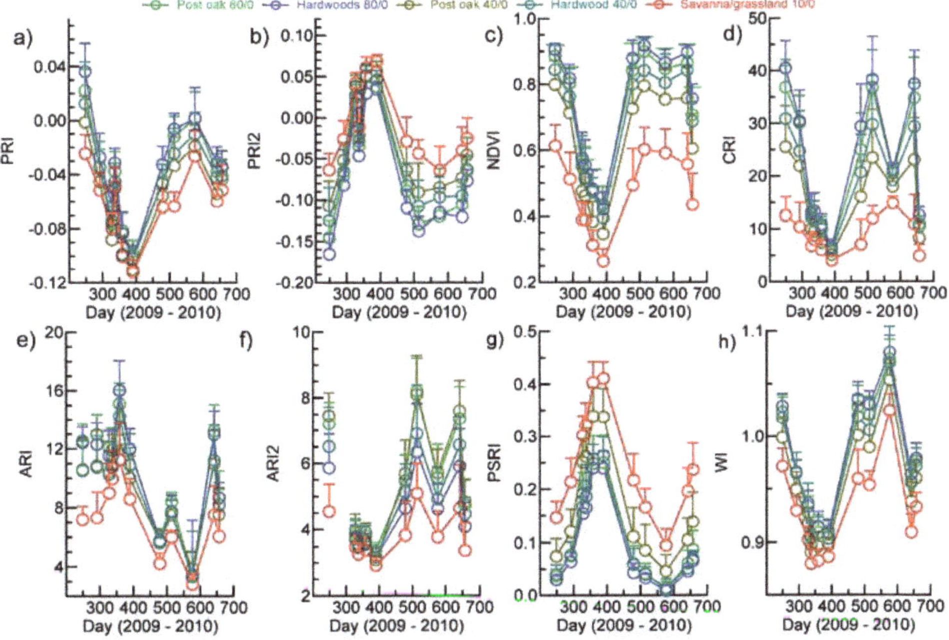

Figure 6. Overview of seasonal phenology of vegetation indices that are indicators of photosynthetic activity and pigments averaged across soil polygons with predominant vegetation types of post oak, bottomland hardwoods and grassland at tree cover percentages of 80, 40 and <10 and no evergreens. The bars denote standard deviations. The gap in the ARI2 profile is due to an unrecoverable error in the calculation for 19 October (see Table 1 for land cover association with soil types). (**a**) PRI; (**b**) PRI2; (**c**) NDVI; (**d**) CRI; (**e**) ARI; (**f**) ARI2; (**g**) PSRI; and (**h**) WI.

3.3. Responses within Tree Density Levels among Soil Types

Available soil water (ASW) is a major variable among dominant soil types on GEWMA, with upland sandy soils having 6–10% ASW and loamy bottomland soils having approximately 14–16% ASW (Table 1). In contrast with the other profile analyses, differences in PRI2 and WI among the different combinations of soil type and tree density mostly exceeded one standard deviation indicating the importance of ASW in canopy processes. The comparisons between selected soils at the extremes of ASW content showed major differences during the leaf-on period among the three soil types predominantly associated with grassland savanna or woodland at tree densities of <10% for PRI2, and smaller differences during the leaf-off period (Figure 8a). There were also differences between the ArD and Na at 80% tree cover in the leaf-on period, but these disappeared in the leaf-off period. There were small differences in WI between ArD and both DkF and DaD soils in the leaf-on and leaf-off periods, and between ArD and Na in the leaf-on period (Figure 8b), but the relative differences were smaller than for PRI2.

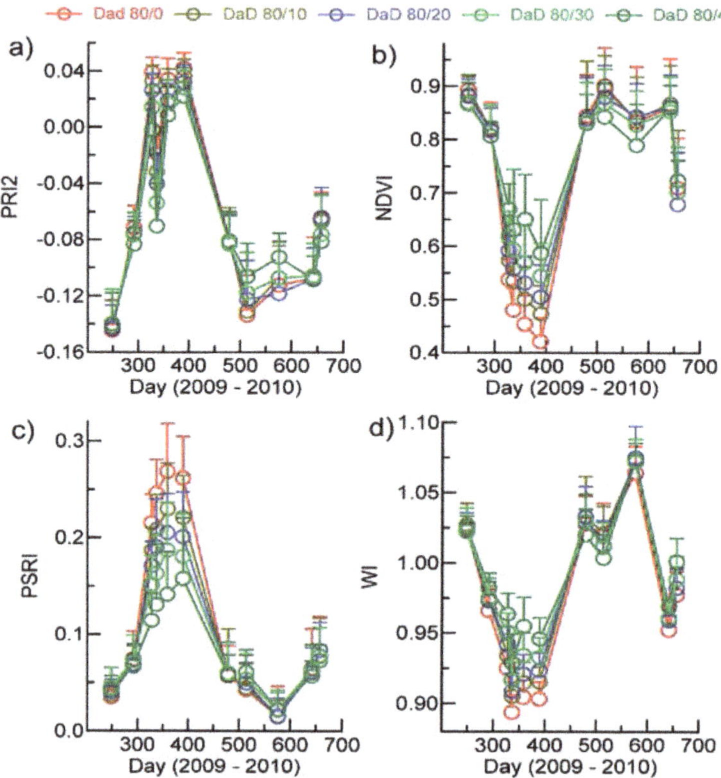

Figure 7. Variation in responses of the Vegetation Indices (VIs) most sensitive to increasing levels of evergreen tree cover within the DaD soil type dominated by post oak woodland. The bars denote standard deviations. (**a**) PRI2; (**b**) NDVI; (**c**) PSRI; and (**d**) WI.

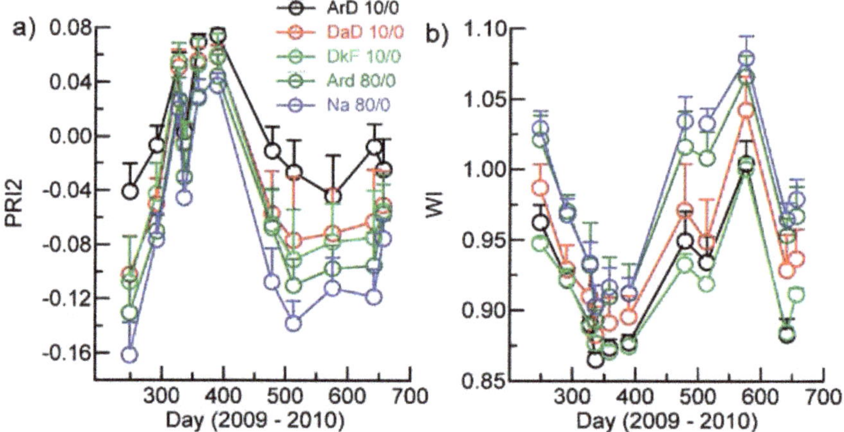

Figure 8. Variation in VI indicators across extremes of deciduous tree cover (<10% and >80%) without evergreens, and across soil types in terms of available soil water (see Table 1). The bars denote standard deviations. (**a**) Photosynthetic activity, PRI2; and (**b**) vegetation water content, WI.

3.4. Sensitivity of VIs to Spatial Variability in Leaf-on and Leaf-off Periods

Images from 6 September 2009 and 25 January 2010—peak leaf-on and leaf-off conditions, respectively—were chosen to examine the sensitivity to vegetation and canopy variation and underlying soil differences that could be obtained by combining the VIs by classification. The 2009 date was chosen because site managers commenced clearing parts of the upland woodland in mid-2010 (see Figure 1) and consistent vegetation cover between leaf-on and leaf-off was required. The images from these dates clearly reflect the differences in leaf-on and leaf-off canopies (Figure 9) and highlight the presence of evergreen vegetation in January (Figure 9b,d,f,h). Whilst the NDVI shows limited within-canopy sensitivity, PRI2 shows a marked difference between evergreen-dominated and deciduous oak-dominated areas in September (Figure 9a,c). The upland post oak woodland and savanna environment associated with drier soil types (Table 1) is sharply delineated in January NDVI and ARI2 (Figure 9b,f). The PSRI and WI show the same patterns as the pigment-specific indices with a similar sharp delineation of the upland post oak woodland and savanna in January in both VIs (Figure 10) and the presence of evergreen vegetation in January (Figure 10b,d).

The image classification results for September and January produced distinctly different class patterns, with the leaf-on September image highlighting the grassland/savanna areas, variation in the post oak woodland canopy, and no clear distinction of hardwoods on bottomland (Figure 11a). By contrast, the results for January clearly distinguished the post oak woodland and savanna grassland on ArD and DaD soil types, areas of evergreen vegetation and the flooded stream bottomland associated with the Na soil type (Figure 11b).

The extracted profiles for each classification illustrate the difference in VI sensitivity, class association with tree and evergreen density, and class association with soil type (Figure 12). Available soil water is a major variable among dominant soil types on GEWMA, with upland sandy soils having 6–10% ASW and loamy bottomland soils having approximately 14–16% ASW (Table 1). Within the image, the dominant soil types were DaD > ArD > DkF > Na, thus representing a contrast between drier soils supporting post oak woodland and savanna grassland, and bottomland loams supporting hardwood forest (Figure 12). In September, most of the area was occupied by classes 1, 2, 6, 7 and 8. Classes 1 (78%) and 2 (61%) were strongly associated with elevations between 100 and 129 m, while classes 7 (55%) and 8 (76%) were strongly associated with elevations between 78 and 100 m. Neither the evergreen vegetation nor the understory fraction varied much between classes, but bare soil was most associated with class 1 and least associated with class 8. In all cases, VI class values showed either a regular positive progression from class 1 to class 8 for ARI2, CRI, NDVI, PRI2 (negative PRI2 represents a higher value) and WI, and a negative progression for PSRI (Figure 12). These progressions represented a regular increase in pigments, photosynthetic activity and canopy water content, and a regular decrease in brownness/yellowness with class number. Class 1 occupied 40% of soil type ArD, while class 2 occupied 20% of each of ArD, DkF and FuB.

In January, most of the area was occupied by classes 1, 3, 4, 5, 6, and 7 (Figure 12b). Bare areas were associated with classes 1 and 3, and evergreen vegetation was associated with classes 7 and 8. The class association with elevation was more marked than in September, with between 88–100% of classes 5–8 being associated with elevations between 78–115 m, and 63% of class 2 being associated with elevations between 78 and 85 m. Classes 7 and 8 had the highest NDVI, CRI and ARI2 and the lowest PRI2 values as well as the highest WI and lowest PSRI. Among classes 1, 3 and 4, associated with the deciduous post oak woodland and savanna grassland, there were lower values for WI for classes 3 and 4 with greater understory cover than class 1 with more bare cover. Class 1 occupied >40% of the Na soil type, class 3 occupied between 30 and 40% of soil types ArD, DaD and FuB, and class 6 occupied around 30% of the minor soil types KnE, DkF and PeC (Figure 12b).

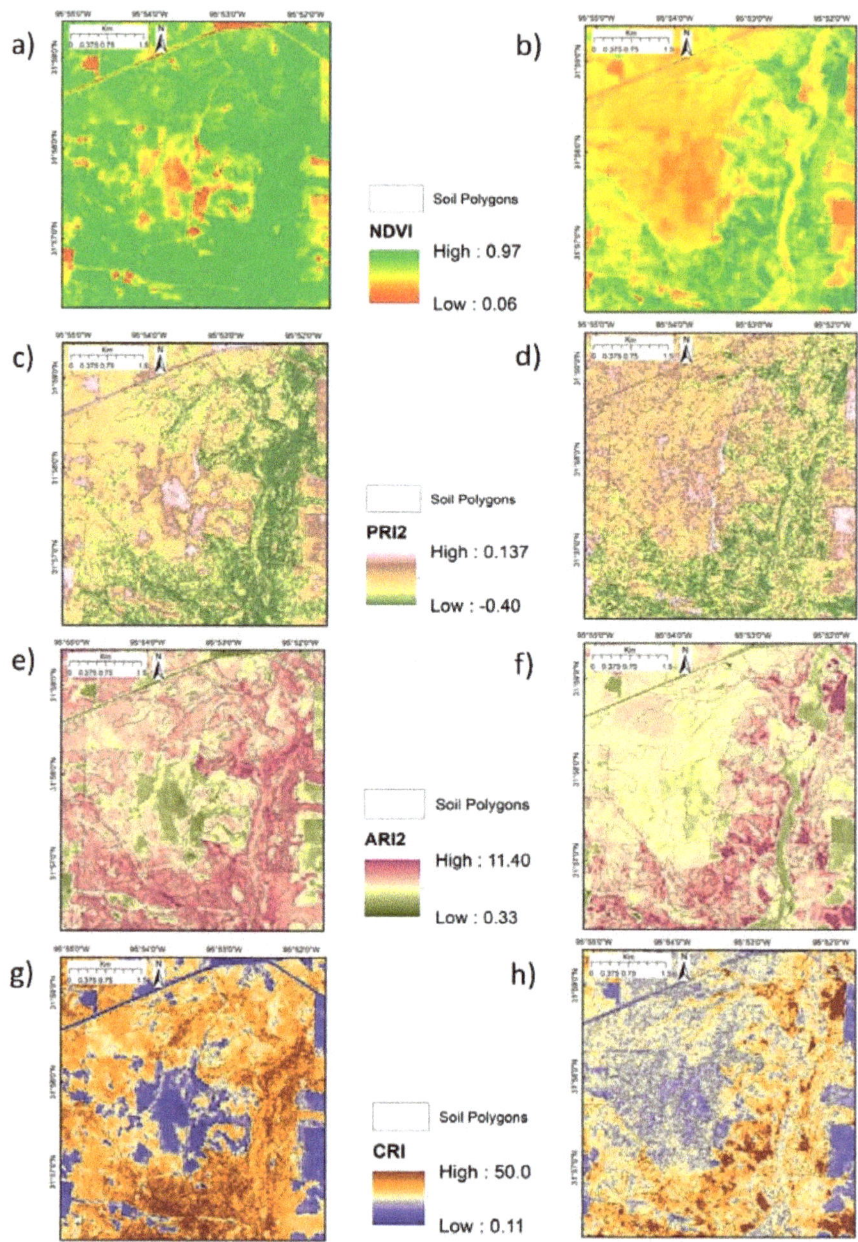

Figure 9. Photosynthesis and pigment-sensitive VIs at leaf-on and leaf-off for deciduous trees. (**a**) NDVI 6 September 2009; (**b**) NDVI 25 January 2010; (**c**) PRI2 6 September 2009; (**d**) PRI1 25 January 2010; (**e**) ARI2 6 September 2009; (**f**) ARI2 25 January 2010; (**g**) CRI 6 September 2009; (**h**) CRI 25 January 2010.

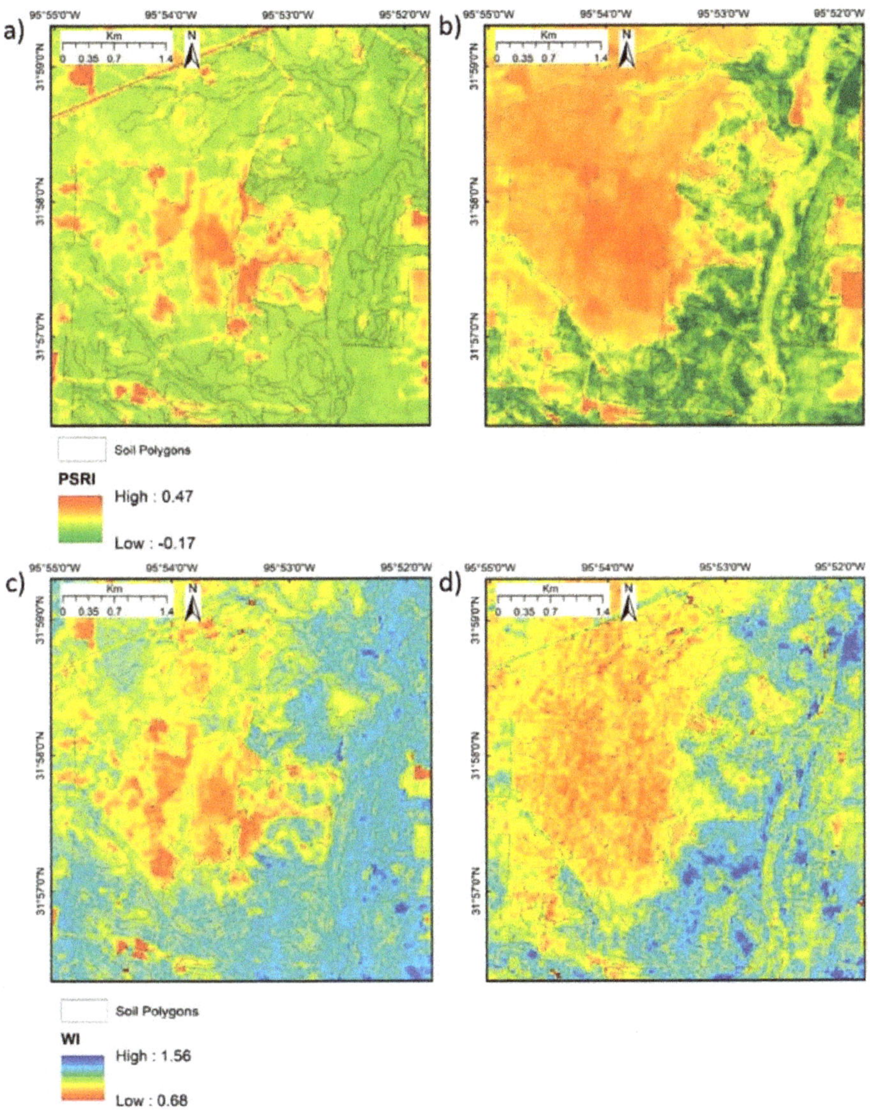

Figure 10. Dryness and moisture-sensitive VIs at leaf-on and leaf-off for deciduous trees. (**a**) PSRI 6 September 2009; (**b**) PSRI 25 January 2010; (**c**) WI 6 September 2009; (**d**) WI 25 January 2010.

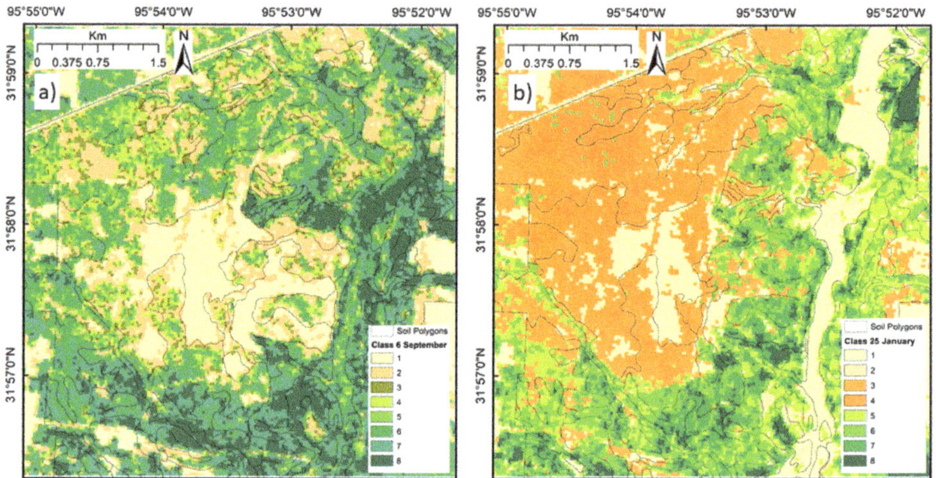

Figure 11. Classification of the northern section of GEWMA based on six VIs aggregated to eight classes. (**a**) September 6, 2009: deciduous leaf-on; (**b**) January 25, 2010: deciduous leaf-off.

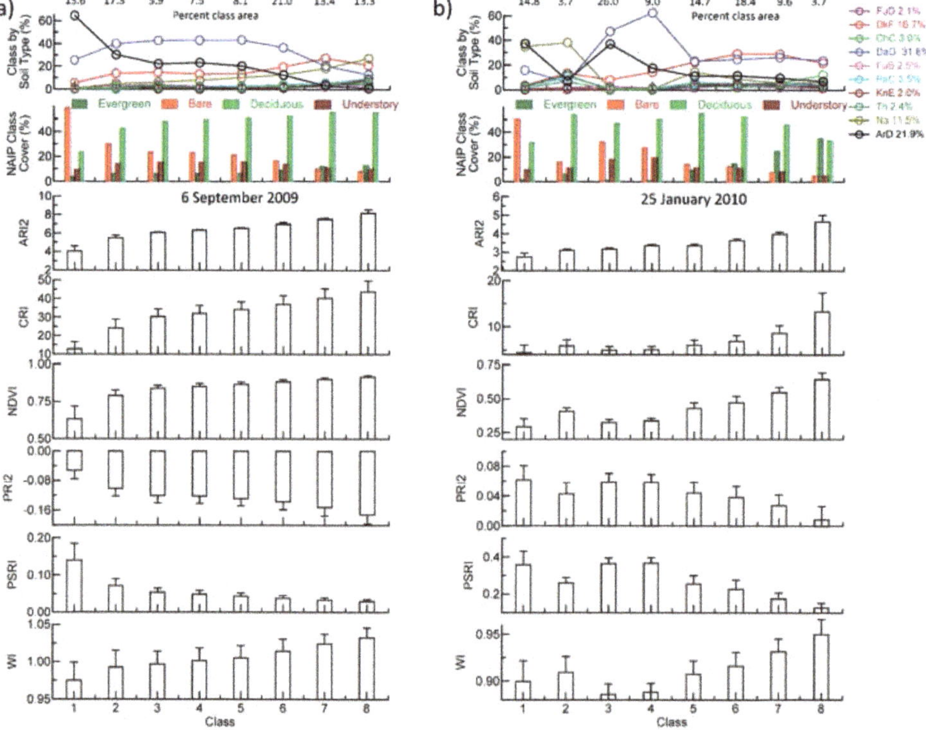

Figure 12. Vegetation Index profiles of cover types from classified NAIP imagery and proportion of NRCS Web Soil Survey soil types associated with (**a**) September 2009 and (**b**) January 2010 classifications shown in Figure 11. Numbers at the top indicate the percentage of the area occupied by each class. Numbers on the soil type legend indicate the percentage of the area occupied by each major soil type (minor soil types were excluded).

4. Discussion

This study on a remnant post oak woodland area has shown that the major differences in spectral phenology as defined by VIs are associated with deciduous tree density, the density of evergreen trees and shrubs—especially during deciduous leaf-off periods—broad vegetation types, and soil types. Although major differences in VIs at peak leaf-on and leaf-off were clearly associated with high and low evergreen density and extremes of ASW associated with soils, combining VIs in a simple classification also revealed variation in canopy water content and photosynthetic capacity within ostensibly uniform post oak savanna on the sandy soils with low ASW. The photosynthetic capacity of evergreens during deciduous leaf-off was clearly indicated by NDVI, PRI1 and PRI2 profiles with obvious growth advantages for further encroachment. Although tree species vary between post oak savanna and bottomland hardwoods, VI phenology was remarkably similar at the same tree densities. Heterogeneous and shared species composition post oak woodland and hardwood forests limited patch-scale discrimination with VI profiles.

The temporal pattern of spectral VIs was dominated by the deciduous tree leaf-on/leaf-off cycle of post oak woodland and hardwoods, and spatial variation in amplitude was primarily determined by tree density from >80% cover down to <10%. There was a greater range in the amplitude of NDVI and CRI than PSRI—an indicator of the chlorophyll/carotenoid ratio [30]—or ARI and ARI2 during leaf-on in response to the tree densities and major vegetation types. The precipitation records from September 2009 to October 2010 indicate that the site did not undergo any unusual moisture stress during the study. In addition, due to a flooding event in May of 2010 with substantial cloud in the preceding couple of months, 2010 spring green-up was not captured. There are significant differences in pigment magnitudes [42], anthocyanin synthesis [43] and rates of decay during senescence associated with stress or different mixes of deciduous species [44,45]. Such differences were either minimal or undetectable in this study. Since deciduous species are highly intermingled in both post oak savanna and bottomland hardwood forest (see Table 2 [3,5]; the pixel resolution of CHRIS and the 90 m resolution of the fishnet used in the analysis meant that patch-scale species differences were likely to be distinguishable).

At the landscape scale, the upland post oak savanna was clearly delineated in the leaf-off VI images, and there was a strong association with soil type and elevation. However, in leaf-on images, PRI2 and WI showed most sensitivity to the differences between upland and bottomland in soil properties, elevation and vegetation type, and to variation within the upland post oak savanna woodland and savanna grassland. The PRI and PRI2 are sensitive to photosynthetic RUE [28,29] and evergreens exhibit reduced midday PRI levels compared to deciduous species [29]. In this study, lower values for PRI and higher values for PRI2 were observed for 40% evergreen tree cover compared to 0–30% evergreen cover at 80% total tree cover in the peak leaf-on period of 2010. This suggests that evergreens suffer some disadvantage in photosynthesis when competing with deciduous trees during the leaf-on period. However, this is a minor factor when compared with the opportunity for photosynthesis and growth in the 3–4 month deciduous leaf-off period. An examination of the weather during the study period (Figure 2) and the long-term climate indicates that the temperature and precipitation are favorable for the growth of evergreens, especially coniferous *Juniperus* spp., while deciduous trees are in a leaf-off state.

The largest differences in pigment/senescence and photosynthetic-sensitive VIs occurred when comparing tree canopies among the sandy soil types with the lowest ASW at tree densities of <10%, and between the soil types with the lowest and highest ASW for tree densities >80%. This suggests that water relations are not only the most important driver of the vegetation types between upland woodland and bottomland hardwood forest at the landscape scale, but also potentially influence the vigor and location of savanna tree species at the patch scale. While differences in leaf-off classes derived from combined VIs were obviously associated with the presence or absence of an evergreen canopy, leaf-on classes associated with deciduous cover of >50% and ≤20% bare (classes 5–8) showed a significant variation in WI that is indicative of different canopy water contents. Studies have shown significant differences in water use attributes such as access to soil water, stomatal conductance, water use efficiency, and leaf-specific hydraulic conductance among *Quercus* spp. [46,47]. The post oak woodland contains a mix of *Quercus* spp., but tree-by-tree identification for any local site was not available or collected for this study. However, spatial variation in class distribution in the water-limited upland savanna environment in September 2010 may be indicative of both local variation in *Quercus* species composition and the vigor of trees associated with differences in water relations attributes.

The observation and analysis of woody encroachment across North America suggest that the potential for "release" of woody encroachers favored by fire suppression is greater at sites with higher precipitation [48], and by inference might be better on bottomland soils with higher ASW rather than on upland soils with lower ASW. Certain *Quercus* species have exhibited higher pre-dawn and midday water potentials than eastern red cedar [49,50] which is known to use water year-round, and almost 100% of precipitation reaches the soil surface [51]. In this study, higher densities of evergreens were clearly negatively associated with the soils with the lowest ASW, higher elevation (compare Figures 3 and 5) and those most dominated by thickened post oak woodland. Evergreens were mainly distributed among a range of soils in the bottomlands as an understory to hardwood canopies and along small tributaries of the main stream, Catfish Creek, at relatively lower elevations than the adjacent terrain. Among the native evergreens, eastern red cedar may have the most potential to encroach on low ASW soils since it has a record of highly successful encroachment in grasslands where it has an advantage in terms of rooting depth [52] and in upland oak forests in Oklahoma [53]. At the time of this study, the upland post oak savanna area was densely thickened with no grasses and a dense thatch of leaf litter (Figure 4a). The combination of high oak density, low ASW, rapid soil drainage at higher elevations, and the advantage in water relations attributes over eastern red cedar may explain its almost complete absence from the upland post oak woodlands.

Close to the end of the study period, a section of the northern upland post oak savanna was partially cleared to enable the planting of grasses and regeneration of a savanna landscape as part of an on-going management plan by the Texas Parks and Wildlife Department (Figure 1b). Small areas of GEWMA had been replanted with *Shizachyrium scoparium* prior to the commencement of this study. Airborne and satellite remote sensing could play a key role in the monitoring and management of GEWMA by tracking the development of replanted areas, identifying evergreen woody encroachment in newly planted areas, selecting areas for spraying, burning and/or mechanical control of eastern red cedar, and monitoring the long-term health of the important remnants of original savanna grassland, post oak woodland and hardwood forest (e.g., [54]). Such monitoring could be carried out with relatively frequent imagery at a 10–20 m resolution from Sentinel 2a and 2b. A prior study simulated Sentinel 2 with Hyperion imagery across a range of sites in North America and showed that suites of VIs could provide clear insights into differences among vegetation states and processes [34]. Various baseline conditions and approaches for the restoration of oak savannas [55,56] could be informed with suitable passive and active remote sensing data (e.g., [34,57]). This study used a now almost 20-year-old sensor with relatively low signal to noise and 400–1000 nm spectral coverage to detect a variety of differences seasonal phenology due to plant functional type, vegetation type, soils, and tree density. Advances in the hyperspectral detection of plant species diversity and function [58–62] provide exciting potential for monitoring important repositories of biodiversity with heterogeneous species composition from space or airborne platforms.

5. Conclusions

This study showed that functional VIs can identify differences in vegetation types that are strongly associated with soil type and landform and indicative of differences in physiological function and plant functional type. The study applied a relatively rare time series of CHRIS imagery at an acquisition frequency that captured variation in canopy dynamics both between and within major vegetation types and tree densities. Although the discrimination of known species-level differences in pigment dynamics and soil water access could not be detected with precision using the CHRIS data due to heterogeneous species mixtures and the patch-scale analysis applied, the potential for this with airborne sensors or new spaceborne hyperspectral sensors was clearly indicated. Although no explicit difference in photosynthetic function among deciduous canopies was detected in this study, nevertheless, the VIs sensitive to carotenoids, anthocyanins and photosynthetic efficiency can discriminate among species and landscape components when retrieved with better-quality sensors. In the meantime, Sentinel 2 with its additional bands in the pigment-sensitive wavelengths at a 20 m pixel resolution, may be used effectively for current monitoring. Although evergreen encroachment on rare oak savanna conservation lands may be easily monitored with leaf-off aerial photography, hyperspectral imagery can provide additional insights into the phenological behaviors that relate to physiological function and competitive advantage. This information could be helpful in the management and control of evergreen shrubs on bottomland areas [63] that present a substantial challenge due to their competitive advantages over deciduous oaks and hardwoods in the leaf-off period.

Author Contributions: Conceptualization, M.J.H. and A.M.; methodology, M.J.H.; field observations and photos A.M. and M.J.H.; formal analysis, M.J.H., R.L. and C.N.; data curation, M.J.H.; writing—original draft preparation, M.J.H.; writing—review and editing, A.M.; project administration, A.M. and M.J.H.

Funding: The involvement of the senior author was partially supported by grants from NASA (Contract #NNX09AQ81G and #NNX10AH20G).

Acknowledgments: We thank the managers of GEWMA for providing access to the site for field observations. Tony Fillipi assisted with field work. Field work costs were covered by funds from Texas Agriculture Experiment Station, Texas A&M University to A.M.

Conflicts of Interest: The authors declare no conflict of interest. The funders had no role in the design of the study; in the collection, analyses, or interpretation of data; in the writing of the manuscript, or in the decision to publish the results.

References

1. McPherson, G.R. *Ecology and Management of North American Savannas*; University of Arizona Press: Tuscon, AZ, USA, 1997; 208p.
2. Srinath, I.; Millington, A.C. Evaluating the potential of the original Texas Land Survey for mapping historical land and vegetation cover. *Land* **2016**, *5*, 4. [CrossRef]
3. Ludeke, K.; German, D.; Scott, J. Texas Vegetation Classification Project: Interpretive Booklet for Phase II. Texas Parks and Wildlife Department and Texas Natural Resources Information System. 2009. Available online: https://morap.missouri.edu/wp-content/uploads/2019/02/Texas_Vegetation_Classification_Phase_2_Interpretive_Booklet.pdf (accessed on 2 December 2009).
4. Gould, F.W. *Texas Plants. A Checklist and Ecological Summary*; Texas Agriculture Experiment Station, Texas A&M University: College Station, TX, USA, 1962.
5. Singhurst, J.R.; Cathy, J.C.; Prochaska, D.; Haucke, H.; Kroh, G.C.; Holmes, W. The vascular flora of Gus Engeling Wildlife Management Area, Anderson County, Texas. *Southeast. Nat.* **2009**, *2*, 347–368. [CrossRef]
6. Smith, S. *Eastern Red-Cedar: Positives, Negatives and Management*; Samuel Roberts Noble Foundation: Ardmore, OK, USA, 2011; 8p.
7. Barnsley, M.J.; Settle, J.J.; Cutter, M.A.; Lobb, D.R.; Teston, F. The PROBA/CHRIS mission: A low-cost smallsat for hyperspectral multiangle observations of the earth surface and atmosphere. *IEEE Trans. Geosci. Remote Sens.* **2004**, *42*, 1512–1520. [CrossRef]
8. Barnsley, M.J.; Lewis, P.; O'Dwyer, S.; Disney, M.I.; Hobson, P.; Cutter, M.; Lobb, D. On the potential of CHRIS/PROBA for estimating vegetation canopy properties from space. *Remote Sens. Rev.* **2000**, *19*, 171–189. [CrossRef]
9. Green, R.O.; Pavri, B.E.; Chrien, T.G. On-orbit radiometric and spectral calibration characteristics of EO-1 Hyperion derived with an underflight of AVIRIS and in situ measurements at Salar de Arizaro, Argentina. *IEEE Trans. Geosci. Remote Sens.* **2003**, *41*, 1194–1203. [CrossRef]
10. Pearlman, J.S.; Barry, P.S.; Segal, C.C.; Shepanski, J.; Beiso, D.; Carman, S.L. Hyperion, a space-based imaging spectrometer. *IEEE Trans. Geosci. Remote Sens.* **2003**, *41*, 1160–1173. [CrossRef]
11. Middleton, E.M.; Ungar, S.G.; Mandl, D.; Ong, L.; Frye, S.; Campbell, P.E.; Landis, D.R.; Young, J.P.; Pollack, N.H. The Earth Observing One (EO-1) satellite mission: Over a decade in space. *IEEE J. Sel. Top. Appl. Earth Observ. Remote Sens. (JSTARS)* **2013**, *6*, 243–256. [CrossRef]

12. Duca, R.; Del Frate, F. Hyperspectral and multiangle CHRIS-PROBA images for the generation of land cover maps. *IEEE Trans. Geosci. Remote Sens.* **2008**, *46*, 2857–2866. [CrossRef]

13. Stagakis, S.; Markos, N.; Sykioti, O.; Kyparissis, A. Monitoring canopy biophysical and biochemical parameters in ecosystem scale using satellite hyperspectral imagery: An application on a *Phlomis fruticosa* Mediterranean ecosystem using multiangular CHRIS/PROBA observations. *Remote Sens. Environ.* **2010**, *114*, 977–994. [CrossRef]

14. Mousivand, A.; Menenti, M.; Gorte, B.; Verhoef, W. Multi-temporal, multi-sensor retrieval of terrestrial vegetation properties from spectral-directional radiometric data. *Remote Sens. Environ.* **2015**, *158*, 311–330. [CrossRef]

15. Verrelst, J.; Romijn, E.; Kooistra, L. Mapping vegetation density in a heterogeneous river floodplain ecosystem using pointable CHRIS/PROBA data. *Remote Sens.* **2012**, *4*, 2866–2889. [CrossRef]

16. García Millán, V.E.; Sánchez-Azofeifa, A.; Málvarez García, G.C.; Rivard, B. Quantifying tropical dry forest succession in the Americas using CHRIS/PROBA. *Remote Sens. Environ.* **2014**, *144*, 120–136. [CrossRef]

17. Torbick, N.; Corbiere, M.A. Multiscale mapping assessment of Lake Champlain cyanobacterial harmful algal blooms. *Int. J. Environ. Res. Pubic Health* **2015**, *12*, 11560–11578. [CrossRef] [PubMed]

18. Gürsoy, Ö.; Birdal, A.C.; Özyonar, F.; Kasaka, E. Determining and monitoring the water quality of Kizilirmak River of Turkey: First results. *Int. Arch. Photogram. Remote Sens. Spat. Inf. Sci.* **2015**, *40*, 1469–1474. [CrossRef]

19. Tian, Y.; Guo, Z.Q.; Qiao, Y.C.; Lei, X.; Xie, F. Remote sensing of water quality monitoring in Guanting Reservoir. *Shengtai Xuebao* **2015**, *35*, 2217–2226. [CrossRef]

20. Latorre-Carmona, P.; Knyazikhin, Y.; Alonso, L.; Moreno, J.F.; Pla, F.; Yan, Y. On hyperspectral remote sensing of leaf biophysical constituents: Decoupling vegetation structure and leaf optics using CHRIS-PROBA data over crops in barrax. *IEEE Geosci. Remote Sens. Lett.* **2014**, *11*, 1579–1583. [CrossRef]

21. Laurent, V.C.E.; Verhoef, W.; Clevers, J.G.P.W.; Schaepman, M.E. Inversion of a coupled canopy-atmosphere model using multi-angular top-of-atmosphere radiance data: A forest case study. *Remote Sens. Environ.* **2011**, *115*, 2603–2612. [CrossRef]

22. Hilker, T.; Coops, N.C.; Hall, F.G.; Nichol, C.J.; Lyapustin, A.; Black, T.A.; Wulder, M.A.; Leuning, R.; Barr, A.; Hollinger, D.Y.; et al. Inferring terrestrial photosynthetic light use efficiency of temperate ecosystems from space. *J. Geophys. Res. Biogeosci.* **2011**, *116*, G03014. [CrossRef]

23. Carmona, F.; Rivas, R.; Fonnegra, D.C. Vegetation index to estimate chlorophyll content from multispectral remote sensing data. *Eur. J. Remote Sens.* **2015**, *48*, 319–326. [CrossRef]

24. Agapiou, A.; Hadjimitsis, D.G.; Alexakis, D.D. Evaluation of Broadband and Narrowband Vegetation Indices for the Identification of Archaeological Crop Marks. *Remote Sens.* **2012**, *4*, 3892–3919. [CrossRef]

25. Datt, B. Remote sensing of chlorophyll a, chlorophyll b, chlorophyll a+b, and total carotenoid content in eucalyptus leaves. *Remote Sens. Environ.* **1998**, *66*, 111–121. [CrossRef]

26. Gitelson, A.A.; Merzlyak, M.N.; Chivkunova, O.B. Optical Properties and Nondestructive Estimation of Anthocyanin Content in Plant Leaves. *Photochem. Photobiol.* **2001**, *71*, 38–45. [CrossRef]

27. Gitelson, A.A.; Zur, Y.; Chivkunova, O.B.; Merzlyak, M.N. Assessing Carotenoid Content in Plant Leaves with Reflectance Spectroscopy. *Photochem. Photobiol.* **2002**, *75*, 272–281. [CrossRef]

28. Filella, I.; Amaro, T.; Araus, J.L.; Peñuelas, J. Relationship between photosynthetic radiation-use efficiency of barley canopies and the photochemical reflectance index (PRI). *Physiol. Plant.* **1996**, *96*, 211–216. [CrossRef]

29. Gamon, J.A.; Serrano, L.; Surfus, J.S. The photochemical reflectance index: An optical indicator of photosynthetic radiation use efficiency across species, functional types, and nutrient levels. *Oecologia* **1997**, *112*, 492–501. [CrossRef] [PubMed]

30. Merzlyak, M.N.; Gitelson, A.A.; Chivkunova, O.B.; Rakitin, V.Y. Nondestructive optical detection of pigment changes during leaf senescence and fruit ripening. *Physiol. Plant.* **1999**, *106*, 135–141. [CrossRef]

31. Peñuelas, J.; Filella, I.; Biel, C.; Serrano, L.; Savé, R. The reflectance at the 950–970 nm region as an indicator of plant water status. *Int. J. Remote Sens.* **1993**, *14*, 1887–1905. [CrossRef]

32. Guerschman, J.-P.; Hill, M.J.; Barrett, D.J.; Renzullo, L.; Marks, A.; Botha, E. Estimating fractional cover of photosynthetic vegetation, non-photosynthetic vegetation and soil in mixed tree-grass vegetation using the EO-1 and MODIS sensors. *Remote Sens. Environ.* **2009**, *113*, 928–945. [CrossRef]

33. Hill, M.J.; Renzullo, L.J.; Guerschman, J.P.; Marks, A.S.; Barrett, D.J. Use of vegetation index "fingerprints" from Hyperion data to characterize vegetation states within land cover/land use types in an Australian tropical savanna. *IEEE J. Sel. Top. Appl. Earth Observ. Remote Sens. (JSTARS)* **2013**, *6*, 309–319. [CrossRef]

34. Hill, M.J. Vegetation index suites as indicators of vegetation state in grassland and savanna: An analysis with simulated SENTINEL 2 data for a North American transect. *Remote Sens. Environ.* **2013**, *137*, 94–111. [CrossRef]

35. Haucke, H.; Prochaska, D. *Management Plan for Gus Engeling Research and Demonstration Area*; Texas Parks and Wildlife Department: Austin, TX, USA, 1998; 22p.

36. Alonso, L.; Gómez-Chova, L.; Moreno, J.; Guanter, L.; Brockmann, C.; Fomferra, N.; Quast, R.; Regner, P. CHRIS/PROBA Toolbox for hyperspectral and multiangular data exploitations. In Proceedings of the 2009 IEEE International Geoscience and Remote Sensing Symposium, University of Cape Town, Cape Town, South Africa, 12–17 July 2009; Volume 2, p. II-202. [CrossRef]

37. Fomferra, N.; Brockmann, C. BEAM—The ENVISAT MERIS and AATSR toolbox. In Proceedings of the MERIS (A)ATSR Workshop 2005, Frascati, Italy, 26–30 September 2005.

38. Guanter, L.; Alonso, L.; Gomez-Chova, L.; Moreno, J. *PROBA/CHRIS Atmospheric Correction Module Algorithm Theoretical Basis Document*; Contract No. 20442/07/I-LG; ESA ESRIN: Frascati, Italy, 2008; 27p.

39. Berk, A.; Anderson, G.P.; Acharya, P.K.; Hoke, M.L.; Chetwynd, J.H.; Bernstein, L.S.; Shettle, E.P.; Matthew, M.W.; Adler-Golden, S.M. *MODTRAN4 Version 3 Revision 1 User's Manual*; Technical Report; Air Force Research Laboratory: Hanscom Air Force Base, MA, USA, 2003; 91p.

40. Tucker, C.J. Red and photographic infrared linear combinations for monitoring vegetation. *Remote Sens. Environ.* **1979**, *8*, 127–150. [CrossRef]

41. Cutter, M. *CHRIS Data Format*; Surrey Satellite Technology: Guildford, UK, 2008; 37p.

42. Moy, A.; Le, S.; Verhoeven, A. Different strategies for photoprotection during autumn senescence in maple and oak. *Physiol. Plant.* **2015**, *155*, 205–216. [CrossRef] [PubMed]

43. Lee, D.W.; O'Keefe, J.; Holbrook, N.M.; Field, T.S. Pigment dynamics and autumn leaf senescence in a New England deciduous forest, eastern USA. *Ecol. Res.* **2003**, *18*, 677–694. [CrossRef]

44. Keskitalo, J.; Bergquist, G.; Gardeström, P.; Jansson, S. A cellular timetable of autumn senescence. *Plant Phys.* **2005**, *139*, 1635–1648. [CrossRef] [PubMed]

45. García-Plazaola, J.I.; Hernández, A.; Becerril, J.M. Antioxidant and pigment composition during autumnal leaf senescence in woody deciduous species differing in their ecological traits. *Plant Biol.* **2003**, *5*, 557–566. [CrossRef]

46. Abrams, M.D. Adaptations and responses to drought in Quercus species of North America. *Tree Phys.* **1990**, *7*, 227–238. [CrossRef]

47. Donovan, L.A.; West, J.B.; McLeod, K.W. Quercus species differ in water and nutrient characteristics in a resource-limited fall-line sandhill habitat. *Tree Phys.* **2000**, *20*, 929–936. [CrossRef]

48. Ratajczak, Z.; Nippert, J.B.; Collins, S.L. Woody encroachment decreases diversity across North American grasslands and savannas. *Ecology* **2012**, *93*, 697–703. [CrossRef]

49. Hinckley, T.M.; Dougherty, P.M.; Lassoie, J.P.; Roberts, J.E.; Teskey, R.O. A severe drought: Impact on tree growth, phenology, net photosynthetic rate and water relations. *Am. Midland Nat.* **1979**, *102*, 307–316. Available online: https://www.jstor.org/stable/2424658 (accessed on 15 August 2019). [CrossRef]

50. Ginter-Whitehouse, D.L.; Hinckley, T.M.; Pallardy, S.G. Spatial and temporal aspects of water relations of three tree species with different vascular anatomy. *For. Sci.* **1983**, *29*, 317–329.

51. Caterina, G.L.; Will, R.E.; Turton, D.J.; Wilson, D.S.; Zou, C.B. Water use of *Juniperus virginiana* trees encroached into mesic prairies in Oklahoma, USA. *Ecohydrology* **2014**, *7*, 1124–1134. [CrossRef]

52. Eggemeyer, K.D.; Awada, T.; Harvey, F.E.; Wedin, D.A.; Zhou, X.; Zanner, C.W. Seasonal changes in depth of water uptake for encroaching trees *Juniperus virginiana* and *Pinus ponderosa* and two dominant C4 grasses in a semiarid grassland. *Tree Phys.* **2009**, *29*, 157–169. [CrossRef] [PubMed]

53. DeSantis, R.D.; Hallgren, S.W.; Lynch, T.B.; Burton, J.A.; Palmer, M.W. Long-term directional changes in upland Quercus forests throughout Oklahoma, USA. *J. Veg. Sci.* **2010**, *21*, 606–615. [CrossRef]

54. Chastain, R.A., Jr.; Struckhoff, M.A.; He, H.S.; Larsen, D.R. Mapping Vegetation Communities Using Statistical Data Fusion in the Ozark National Scenic Riverways, Missouri, USA. *Photogramm. Eng. Remote Sens.* **2008**, *74*, 247–264. [CrossRef]

55. Asbjornsen, H.; Brudvig, L.A.; Mabry, C.M.; Evans, C.W.; Karnitz, H.M. Defining reference information for restoring ecologically rare tallgrass oak savannas in the Midwestern United States. *J. For.* **2005**, *103*, 345–350. [CrossRef]

56. Grundel, R.; Pavlovic, N.B. Using conservation value to assess land restoration and management alternatives across a degraded oak savanna landscape. *J. Appl. Ecol.* **2008**, *45*, 315–324. [CrossRef]

57. Chen, Q.; Baldocchi, D.; Gong, P.; Kelly, M. Isolating individual trees in a savanna woodland using small footprint LiDAR data. *Photogram. Eng. Remote Sens.* **2006**, *72*, 923–932. [CrossRef]

58. Asner, G.P.; Jones, M.O.; Martin, R.E.; Knapp, D.E.; Hughes, R.F. Remote sensing of native and invasive species in Hawaiian forests. *Remote Sens. Environ.* **2008**, *112*, 1912–1926. [CrossRef]

59. Asner, G.P.; Martin, R.E.; Anderson, C.B.; Knapp, D.E. Quantifying forest canopy traits: Imaging spectroscopy versus field survey. *Remote Sens. Environ.* **2015**, *158*, 15–27. [CrossRef]

60. Feilhauer, H.; Asner, G.P.; Martin, R.E. Multi-method ensemble selection of spectral bands related to leaf biochemistry. *Remote Sens. Environ.* **2015**, *164*, 57–65. [CrossRef]

61. Chadwick, K.; Asner, G. Organismic-scale remote sensing of canopy foliar traits in lowland tropical forests. *Remote Sens.* **2016**, *8*, 87. [CrossRef]

62. McManus, K.; Asner, G.; Martin, R.; Dexter, K.; Kress, W.; Field, C. Phylogenetic structure of foliar spectral traits in tropical forest canopies. *Remote Sens.* **2016**, *8*, 196. [CrossRef]

63. Mitchell, R.; Cathey, J.C.; Drabbert, B.; Prochaska, D.; Dupree, S.; Sosebee, R. Managing Yaupon with fire and herbicodes in the Texas post oak savannah. *Rangelands* **2005**, *27*, 17–19. [CrossRef]

Article

Monitoring Grass Phenology and Hydrological Dynamics of an Oak–Grass Savanna Ecosystem Using Sentinel-2 and Terrestrial Photography

Pedro J. Gómez-Giráldez [1,*], María J. Pérez-Palazón [2], María J. Polo [2] and María P. González-Dugo [1]

[1] IFAPA. Institute of Agricultural and Fisheries Research and Training of Andalusia. Avd. Menéndez Pidal s/n, 14071 Córdoba, Spain; mariap.gonzalez.d@juntadeandalucia.es
[2] Fluvial Dynamics and Hydrology Research Group. Andalusian Institute for Earth System Research. University of Cordoba. Campus Rabanales, Edificio Leonardo Da Vinci, Área de Ingeniería Hidráulica, 14014 Córdoba, Spain; mjppalazon@gmail.com (M.J.P.-P.); mjpolo@uco.es (M.J.P.)
* Correspondence: pjgomezgiraldez@gmail.com

Received: 26 December 2019; Accepted: 8 February 2020; Published: 11 February 2020

Abstract: Annual grasslands are an essential component of oak savanna ecosystems as the primary source of fodder for livestock and wildlife. Drought resistance adaptation has led them to complete their life cycle before serious soil and plant water deficits develop, resulting in a close link between grass phenology and soil water dynamics. In this work, these links were explored using a combination of terrestrial photography, satellite imagery and hydrological ground measurements. We obtained key phenological parameters of the grass cycle from terrestrial camera data using the Green Chromatic Coordinate (GCCc) index. These parameters were compared with those provided by time-series of vegetation indices (VI) obtained from Sentinel-2 (S2) satellites and time-series of abiotic variables, which defined the hydrology of the system. The results showed that the phenological parameters estimated by the S2 Normalized Difference Vegetation Index (NDVI) (r = 0.83, p < 0.001) and soil moisture (SM) (r = 0.75, p < 0.001) presented the best agreement with ground-derived observations compared to those provided by other vegetation indices and abiotic variables. The study of NDVI and SM dynamics, that was extended over four growing seasons (July 2015–May 2019), showed that the seasonality of both variables was highly synchronized, with the best agreements at the beginning and at the end of the dry seasons. However, stage changes were estimated first by SM, followed by NDVI, with a delay of between 3 and 10 days. These results support the use of a multi-approach method to monitor the phenology and the influence of the soil moisture dynamic under the study conditions.

Keywords: vegetation indices; oak-grass savanna; phenology; hydrology; Sentinel-2

1. Introduction

Monitoring the phenology of Mediterranean ecosystems is key to adequately assessing the impacts of global warming on different time scales, identifying pre-critical states in the framework of early warning decision-making systems, and establishing adaptation planning and management in the medium- to long-term time horizons. The natural variability of the climatic–hydrological regime in these areas, usually with complex spatial patterns of the vegetation that is sometimes difficult to access, makes it necessary to exploit the available data from remote sensing sources.

The holm oak savanna ecosystem of the Iberian Peninsula (known as *dehesa* in Spain and *montado* in Portugal) is an ancient system emerging as the only productive and sustainable structure able to deal with the combination of low soil fertility and the high variability of the Mediterranean climate [1]. This landscape is formed by a low density of holm oak trees, an understory of grasslands and, occasionally, shrubs and crops of cereals and legumes. The result is a mixture of natural facto and management, making it difficult to differentiate the absolute determinants of its structure [2]. However, water availability is known to be important in the conformation of this mixture of grasses and woody plants which, as stated in [3], is the only stable state of equilibrium when a marked seasonality in water availability occurs. In this state, canopy density maximizes biomass, thus minimizing water stress [4]. Annual grasslands are an essential component of this system as the primary source of fodder for livestock, and is the main economic activity in these areas, but also contains a high richness of vascular plants supporting a wide diversity of habitats [5]. The escape mechanism—i.e., the ability for the plant to complete its life cycle before serious soil and plant water deficits develop [6]—is used by these grasses to cope with the long summer dry season and the recurrent water scarcity events of the Mediterranean climate. This results in a close link between grass phenology and soil water dynamics.

Plant phenological shifts have been observed as a result of changes in the climate [7–10]. Phenological research has advanced in developing climate–phenology models derived from ground observations [11,12] and also in improving the monitoring methods constructed from satellite information [13]. In recent years, the use of remote sensing in the study of phenology has increased thanks to a greater availability of global products, which allow one to extend these studies in space and time. In particular, the use of vegetation indices (VIs) has been generalized to determine changes in vegetation on different spatial scales, ranging from centimeters to tens of kilometers. Some examples are the use of the AVHRR (Advanced Very-High-Resolution radiometer) sensor for global studies [14–16]; broad scale analysis using MERIS (Medium Resolution Imaging Spectrometer) and MODIS (Moderate-Resolution Imaging Spectroradiometer) data for maize crops and forests [17,18]; studies on a detailed scale using Landsat or ASTER (Advanced Spaceborne Thermal Emission and Reflection Radiometer) for multiple crops monitoring [19,20]; and at a very high spatial resolution with UAVs (unmanned aerial vehicles) in vineyards [21]. The Sentinel-2 (S2) sensor, part of the European Union Copernicus Program, has provided the renewed possibility of addressing canopy phenological changes on a local scale with high spatial and temporal resolutions (10–20 m or 5 days), and an improved spectral resolution providing narrow bands in the red-edge region. Despite the short time series available (from July 2015 to present), S2 has been used for the monitoring of vegetation, addressing aspects such as phenology [22], complementing other sensor datasets such as MODIS [23] or Landsat [24], and monitoring biophysical variables through the proposal of new indices making use of the red-edge bands [25]. However, its potential for improving the accuracy of estimating the phenological transition stages has not been fully explored yet.

Remote sensing studies rely on field data of phenological parameters to explain their results. However, field monitoring requires frequent and time-consuming observations, especially over areas which are difficult to access. The use of digital terrestrial cameras offers a cost-effective solution, providing real-time and reliable information to track vegetation phenology. Previous studies have used cameras for phenological monitoring [26,27], and the creation of networks of these cameras for a more global phenological follow-up is being increasingly demanded [28]. In addition, these cameras can complement the information provided by other equipment, such as weather stations or eddy covariance systems, in a wide range of applications [29–32]. The combination of these ground observations with satellite images has also been explored [33–35], and the integration of highly detailed terrestrial photography with large-scale satellite images is a promising tool that could reduce the scale gap between phenological field and remote sensing studies.

The Mediterranean climate of this area presents a high vulnerability to global warming, with an increasing occurrence of extreme droughts and torrential rainfall. This region is projected to warm more and experience greater changes in the precipitation regime over the course of the century [36–38]. An improved knowledge of how climate and hydrology influence plant phenology is key to making appropriate decisions to cope with these threats [39]. In addition, changes in values of satellite VIs have proved to be able to reflect the heterogeneity of the hydrological behavior in this region [40]. In the same line, phenological variations, assessed by using satellite data, could be useful for evaluating the cumulative effect of the hydrological regime over large areas, in a capacity that would not be achievable with a field observation network.

In this work, images provided by a terrestrial camera were combined with meteorological information and satellite data to provide an insight into the dynamics of phenological processes and their links with the hydrological behavior of the grassland of an oak–grass savanna ecosystem. On this basis, the objectives of the present study were: (1) to explore the potential of the S2 satellites to monitor phenological changes over grasslands using a variety of vegetation indices derived from different band combinations, both broadband and narrowband indices, taking advantage in the latter case of the new possibilities offered by the red-edge bands of these satellites, (2) identify the links between hydrology and vegetation phenology derived from terrestrial photography, evaluating the response of grass vegetation and its life cycle to changes in the main abiotic variables controlling this system, and (3) study the relationship between satellite VIs and the hydrological state of the system regarding their ability to monitor grassland phenology.

2. Material and Methods

2.1. Study Site and Datasets

The study was conducted in a *dehesa* farm, called "Santa Clotilde" (Figure 1a), located in Southern Spain (Córdoba, 38° 12' 37'' N, 4° 17' 24'' W, 734 m.a.s.l.) [41]. It is part of an environmentally protected area (*Sierra de Cardeña y Montoro Natural Park*) with a landscape composed of sparse holm oak trees and natural grasses. The semiarid climate in this region is characterized by cold winters and long dry summers, with periodic severe droughts. Beef cattle and pigs are raised extensively on grasslands and acorns.

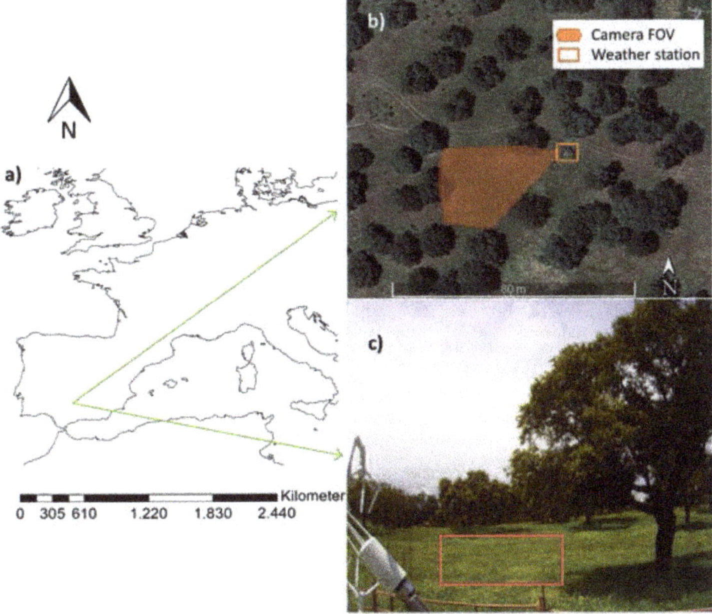

Figure 1. (**a**) Farm location. (**b**) Aerial photography of the experimental site, including the location of the weather station and the terrestrial digital camera and its field of view (FOV). (**c**) Image from the terrestrial digital camera and region of interest (ROI) for natural grass used in the analysis (red polygon).

Field measurements were taken at a grass-dominated site (Figure 1b), in which the grasses were annual species of a medium size (with a maximum plant height between 0.5–1 m) and a variable coverage. The field has a great variety of species, chiefly grams such as *Vulpia spp*, *Bromus spp* or *Aegilops spp*, although legumes such as *Trifolium spp*, *Ornithopus spp*, *Uncaria spp* or *Medicago spp* are also found [42]. The soil at the site is dominated by Eutric Cambisols, with a loamy sand texture in the grass root zone (0–30 cm). The soil moisture content, modeled at field capacity (FC) and wilting point (WP) [43], was equal to 0.18 and 0.08 m^3/m^3, respectively.

The site's instrumentation included a digital camera, micrometeorological equipment and soil moisture probes. Terrestrial photography was acquired using a CC5MPX Digital Camera (Campbell Sci. Inc) installed at a height of 1.65 m above the ground surface. This is a programmable camera of 5 megapixels which has been specially prepared to be placed outdoors (at a thermal amplitude −40 to 60 °C). Its field of view (FOV) is 790 square meters (Figure 1b), and images were obtained from the installation every hour over the daylight period, generally from 8 a.m. to 6 p.m., from the beginning of December 2017 to the end of May 2019.

According to data availability, two study periods were defined: a first period, herein called A, from December 2017 to May 2019, with available terrestrial images on a daily basis measuring field grassland greenness, and an extended period of analysis (study period B), from July 2015 to May 2019, which covered the whole period with Sentinel-2 availability (since first acquisition time).

The set of abiotic variables measured at the site during the whole study period (Figure 1b) included the following:

1. Air temperature, recorded using an HMP45A probe (Vaisala OyJ); daily minimum (T_{min}), maximum (T_{max}) and mean temperature (T_{med}) values were computed from half-hourly records.
2. Incoming and outgoing solar radiation (Rad), measured with a four-way radiometer NR-01 (Campbell Sci. Inc); daily cumulative radiation fluxes were computed from half-hourly records.
3. Vapor pressure deficit (VPD) was derived from atmospheric pressure and relative humidity (HMP45A probe (Vaisala OyJ); daily values were computed from half-hourly records.
4. Precipitation (R) was measured with a weighing-type recording rain-gauge ARG100 (Campbell Sci. Inc); daily values were computed from the aggregation of 30 min cumulative values.
5. Volumetric soil moisture (SM) was measured at two depths (10 and 30 cm) with an ENVIROSCAN (Campbell Sci. Inc) probe at 10 min intervals. Daily values were computed from these data and averaged between both depths.

Figure 2 shows the general meteorological characterization of the study period with monthly values of R and T_{med} and the monthly average of the province of Córdoba with historical data from 1971–2000 [44]. A very irregular annual distribution of rainfall was observed, as expected, as well as a great thermal amplitude on different time scales. Compared with the historical values, the four hydrological years of the study period could be considered within the average regime in terms of mean temperature, with values in summer close to 30 °C that may drop below 10 °C in winter, and mean annual temperature of 16 °C. In terms of total precipitation, the historical annual value in the area was about 650 mm: 2015/2016 and 2016/2017 were normal periods (close to 700 mm) but with different temporal distributions; 2017/2018 was an anomalous year in the seasonal distribution of precipitation in that it was drier than average until the month of March, in which 356 mm of the final 820 mm of annual precipitation was concentrated; and 2018/2019 was the driest period (500 mm).

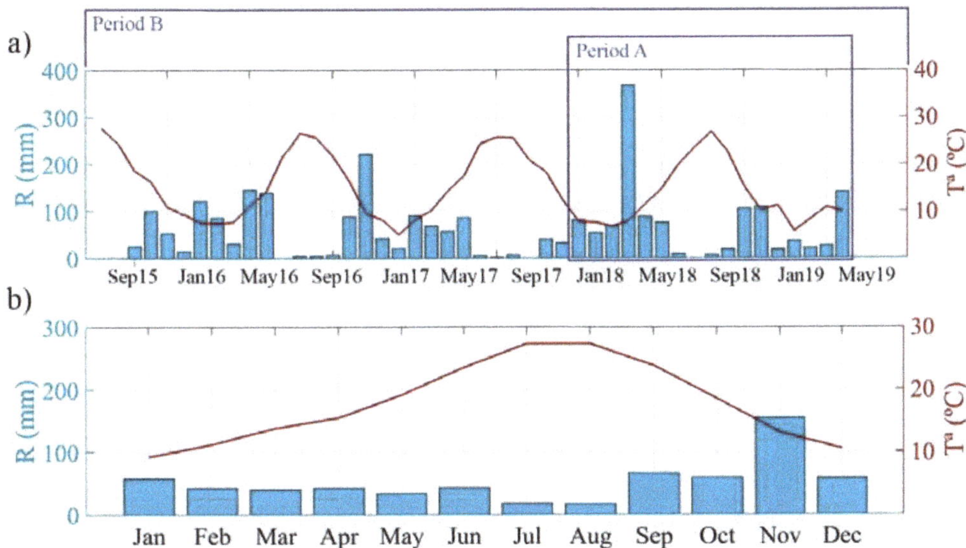

Figure 2. Monthly values of precipitation and average temperature (**a**) during study period B from the weather station in Santa Clotilde, (**b**) from historic data 1971–2000 in Córdoba.

The satellite image dataset consisted of 242 granules of Sentinel-2 images. It included all available images with less than 15% of cloud coverage from July 2015 to May 2019. From March 2017 to May 2019, the atmospherically corrected product Level 2A available from ESA (European Space Agency) was used. From July 2015 to March 2017, the images were atmospherically corrected using the module Sen2cor included in SNAP (*ESA Sentinel Application Platform v2.3.0*). Seven bands were processed to compute the different vegetation indices used in this study (the central wavelength and the bandwidth are indicated in parentheses): B2 = blue (492 nm; 66 nm), B3 = green (559 nm; 36 nm)), B4 = red (665 nm; 31 nm), B5 = red edge 1 (704 nm; 15 nm), B6 = red edge 2 (740 nm; 15 nm), B7 = red edge 3 (783 nm; 20 nm) and B8 = near infrared (NIR) (833 nm; 106 nm).

2.2. Phenological Parameters From Terrestrial Photography

The greenness of the grass, observed by terrestrial photography, was used as a field indicator of grass phenology and used to evaluate the phenological parameters retrieved from the satellite images. A quantitative value of this greenness was computed using the Green Chromatic Coordinate (GCC) (Equation (1)) [45]. The GCC index is not a direct measurement of chlorophyll content, and previous research [35,46,47] has addressed its sensibility to changing levels of green pigmentation in vegetation and how canopy phenology can be monitored and quantified without the need for a human observer. This index is a non-linear transformation of the digital levels of the green color of the picture, and it is known to minimize the effects of different lighting between images taken on different days. Their characterization of greenness has proved to be better than those of other camera indices [48]:

$$GCC = \frac{G}{R + G + B} \tag{1}$$

where

GCC: Green Chromatic Coordinate index;
R: digital level in red;
G: digital level in green;

B: digital level in blue.

This index was calculated as the average value of the pixels of a homogeneous grass region of interest (ROI) (Figure 1c). Daily index values were computed for period A (December 2017 to May 2019) using the 90th percentile, in which all the GCC values were used in a 3-day moving window, assigning the average value to the central day. This methodology is assumed to reduce the error in the estimates by more than 30% [49].

To calculate the phenological parameters from GCC values, the 50% amplitude method [50] was used. In this method, the temporal evolution of a variable is considered as a function, and the state changes are supposed to be within 50% of the amplitude of this function. The amplitude was fixed for each annual growing cycle as the difference of the minimum GCC (baseline value) and the maximum value (peak value). The parameters calculated were as follows:

Start of season (SOS): time when 50% of the amplitude is reached.

Peak of season (POS): time when the data reach the maximum value of the cycle.

End of season (EOS): time when 50% of the amplitude is reached to the right of the peak value.

To homogenize the extraction of values, the data were fitted using a double logistic function (Equation (2)), which is commonly employed to process remotely sensed data for phenology monitoring [51–53]. This function has proved to be useful for reducing the noise in a satellite time-series and enforcing a general seasonal shape on noisy data [54].

$$v(t) = v_{min} + v_{max}\left(\frac{1}{1 + e^{x_1 y_1 t}} - \frac{1}{1 + e^{x_2 y_2 t}}\right)v(t) = v \tag{2}$$

where

$v(t)$: value of the function at time t;

v_{min}: minimum value of the amplitude;

v_{max}: maximum value of the amplitude;

x and y: parameters that control the shape of the curve. x_1 and x_2 control the left and right inflexion points, respectively, and y_1 and y_2 represent the rate of change at time t.

The TIMESAT program [51,54,55] was used to fit the double logistic function to the time series and extract the phenological parameters.

2.3. Selection of Satellite Vegetation Indices

The time series of various vegetation indices, derived from Sentinel-2 images, were studied to evaluate their ability to capture the phenology of the vegetation observed in the field using terrestrial photography. An initial set of VIs, listed in Table 1, were calculated and spatially averaged in the FOV given in Figure 1b. The first group of broad-band indices included the Normalized Difference Vegetation Index (NDVI) [56], and a version of this which changed the red reflectance for the green one—the Green Normalized Difference Vegetation Index (GNDVI) [57]; the Soil Adjusted Vegetation Index (SAVI) [58]; the Enhanced Vegetation Index (EVI) [59] and a simplified version, EVI2 [60]. The new possibilities of Sentinel-2 red edge bands made a second group of narrowband indices possible, including an adaptation of the Meris Terrestrial Chlorophyll Index (MTCI) [61] and two indices developed specifically for Sentinel-2 data: the Inverted Red Edge Chlorophyll Index (IRECI) and Sentinel-2 Red Edge Position (S2REP) [25]. Finally, GCC was computed using Sentinel-2 bands and was named as GCCs to distinguish it from the GCC from the camera, herein called GCCc. All VIs were calculated for the whole study period and spatially averaged at the camera FOV.

Table 1. Vegetation indices used in the study (listed alphabetically), formulation with Sentinel-2 bands and the available spatial resolution. Bands of Sentinel-2: B2 = blue, B3 = green, B4 = red, B5 = red edge 1, B6 = red edge 2, B7 = red edge 3, B8 = near infrared (NIR). EVI: Enhanced Vegetation Index; GCC: Green Chromatic Coordinate; GNDVI: Green Normalized Difference Vegetation Index; IRECI: Inverted Red Edge Chlorophyll Index: MTCI: Meris Terrestrial Chlorophyll Index; S2REP: Sentinel-2 Red Edge Position; SAVI: Soil Adjusted Vegetation Index.

Index	Sentinel-2 Formulation	Spatial Resolution
EVI	$2.5(B8 - B4)/(B8 + 6B4 - 7.5B2 + 1)$	10 m
EVI2	$2.5(B8 - B4)/(B8 + 2.4B4 + 1)$	10 m
GCCs	$B3/(B2 + B3 + B4)$	10 m
GNDVI	$(B8 - B3)/(B8 + B3)$	10 m
IRECI	$(B7 - B4)/(B5/B6)$	20 m
MTCI	$(B6 - B5)/(B5 - B4)$	20 m
NDVI	$(B8 - B4)/(B8 + B4)$	10 m
S2REP	$705 + 35(((B7 + B4)/2) - B5)/(B6 - B5))$	20 m
SAVI	$1.5(B8 - B4)/(B8 + B4 + 0.5)$	10 m

Once the indices values were obtained for Sentinel-2 acquisition days, a daily time series for period A was computed using a linear interpolation between every two consecutive images. To minimize the effect of possible spikes due to clouds or poor atmospheric conditions, daily index values were smoothed using the Savitzky–Golay filter [62]. This method replaces each data value by a linear combination of nearby values in a moving window, helping to reduce the noise and producing a higher quality VI time series [63]. These satellite-derived time series were compared with GCCc values for the same period; the statistical analysis was able to identify those of satellite VI better on reproducing the digital camera observations. In order to compare the performance of the different indices, their values were normalized.

Similar to GCCc, satellite VI time-series were fitted to a double logistic function, and the 50% amplitude method was applied to estimate the phenological parameters, comparing their results with those obtained from GCCc data.

2.4. Selection of Abiotic Variables

The daily evolutions of different weather variables together with the water content of the root-depth soil layer were studied to locally evaluate their links with vegetation phenology and to obtain information on the main factors influencing their behavior.

The set of abiotic variables initially selected were those commonly identified in phenological studies (T_{max}, T_{med} and R), the meteorological variables often used in crop growth models (T_{min}, Rad and VPD), and the water content in the root depth of the soil layer of the grasses (SM). Daily data were obtained from half-hourly values of the weather station for the period December 2017 to May 2019, as described in Section 2.1.

The relationship between these variables and the daily series of GCCc were analyzed throughout period A. The principal explanatory variable was also employed to determine phenological parameters by fitting its values to a double logistic function, as described previously, and applying the 50% amplitude method.

2.5. Satellite VI and Abiotic Variable Relationship

The dynamic of the selected satellite VI and the most significant abiotic variable were compared during period B (July 2015 to May 2019). For this purpose, data from both of them for this period were fitted to a double logistic function, and the phenological parameters were extracted in the same way as period A.

2.6. Statistical Analysis

To evaluate the relationships between the GCCc, satellite VIs, and the abiotic variables, a Pearson correlation matrix and principal component analysis (PCA) were carried out to detect the group of variables that similarly governed the behavior of the system [64]. This method generates a new set of variables, called principal components (PCs), which are linear combinations of the original ones. The variables are related by coefficients that range between −1 and 1. The closer to 1 or −1 they are in a certain PC, the higher the direct or inverse influence they exert, respectively, on the PC. All the PCs are orthogonal to each other, so there is no redundant information. Once the PCs are obtained, all the variables of the system are grouped by processes. Since the original variables have different dimensions, standardization was applied prior to the PCA by dividing each one by its standard deviation. This process was applied firstly to GCCc with satellite VIs in order to select the representative ones, and, secondly, to GCCc with the abiotic variables.

The TIMESAT program was used to fit a double logistic function to all the variables' time series and extract the phenological parameters. Finally, the coefficient of determination (R^2) and the root mean square error (RMSE) were used to evaluate the results.

All these calculations and statistical treatments were conducted by programming in MATLAB (The Mathworks Inc., MA). The algorithms used were *corcoeff* for Pearson correlation matrices, *pca* for PCA and the *Curve fitting tool* for R^2 and RMSE.

To assess the relationship between the phenological parameters estimated using the indices, the differences in days i.e., the biases with respect to the values obtained by the GCCc were computed, and the percentage of those biases was calculated with respect to the total length of the phenological cycle, with an estimated value of 210 days (seven months) as an average [1]. In the case of the joint assessment of satellite VI and the abiotic variable, the bias in days was also calculated.

3. Results

3.1. Deriving Phenological Parameters From Terrestrial Photography

Figure 3 shows the evolution of the GCCc values throughout period A consisting of one and a half growing seasons and the function fitted to those data. The days when the baseline, SOS, EOS and the two POS for the period were reached have been identified, and the pictures (Figure 3a–e) illustrate the state of the grass at each phenological stage.

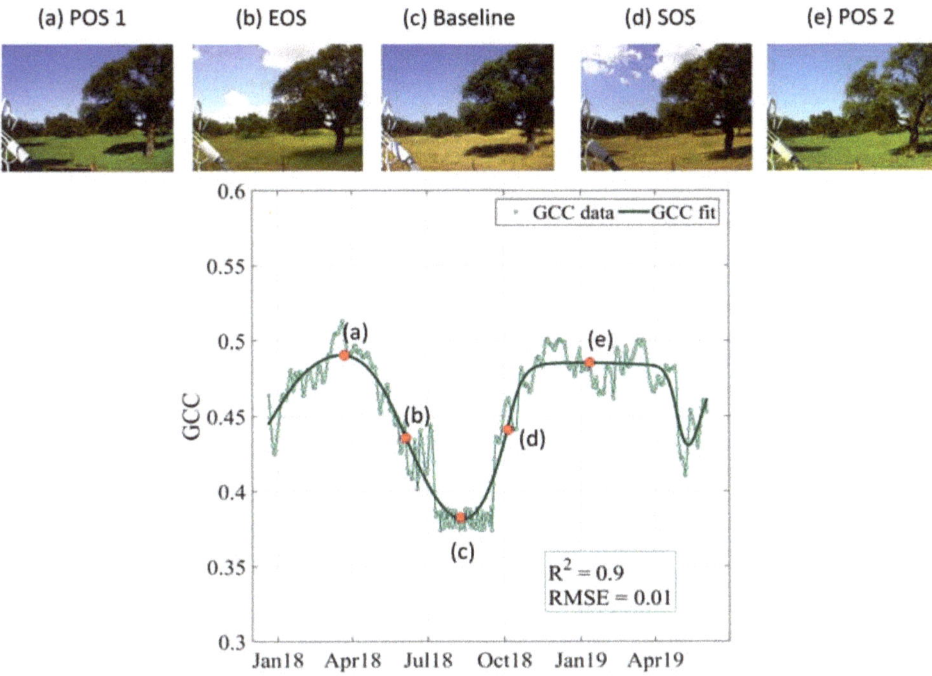

Figure 3. Evolution of GCCc (raw data, fitted function and error) throughout period A and corresponding images of the phenological dates (red dots): (**a**) peak of season (POS) 1 (03/22); (**b**) end of season (EOS (06/05); (**c**) Baseline (08/17); (**d**) start of season (SOS) (10/17); (**e**) POS 2 (01/15).

The lack of a first cycle starting point and the ending of the second one prevented the analysis of a complete single cycle. The start of the period coincided with the installation of the camera; the premature end of the second cycle was artificial and due to common farm management practices, i.e., plowing and preparing the soil for a future cereal crop. Thus, the parameters to be studied were the EOS of the first cycle, the SOS of the second and the peaks of both cycles. The baseline value used was the summer minimum of 2018.

The fitted function presented a high correlation of $R^2 = 0.9$ and a very low error of RMSE = 0.01. It can be seen that the dates of the phenological parameters given by the 50% amplitude method correctly matched the observations derived from the visual inspection of the digital camera photographs. Both POSs showed a green full cover; in the baseline, the grass was totally dry, and EOS and SOS showed the beginning of senescence and the grass becoming green, respectively.

3.2. Terrestrial Camera vs. Satellite-derived Indices

3.2.1. Satellite and Ground-based Indices Comparison

The behavior of the different satellite indices and GCCs in the study area was analyzed using the histograms of the values obtained during a complete growing cycle (year 2018). Figure 4 presents the distribution of the normalized VI values throughout this year. It can be observed that some variability was derived from the different algorithms and band combinations. All indices presented a first peak corresponding to low/very low values. This peak was more pronounced for EVI2, SAVI, IRECI and GCCs. Most indices also presented a second peak corresponding to intermediate/high index values. NDVI, EVI, GNDVI showed a more balanced distribution between both peaks than the previous group. However, MTCI and S2REP shapes do not completely follow this general pattern.

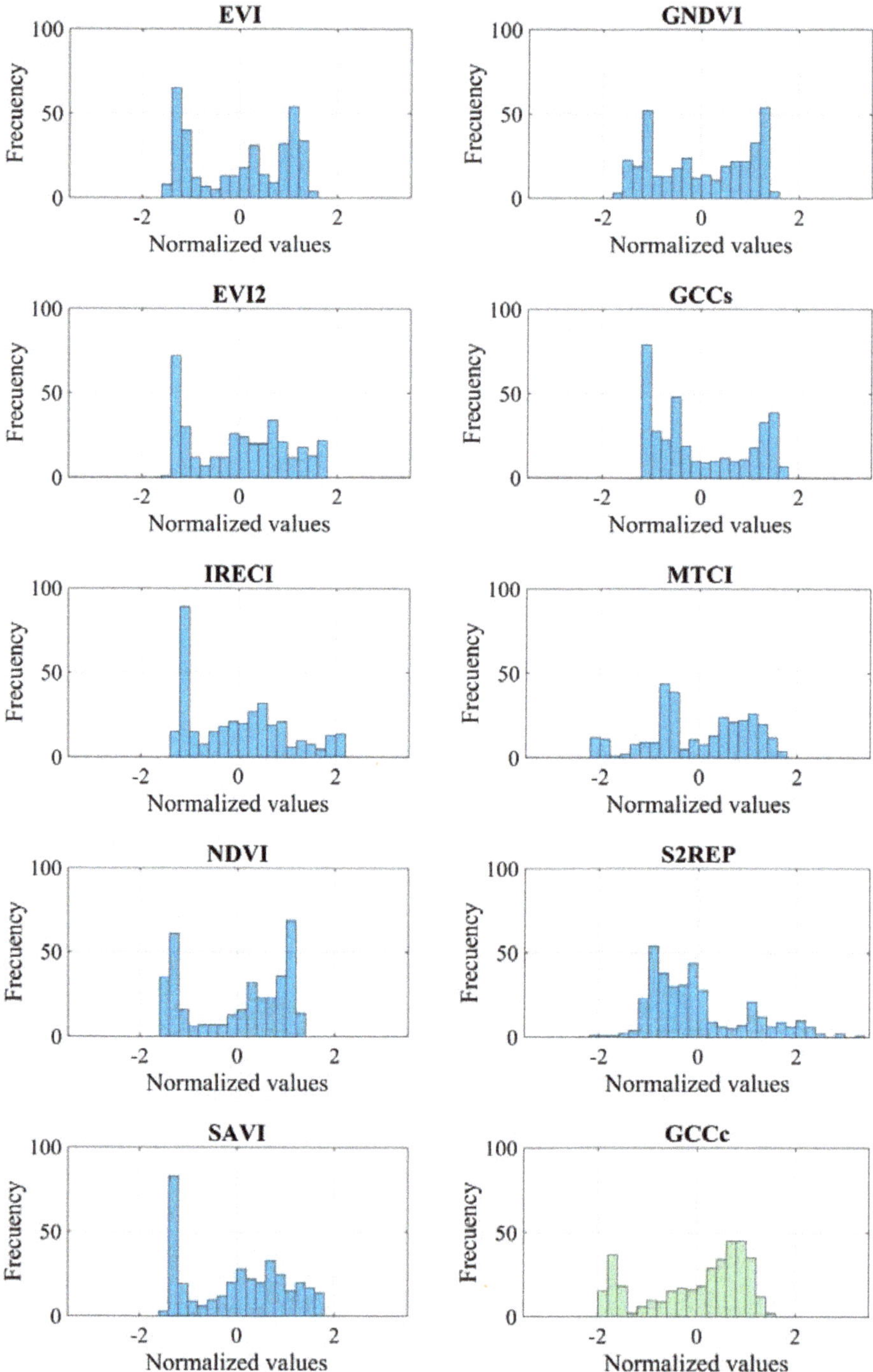

Figure 4. Histograms of vegetation index (VI) normalized values in the study area during 2018. Blue: satellite VIs; Green: Digital camera index.

Table 2 shows the Pearson correlation coefficients of daily time series of GCCc with the daily time series of a selection of satellite-derived VIs.

Table 2. Pearson correlation matrix between GCC data and the satellite VI on a daily scale. * p < 0.001.

Variable	r (GCC)
EVI	0.72 *
EVI2	0.77 *
GCCs	0.79 *
GNDVI	0.82 *
IRECI	0.71 *
MTCI	0.52 *
NDVI	0.83 *
S2REP	−0.44 *
SAVI	0.78 *

The correlation of all satellite indices with GCCc was significant (p < 0.001) and had a high correlation value (r > 0.7), except for MTCI and S2REP (r = 0.52 and –0.44, respectively).

The results for PCA are shown in Figure 5. Most of the variance (79%) was explained by the PC 1, in which two groups of indices could be distinguished: a first group of higher coefficients (equal to or greater than 0.3) included GCCc and GCCs with the rest of the satellite indices except the MTCI and S2REP, which would together form a second group with lower or negative coefficients. In turn, these two indices presented higher values (> 0.5) in PC 2, which explained 11% of the total variance.

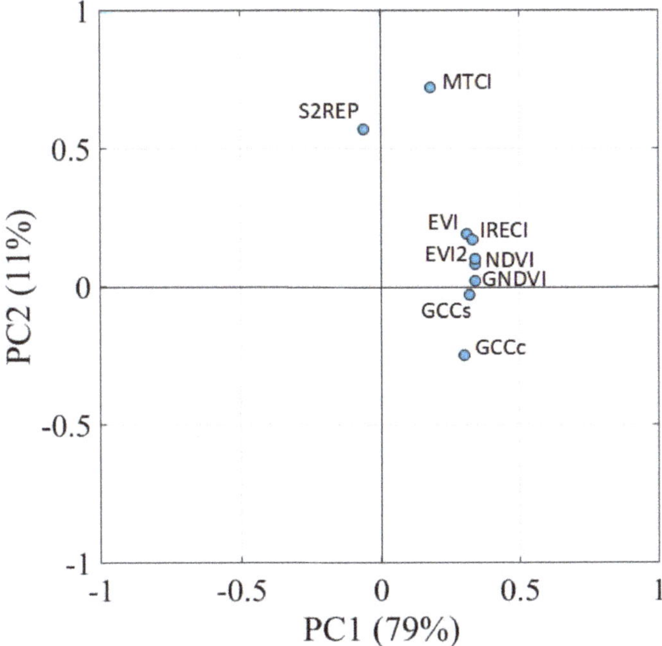

Figure 5. Principal component analysis (PCA) of GCCc and vegetation indices. The two first principal components (PCs) are shown. Between parenthesis: percentage of variance explained.

3.2.2. Satellite-derived Phenology

Based on the results of the previous section, VIs with a correlation higher than 0.7 and placed in the same group of data in PCA were selected for the subsequent analysis. Thus, seven satellite VIs were selected for the phenological analysis: EVI, EVI2, GCCs, GNDVI, IRECI, NDVI, and SAVI.

The smoothed daily data from the fitted functions are shown in Figure 6 together with the phenological parameters obtained from those VIs.

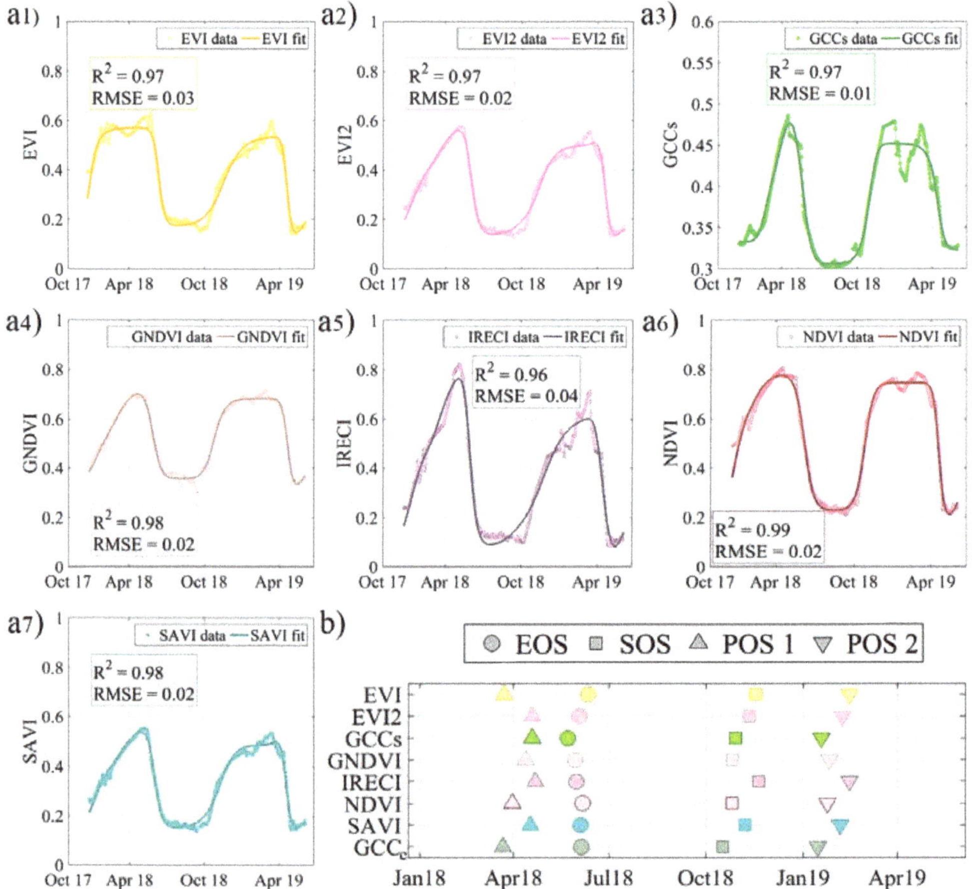

Figure 6. (a) Evolution of vegetation indices (raw data, fitted function and error) throughout the study period (a1 = EVI, a2 = EVI2, a3 = GCCs, a4 = GNDVI, a5 = IRECI, a6 = NDVI, a7 = SAVI). (b) Comparison of the phenological dates of GCCc and satellite VIs in period A.

All fitted functions showed very high correlation values ($R^2 > 0.95$) and low RMSEs (<0.03).

It can be observed (Figure 6) that the EOS was the parameter which was most similarly estimated by all indices and had the lowest bias with respect to the GCCc. The estimation of both POSs presented a higher dispersion. For this reason, POS 1 and POS 2 derived from the original raw data have been also included in Table 3 in addition to the values derived from the adjusted functions. These data are not presented for the EOS and SOS parameters due to the observed existence of erratic spikes that may cause erroneous estimations of green-up or real senescence shifts. Table 3 summarizes the biases of the phenological dates of the satellite VI with respect to GCCc (Figure 6b).

Table 3. Biases in days of the phenological parameters extracted from satellite vegetation indices with respect to GCCc, for POS 1, EOS, SOS and POS 2. The percentage of error with respect to the whole cycle is displayed in parentheses.

	POS 1 (raw data)	POS 1	EOS	SOS	POS 2	POS 2 (raw data)
EVI	−8 (3.7)	1 (0.5)	6 (2.9)	31 (14.8)	30 (14.3)	36 (17.2)
EVI2	20 (9.3)	28 (13.3)	−2 (9)	25 (11.9)	23 (11)	23 (11)
GCCs	18 (8.7)	28 (13.3)	−13 (6.2)	12 (5.7)	3 (1.4)	−3 (1.4)
GNDVI	13 (6)	22 (10.5)	−6 (2.9)	9 (4.3)	11 (5.2)	6 (2.9)
IRECI	26 (12.4)	31 (14.8)	−5 (2.4)	34 (16.2)	30 (14.3)	37 (17.7)
NDVI	4 (1.9)	9 (4.3)	1 (0.5)	9 (4.3)	9 (4.3)	2 (0.9)
SAVI	17 (8.2)	26 (12.4)	−1 (0.5)	21 (10)	21 (10)	23 (11)

The estimation of the green up (SOS) was delayed by all satellite VIs, with GCCs, GNDVI and NDVI presenting the lowest errors of around ten days (less than 10% of the cycle). The agreement was higher for the end of the season, with errors below 10% in all cases and biases of only one day using NDVI and SAVI. For both peaks of the season, there was a general delay in the estimations, with lower errors in POS 1 than in POS 2 in most cases. EVI and NDVI produced the best results for POS1, and GCCs and NDVI for POS 2. However, in the case of POS, the dates obtained using the original raw data generally improved the estimates. It could be of interest to evaluate the impact of specific weather events. In general, NDVI obtained the most consistent results and lowest errors; considering this, NDVI was selected for further analysis in this work.

3.3. Analysis of Abiotic Variables and Greenness Dynamics

The results of the Pearson correlation matrix between the daily time series of GCCc and the daily time series of selected abiotic variables for period A are shown in Table 4. All correlations were significant ($p<0.001$), and soil moisture (SM) presented the highest correlation ($r = 0.75$). Some variables, including VPD, Rad, T_{min}, T_{med}, and T_{max}, were inversely correlated with greenness. Rainfall, although positively correlated, presented the lowest coefficient of all analyzed variables. In the PCA results (Figure 7), the first PC explained 60% of the total variance, showing high coefficients (greater than 0.3) for most variables. Similar to the correlation matrix results, a first group with negative coefficients included GCCc and variables favoring an increase of soil water content (SM and R); a second group with positive coefficients was formed by variables with high values during the summer when the grass canopy was dry.

Table 4. Pearson correlation matrix between GCCc data and the abiotic variables at a daily scale. * $p < 0.001$.

Variable	r (GCCc)
SM	0.75 *
VPD	−0.68 *
R	0.17 *
Rad	−0.56 *
T_{min}	−0.68 *
T_{med}	−0.72 *
T_{max}	−0.69 *

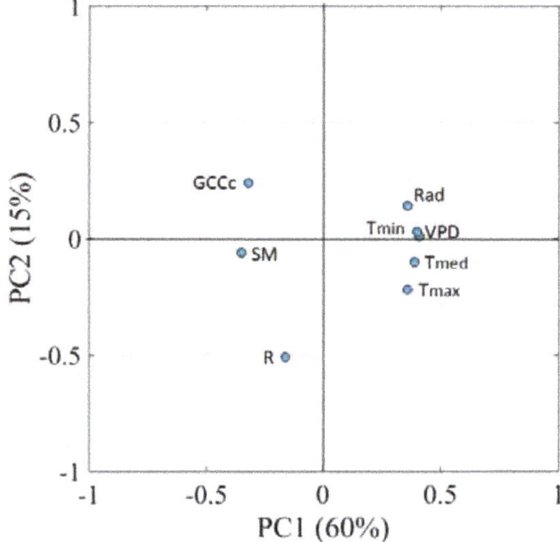

Figure 7. Principal component analysis (PCA) of GCCc and abiotic variables. The two first principal components (PCs) are shown. Between parenthesis: percentage of variance explained.

Grassland Phenology and Soil Moisture Relationship

Considering the previous results, SM has been selected to evaluate its capability to predict the phenology of the grass canopy of oak–grass savanna ecosystems. Figure 8a shows the distribution of SM data and the fitted function. This function presented a correlation of $R^2 = 0.9$ and an error of RMSE = 0.02, indicating a good agreement, although to a lesser extent than those obtained for GCCc and satellite VIs.

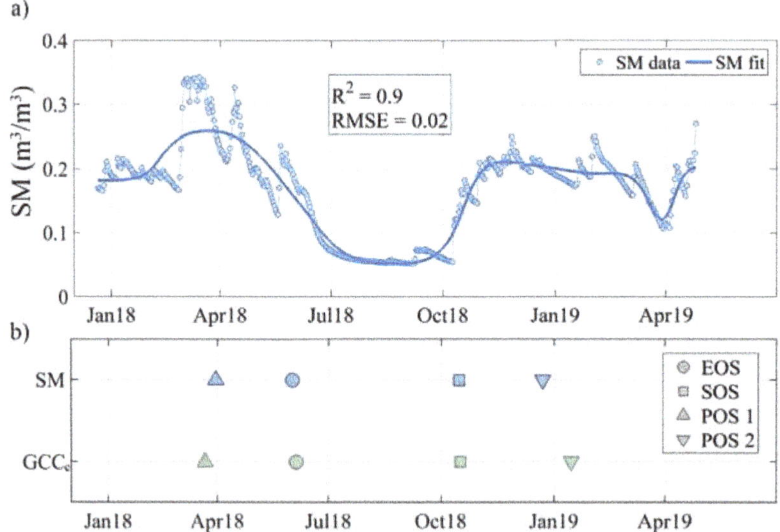

Figure 8. (a) Evolution of SM (raw data, fitted function and error) for the study period. (b) Comparison of the phenological dates of GCCc (green) and SM (blue) in period A.

The phenological parameters predicted by this function were compared to those derived from GCCc (Figure 8b). The SOS and EOS were quite similar to the timings estimated by GCCc, with a slight advance in both dates according to SM estimations. The 50% amplitude method resulted in similar values of SM, around 0.14 m^3 /m^3 on average, marking the beginning and the end of the growing season.

3.4. Relationships Between Soil Moisture and NDVI

The relationship between NDVI and SM—i.e., the variables that best suited the phenological monitoring of the grassland according to the previous results—were analyzed during period B (July 2015 to May 2019) comprising a total of four growing seasons.

Figure 9 shows the data for SM, the smoothed data for NDVI, the functions fitted to both series, and the phenological parameters extracted from them. The marked and mostly synchronized seasonality of both variables can be observed. The highest differences were found during the second year, corresponding to the growing season 2016/2017, with differences of 9 days (4.3%) for SOS, 28 days (13.3%) for POS and 16 days (7.6%) for EOS. Both fitted functions showed a high correlation (R^2 = 0.9 for SM and R^2 = 0.93 for NDVI) and low errors (RMSE = 0.02 for SM and RMSE = 0.04 for NDVI). Compared to GCCc, NDVI showed a later response to water availability, resulting in lower differences in monitoring the phenology.

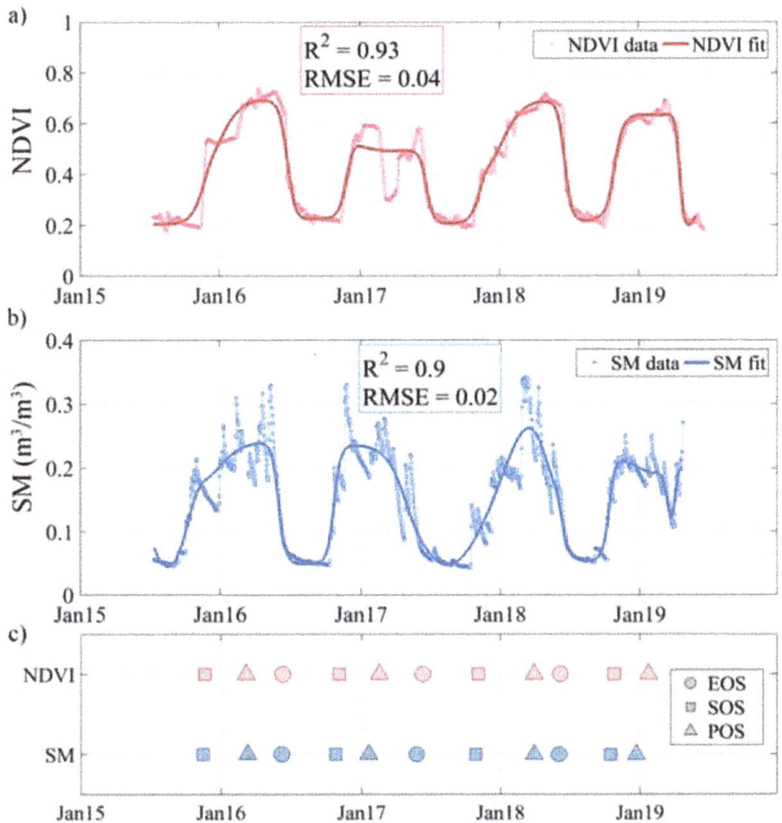

Figure 9. (a) Evolution of NDVI (raw data, fitted function and error) throughout the study period. (b) Evolution of SM (raw data, fitted function and error) during the study period. (c) Comparison of the phenological dates of NDVI and SM in the whole study period

4. Discussion

4.1. Capability of the Terrestrial Photography to Provide Phenological Parameters

The two hydrological years of study period A are an example of Mediterranean climate variability (Figure 2), and therefore of phenology variability. In 2017/2018, most of the precipitation was concentrated in a short period of time at the beginning of March, and POS 1 occurred shortly after those intense events. On the other hand, 2018/2019 was a drier year, but with a more regular rainfall distribution in the fall and spring. A lower GCCc maximum value was reached, but the growing season was longer, producing a plateau shape in which an intermediate value was identified as POS 2.

GCCc values ranged between 0.35 and 0.5, reaching their maximum values between November and April, eventually presenting drops of varying duration and intensity due to the particular weather conditions of the year. This temporal trend of the GCCc time series was similar to the one observed by other studies in Mediterranean grasslands [65,66]. The reasonable behavior of the CGGc of canopy status in the field according to the weather conditions, the previous studies and the visual observations provided by the terrestrial photography supported the use of GCCc time series as a reference to evaluate the performance of satellite vegetation indices.

4.2. Comparison of Ground and Satellite-based Indices

The first peak with a low value observed in Figure 4 corresponds to a predominance of dry grass/bare soil coverage during the dry season. Indices particularly sensitive to soil contribution, such as EVI, EVI2 and SAVI [58], present a higher concentration around these low values. The second peak may correspond to the central time of the spring, when the greening is widespread, while intermediate values reflect periods of green-up and senescence. The narrow band indices (IRECI, MTCI and S2REP) showed a wider distribution in values, probably due to the greater sensitivity to the changes that the red edge presented. It is worth mentioning that NDVI and GNDVI presented a sharpened ending at high values compared to other indices, with an accumulation of intermediate–high values that did not reach its maximum. This might indicate a certain degree of saturation of both indices under high grass cover conditions. This issue, described in previous studies for forest or high biomass crops [67,68], appears to occur here also to some extent. It is also worth noting the difference in the distribution of GCCs and GCCc; with the same formulation, the satellite version presented lower values than those calculated using the field camera. This result could have been due to the differing geometry of the image acquisition.

The low correlation for MTCI shown in Table 2 could be explained by the rapid response of this index to chlorophyll variations [61]. This causes the time series to follow a different trend than the rest of the indices, which is more connected to green foliage density [69] with smoother transitions. MTCI has presented good results in more homogeneous canopies, such as subalpine grasslands [33] or in deciduous trees in Central Europe [70]. In this case, the high diversity of species, typical of Mediterranean grasslands, might explain the high variability observed in these values. The case of S2REP is similar since it was also designed to measure chlorophyll variation using three bands located in the red-edge region, and allegedly provides a better characterization of the red-edge slope than MTCI [25]. IRECI also uses three bands in the red edge, although the high correlation presented with GCCc might be explained by a different formulation. In this case, the index was designed to focus on the chlorophyll content of the canopy instead of the leaf scale addressed by S2REP [25]. The influence of the vegetation ground coverage and foliage density was higher and more similar to the other broad-band indices.

The results for PCA in Figure 5 showed that most of the variance (79%) was explained by the PC 1, in which two groups could be distinguished; a first group of higher coefficients (equal to or greater than 0.3) included GCCc and the rest of the satellite indices except the MTCI and S2REP, which would form a second group with lower or negative coefficients. In turn, these two indices had high values (>0.5) in PC 2, which explained 11% of the total variance. This grouping confirmed the results obtained by the Pearson correlation and suggested that these two red-edge indices might be related to a different process than the rest. In that case, MTCI and S2REP would be more useful for estimating leaf chlorophyll content [61,71], while the rest were more influenced by canopy foliage. Regarding the pixel size, it does not seem to have had an influence on previous results or to determine the selection of the index for the conditions of this study. IRECI, at 20 m in size (like MTCI and S2REP), was included in the subsequent phenological analysis.

The estimation of the green up (SOS) was delayed by all satellite VIs, but GCCs, GNDVI and NDVI presented low errors of around ten days (less than 10% of the cycle). The agreement was higher for EOS, with errors below 10% in all cases and biases of only one day using NDVI and SAVI. However, in the case of POS, the dates obtained using the original raw data generally improved the estimates—something of interest to evaluate the impact of specific weather events—nevertheless, in the case of analyzing long time series, fitted functions provide a useful homogenization procedure, generally leading to a more consistent analysis.

The disagreement in terms of POS estimations between satellite-borne and ground-borne sensors can be explained by the different acquisition angle of both systems. The field camera captures flowering and the presence of non-photosynthetic elements such as stems [46], which might reduce greenness and anticipate peak time. Despite this factor, NDVI exhibited a very similar behavior to GCCc, with lower differences in days than those of previous studies in grasslands [66,72]. A possible reason for the comparatively lower errors could be the higher spatial resolution employed in this case (10 m for Sentinel-2 versus 250 or 500 m for MODIS, as previously used). This factor may become a major issue in mixed grass/tree systems [34]. GNDVI also provided good general results for EOS and SOS, although POS, especially for the first year, was somewhat delayed. GCCs, with the same formulation as GCCc, presented higher errors than NDVI when applied to satellite bands. EVI, EVI2, IRECI and SAVI displayed the highest difference of days in this study, despite EVI2 and NDVI having performed similarly in other types of grasslands [47] and with even better EVI2 results for maize yield [73] and EVI for deciduous tree coverages ([74,75]). In general, NDVI obtained the most consistent results and the lowest errors (less than 10% in all cases). These good results support the use of this multi-scale approach to complement phenology studies in grasslands. Even though the availability of terrestrial camera observations is not widespread, the increasing number of these stations is a promising framework to encourage this combined strategy.

4.3. Analysis of the Dynamics of Abiotic Variables, Greenness and Phenology

The results shown in Table 4 are consistent with the water-limited condition of this ecosystem, where the dynamics of plant development and soil moisture are intimately related. This explains the high correlation of SM with GCCc, and the positive values found for SM and rainfall, with both variables favoring an increase in soil water content. The low correlation coefficient of the rainfall was also reasonable considering the daily time scale and the bulk analysis performed over a period with high rainfall variability. Rainfall has been previously related to the timing of leaf flush in dry forest [76], suggesting that an event-based analysis at specific times of the year might produce different results. High values of solar radiation, air temperature and VPD—characteristics of the dry season—are linked to an increase in atmospheric water demand, thus accelerating the water stress process in the vegetation and explaining the inverse relationships with grass greenness. This general description of the ecosystem behavior is also supported by PCA results (Figure 7). The first PC explained 60% of the total variance, showing high coefficients (greater than 0.3) for most variables. Similar to the correlation matrix results, a first group with negative coefficients included GCCc and variables favoring an increase in soil water content (SM and R); a second group with positive coefficients was formed by variables with high values during the summer when the grass canopy was dry. These variables might negatively influence the plant water status in water scarce conditions. No significant differences in both analyses were obtained when the different growing season stages were analyzed separately.

Analyzing the behavior of SM shown in Figure 8, the value of SM selected by the 50% method for the EOS seems slightly high, as it coincides with a fraction of around 60% of the total available soil water. This depletion fraction is close to the threshold used for crops [77] below which the soil water can no longer be transported quickly enough towards the roots to meet the transpiration demand, and the plant starts to experience stress. However, in this application, this threshold is supposed to indicate the complete senescence of the plant rather than the beginning of plant water stress, which might indicate that both processes—the soil drying and the response of the vegetation—happened very quickly.

The use of SM to compute the POS in this environment makes less sense than for SOS and EOS, and the biases found with respect to GCCc estimations confirm this point. First of all, the data reflected a significant difference between the two POSs reached in the field-calibrating period as a consequence of the natural climatic variability of this region. POS 1 estimation with SM was closer to GCCc estimation, occurring with a bias of 9 days (4.3%). For POS 2, the bias was of −23 days (11%), presenting a difficult identification of the peak due to a flatter distribution of SM over the mid-growing season. Looking at the distribution of GCCc (Figure 3) and SM (Figure 8a), the shape of the first cycle was sharper because high SM values were concentrated in a short period, producing a fast response in GCCc values, with the POS reached before the SM peak, once the plant had no limitation to accessing soil water. In the second cycle, the distribution of SM followed a distribution with two similar low peaks (November 2018 and February 2019) with a slightly drier period around January. This reduction in SM was reflected by both GCCc (Figure 3) and some satellite VIs such as NDVI or GCCs (Figure 6), confirming the close link between SM and greenness and also the fast response of the vegetation to relatively small reductions in SM. This change in SM was small and it was not reproduced by the different fitted functions however, nevertheless, it caused a notable bias in the peak identification. Using the unfitted data for the POS, the results were similar; in POS 1, there was an advance of 9 days (4.3%), and in POS 2, this was 31 (14.7%).

4.4. Relationship Between SM and NDVI

The synchronized seasonality of both variables observed in Figure 9 has already been pointed out for other vegetation types under arid or semi-arid conditions in [78], in which the authors reported an increasing synchronization following a gradient of dry season length in a region. It is worth noting here the similar shapes of both variables following the beginning and the end of the dry periods, supporting the higher control of soil moisture on vegetation development during those periods. In general, an average delay between 3 and 10 days can be noted in the estimation of most phenological parameters with NDVI with respect to SM. The highest differences were found in the second year, corresponding to the growing season 2016/2017, with differences of 9 days (4.3%) for SOS, 28 days (13.3%) for POS and 16 days (7.6%) for EOS. During that year, the grassland was plowed and sown in the middle of February, following a management practice conducted every nine years or so that is currently disappearing in the area. Therefore, these two grass-growing cycles were artificially shaped and were not adequately reproduced by the fitting function. The poor adjustment of the function affects the phenological estimations, increasing their errors. Apart from this specific case, both fitted functions showed a high correlation ($R^2 = 0.9$ for SM and $R^2 = 0.93$ for NDVI) and low errors (RMSE = 0.02 for SM and RMSE = 0.04 for NDVI). Compared to GCCc, NDVI showed a later response to water availability, resulting in lower differences in monitoring the phenology. However, this difference could simply be due to the lower temporal resolution of satellite data, with a minimum of 5 days, compared to the daily availability of the terrestrial photography.

5. Conclusions

The phenology of a semi-arid grassland was monitored using terrestrial photography, satellite images, and abiotic ground measurements. The similarities and discrepancies observed using these datasets supported the capability of satellite vegetation indices to track important phenological parameters of this canopy and highlight the close relationship between these phenological parameters and the soil moisture dynamic under the study conditions.

The methodology to process the ground and satellite time series and extract the phenological information, implemented in TIMESAT, successfully fitted the data, with high correlation coefficients and low errors (R^2 > 0.9 and RMSE < 0.03 in all cases at the two scales), and identified key seasonal changes of the natural grasslands. On the ground, daily time series of the GCCc, computed using digital photography, estimated grassland phenological parameters in accordance with visual inspection and with the observed climatic conditions of the experimental study period. On the S2 satellite platform, NDVI was the index that better reproduced ground GCCc behavior, with the highest correlation (r = 0.83) and less than 10 days of difference for all the phenological parameters studied (representing less than 5% error within a grass cycle). None of the VIs using bands in the red-edge region improved the NDVI results. Two of them, MTCI and S2REP, followed a different trend than the rest of the explored indices. Both indices presented a high temporal variability, which could be explained by chlorophyll content variations caused by the diversity of species of this grassland, different to the more homogeneous canopies of previous studies [33,70]. The rest of the indices were more related to foliage density [69] and presented smoother changes in this canopy.

The abiotic variable that best represented the behavior of GCCc was SM (r = 0.75). The estimation of SOS and EOS phenological parameters using SM time series presented a good agreement with GCCc, with SM reaching threshold values a few days before greenness ones, as measured by GCCc. The values of SM found at the beginning and the end of the growing seasons, when changes between green and senescent stages occurred, were high according to the modeled soil water holding capacity. This might indicate that these modeled values were not accurate enough to assess the quick response of this canopy to the soil drying process. On the other hand, SM was not a good indicator of the POS, presenting significant biases with respect to GCCc estimations.

Finally, the behavior of NDVI and SM during the four growing seasons points out the synchronized seasonality shown in this system by the vegetation greenness, measured here by the NDVI and the SM. Higher agreement was found at the beginning and the end of the dry season, with stage changes estimated first by SM followed by NDVI with a delay of between 3 to 10 days. Taking into account these results, it is possible to monitor the water state of the soil from the phenological parameters obtained with NDVI of S2 and also to determine the phenological state of the cover from the SM with errors below ten days for this type of coverage.

Author Contributions: Conceptualization, P.J.G.-G. and M.P.G.-D.; methodology, P.J.G.-G., M.P.G.-D., M.J.P.-P. and M.J.P.; software, P.J.G.-G. and M.J.P.-P.; data curation, P.J.G.-G.; writing, review, and editing, P.J.G.-G., M.P.G.-D., M.J.P.-P. and M.J.P.; graphical deployment, P.J.G.-G. and M.J.P.-P.; supervision, M.P.G.-D. and M.J.P. All authors have read and agreed to the published version of the manuscript.

Funding: This work was supported by the SensDehesa (PP.PEI.IDF201601.16) project, co-funded at 80% by the European Regional Development Fund (ERDF), as part of the Andalusian operational program 2014–2020; additional support was provided by the project "Control and early warning of critical ecohydrological states in areas of Dehesa and high mountains through terrestrial photography", funded by the Biodiversity Foundation, Spanish Ministry of Agriculture, Fisheries, Food, and Environment.

Acknowledgments: The authors are grateful to the owner of the Santa Clotilde farm for providing access to the site.

Conflicts of Interest: The authors declare no conflicts of interest. The funders had no role in the design of the study; in the collection, analyses, or interpretation of data; in the writing of the manuscript, or in the decision to publish the results.

References

1. Olea, L.; Miguel-ayanz, A.S. The Spanish dehesa, a traditional Mediterranean silvopastoral system. In Proceedings of the 21st General Meeting of the European Grassland Federation, Badajoz, Spain, 3–6 April 2006; pp. 1–15.

2. Scholes, R.J.; Archer, S.R. Tree-Grass Interactions in Savannas. *Annu. Rev. Ecol. Syst.* **1997**, *28*, 517–544. [CrossRef]

3. Eagleson, P.S.; Segarra, R.I. Water-Limited Equilibrium of Savanna Vegetation Systems. *Water Resour. Res.* **1985**, *21*, 1483–1493. [CrossRef]

4. Eagleson, P.S.; Tellers, T.E. Ecological optimality in water-limited natural soil-vegetation systems: 2. Tests and applications. *Water Resour. Res.* **1982**, *18*, 341–354. [CrossRef]

5. Moreno, G.; Gonzalez-Bornay, G.; Pulido, F.; Lopez-Diaz, M.L.; Bertomeu, M.; Juárez, E.; Diaz, M. Exploring the causes of high biodiversity of Iberian dehesas: The importance of wood pastures and marginal habitats. *Agrofor. Syst.* **2016**, *90*, 87–105. [CrossRef]

6. Turner, N.C. Drought resistance and adaptation to water deficits in crop plants. *Stress Physiol. Crop Plants* **1979**, 343–372. Available online: https://ci.nii.ac.jp/naid/10006285376/en/ (accessed on 25 December 2019).

7. Parmesan, C.; Yohe, G. A globally coherent fingerprint of climate change impacts across natural systems. *Nature* **2003**, 37–42. [CrossRef]

8. IPCC 2001. Climate change 2001: Impacts, adaptation, and vulnerability. Contribution of Working Group II to the third assessment report of the Intergovernmental Panel on Climate Change (IPCC). In *Choice Reviews Online (Vol. 39)*; McCarthy, J.J., Canziani, O.F., Leary, N.A., Dokken, D.J., White, K.S., Eds.; Cambridge University Press: Cambridge, UK, 2001. [CrossRef]

9. Parmesan, C. Influences of species, latitudes and methodologies on estimates of phenological response to global warming. *Glob. Chang. Biol.* **2007**, *13*, 1860–1872. [CrossRef]

10. Cleland, E.E.; Chuine, I.; Menzel, A.; Mooney, H.A.; Schwartz, M.D. Shifting plant phenology in response to global change. *Trends Ecol. Evol.* **2007**, *22*, 357–365. [CrossRef]

11. Menzel, A. Phenology: Its importance to the global change community—An editorial comment. *Clim. Chang.* **2002**, *54*, 379–385. [CrossRef]

12. Rafferty, N.E.; CaraDonna, P.J.; Burkle, L.A.; Iller, A.M.; Bronstein, J.L. Phenological overlap of interacting species in a changing climate: An assessment of available approaches. *Ecol. Evol.* **2013**, *3*. [CrossRef]

13. Fisher, J.I.; Richardson, A.D.; Mustard, J.F. Phenology model from surface meteorology does not capture satellite-based greenup estimations. *Glob. Chang. Biol.* **2007**, *13*, 707–721. [CrossRef]

14. Justice, C.O.; Townshend, J.R.G.; Holben, A.N.; Tucker, C.J. Analysis of the phenology of global vegetation using meteorological satellite data. *Int. J. Remote Sens.* **1985**, *6*, 1271–1318. [CrossRef]

15. Reed, B.C.; Brown, J.F.; VanderZee, D.; Loveland, T.R.; Merchant, J.W.; Ohlen, D.O. Measuring phenological variability from satellite imagery. *J. Veg. Sci.* **1994**, *5*, 703–714. [CrossRef]

16. Jin, Z.; Zhuang, Q.; He, J.S.; Luo, T.; Shi, Y. Phenology shift from 1989 to 2008 on the Tibetan Plateau: An analysis with a process-based soil physical model and remote sensing data. *Clim. Chang.* **2013**, *119*, 435–449. [CrossRef]

17. Viña, A.; Gitelson, A.A.; Rundquist, D.C.; Keydan, G.; Leavitt, B.; Schepers, J. Monitoring maize (Zea mays L.) phenology with remote sensing. *Agron. J.* **2004**, *96*, 1139–1147. [CrossRef]

18. Xiao, X.; Hagen, S.; Zhang, Q.; Keller, M.; Moore, B. Detecting leaf phenology of seasonally moist tropical forests in South America with multi-temporal MODIS images. *Remote Sens. Environ.* **2006**, *103*, 465–473. [CrossRef]

19. Duchemin, B.; Hadria, R.; Erraki, S.; Boulet, G.; Maisongrande, P.; Chehbouni, A.; Escadafal, R.; Ezzahar, J.; Hoedjes, J.C.B.; Kharrou, M.H.; et al. Monitoring wheat phenology and irrigation in Central Morocco: On the use of relationships between evapotranspiration, crops coefficients, leaf area index and remotely-sensed vegetation indices. *Agric. Water Manag.* **2006**, *79*, 1–27. [CrossRef]

20. Peña-Barragán, J.M.; Ngugi, M.K.; Plant, R.E.; Six, J. Object-based crop identification using multiple vegetation indices, textural features and crop phenology. *Remote Sens. Environ.* **2011**, *115*, 1301–1316. [CrossRef]

21. Lamb, D.W.; Weedon, M.M.; Bramley, R.G.V. Using remote sensing to predict grape phenolics and colour at harvest in a Cabernet Sauvignon vineyard: Timing observations against vine phenology and optimising image resolution. *Aust. J. Grape Wine* **2004**, *10*, 46–54. [CrossRef]

22. Poenaru, V.; Badea, A.; Dana Negula, I.; Moise, C. Monitoring Vegetation Phenology in the Braila Plain Using Sentinel 2 Data. *Sci. Pap. Ser. E Land R* **2017**, *6*, 175–180. Available online: https://scihub.copernicus.eu/ (accessed on 25 December 2019).

23. Lange, M.; Dechant, B.; Rebmann, C.; Vohland, M.; Cuntz, M.; Doktor, D. Validating MODIS and sentinel-2 NDVI products at a temperate deciduous forest site using two independent ground-based sensors. *Sensors* **2017**, *17*, 1855. [CrossRef] [PubMed]

24. Jönsson, P.; Cai, Z.; Melaas, E.; Friedl, M.A.; Eklundh, L. A method for robust estimation of vegetation seasonality from Landsat and Sentinel-2 time series data. *Remote Sens.* **2018**, *10*, 635. [CrossRef]

25. Frampton, W.J.; Dash, J.; Watmough, G.; Milton, E.J. Evaluating the capabilities of Sentinel-2 for quantitative estimation of biophysical variables in vegetation. *ISPRS J. Photogramm Remote Sens.* **2013**, *82*, 83–92. [CrossRef]

26. Richardson, A.D.; Braswell, B.H.; Hollinger, D.Y.; Jenkins, J.P.; Ollinger, S.V. Near-surface remote sensing of spatial and temporal variation in canopy phenology. *Ecol. Appl.* **2009**, *19*, 1417–1428. [CrossRef] [PubMed]

27. Motohka, T.; Nasahara, K.N.; Oguma, H.; Tsuchida, S. Applicability of Green-Red Vegetation Index for remote sensing of vegetation phenology. *Remote Sens.* **2010**, *2*, 2369–2387. [CrossRef]

28. Brown, T.B.; Hultine, K.R.; Steltzer, H.; Denny, E.G.; Denslow, M.W.; Granados, J.; Richardson, A.D. Using phenocams to monitor our changing earth: Toward a global phenocam network. *Front. Ecol. Environ.* **2016**, *14*, 84–93. [CrossRef]

29. Moore, C.E.; Beringer, J.; Evans, B.; Hutley, L.B.; Tapper, N.J. Tree-grass phenology information improves light use efficiency modelling of gross primary productivity for an Australian tropical savanna. *Biogeosciences* **2017**, *14*, 111–129. [CrossRef]

30. Knox, S.H.; Dronova, I.; Sturtevant, C.; Oikawa, P.Y.; Matthes, J.H.; Verfaillie, J.; Baldocchi, D. Using digital camera and Landsat imagery with eddy covariance data to model gross primary production in restored wetlands. *Agric. For. Meteorol.* **2017**, 237–238. [CrossRef]

31. Pimentel, R.; Herrero, J.; Polo, M.J. Subgrid parameterization of snow distribution at a Mediterranean site using terrestrial photography. *Hydrol. Earth Syst. Sc.* **2017**, *21*, 805–820. [CrossRef]

32. Polo, M.J.; Herrero, J.; Pimentel, R.; Pérez-Palazón, M.J. The Guadalfeo Monitoring Network (Sierra Nevada, Spain): 14 years of measurements to understand the complexity of snow dynamics in semiarid regions. *Earth Syst. Sci. Data* **2019**, *11*, 393–407. [CrossRef]

33. Migliavacca, M.; Galvagno, M.; Cremonesec, E.; Rossini, M.; Meroni, M.; Sonnentage, O.; Cogliati, S.; Mancaf, G.; Diotri, F.; Busetto, L.; et al. Using digital repeat photography and eddy covariance data to model grassland phenology and photosynthetic CO_2 uptake. *Agric. For. Meteorol.* **2011**, *151*, 1325–1337. [CrossRef]

34. Liu, Z.; Wu, C.; Peng, D.; Wang, S.; Gonsamo, A.; Fang, B.; Yuan, W. Improved modeling of gross primary production from a better representation of photosynthetic components in vegetation canopy. *Agric. For. Meteorol.* **2017**, *233*, 222–234. [CrossRef]

35. Moore, C.E.; Brown, T.; Keenan, T.F.; Duursma, R.A.; Van Dijk, A.I.J.M.; Beringer, J.; Liddell, M.J. Reviews and syntheses: Australian vegetation phenology: New insights from satellite remote sensing and digital repeat photography. *Biogeosciences* **2016**, *13*, 5085–5102. [CrossRef]

36. Norrant, C.; Douguédroit, A. Monthly and daily precipitation trends in the Mediterranean (1950–2000). *Theor. Appl. Climatol.* **2006**, *83*, 89–106. [CrossRef]

37. Cortesi, N.; González-Hidalgo, J.C.; Brunetti, M.; Martin-Vide, J. Daily precipitation concentration across Europe 1971–2010. *Nat. Hazards Earth Syst. Sci.* **2012**, *12*, 2799–2810. [CrossRef]

38. Solomon, S.; Quin, D.; Manning, M.; Chen, Z.; Marquis, M.; Averyt, K.; Tignort, M.; Miller, H. *Climate Change: The Physical Science Basis. Contribution of Working Group I to the Fourth Assessment Report of the Intergovernmental Panel on Climate Change*; Cambridge University Press: Cambridge, UK; New York, NY, USA, 2017.

39. Kovats, R.S.; Valentini, R.; Bouwer, L.M.; Georgopoulou, E.; Jacob, D.; Martin, E.; Rounsevell, M.; Soussana, J.F. *Climate Change 2014: Impacts, Adaptation, and Vulnerability. Part B: Regional Aspects. Contribution of Working Group II to the Fifth Assessment Report of the Intergovernmental Panel on Climate Change*; Barros, V.R., Field, C.B., Dokken, D.J., Mastrandrea, M.D., Mach, K.J., Bilir, T.E., Chatterjee, M., Ebi, K.L., Estrada, Y.O., Genova, R.C., et al., Eds.; Cambridge University Press: Cambridge, UK; New York, NY, USA, 2014; pp. 1267–1326.

40. Gómez-Giráldez, P.J.; Aguilar, C.; Polo, M.J. Natural vegetation covers as indicators of the soil water content in a semiarid mountainous watershed. *Ecol. Indic.* **2014**, *46*, 524–535. [CrossRef]

41. Andreu, A.; Kustas, W.P.; Polo, M.J.; Carrara, A.; González-Dugo, M.P. Modeling surface energy fluxes over a dehesa (oak savanna) ecosystem using a thermal based two-source energy balance model (TSEB) I. *Remote Sens.* **2018**, *10*, 567. [CrossRef]

42. García-moreno, A.; Fernández-Rebollo, P.; Muñoz, M.; Carbonero, M. Gestión de los pastos en la dehesa. Instituto de Investigación y Formación Agraria y Pesquera (IFAPA). 2016 Dep. Legal CO-614-2016. Available online: https://www.juntadeandalucia.es/agriculturaypesca/ifapa/servifapa/contenidoAlf?id=1e33ce7b-9a13-4d73-8832-f367be91c551§or=69cc80a0-9a2d-11df-accb-b374239e8181§orf=69cc80a0-9a2d-11df-accb-b374239e8181&l=material (accessed on 25 December 2019).

43. Schaap, M.G.; Leij, F.J.; Van Genuchten, M.T. Rosetta: A computer program for estimating soil hydraulic parameters with hierarchical pedotransfer functions. *J. Hydrol.* **2001**, *251*, 163–176. [CrossRef]

44. Agencial Estatal de Meteorología (Ministerio de Medio Ambiente y Medio Rural y Marino); Instituto de Meteorologia de Portugal. Atlas climático de España y Portugal. 2011. Available online: http://www.aemet.es/es/serviciosclimaticos/datosclimatologicos/atlas_climatico (accessed on 25 December 2019).

45. Gillespie, A.R.; Kahle, A.B.; Walker, R.E. Color enhancement of highly correlated images. II. Channel ratio and "chromaticity" transformation techniques. *Remote Sens.* **1987**, *22*, 343–365. [CrossRef]

46. Vrieling, A.; Meroni, M.; Darvishzadeh, R.; Skidmore, A.K.; Wang, T.; Zurita-Milla, R.; Oosterbeek, K.; O'Connor, B.; Paganini, M. Vegetation phenology from Sentinel-2 and field cameras for a Dutch barrier island. *Remote Sens. Environ.* **2018**, *215*, 517–529. [CrossRef]

47. Zhang, X.; Jayavelu, S.; Liu, L.; Friedl, M.A.; Henebry, G.M.; Liu, Y.; Schaaf, C.-B.; Richardson, A.D.; Gray, J. Evaluation of land surface phenology from VIIRS data using time series of PhenoCam imagery. *Agric. For. Meteorol.* **2018**, *256–257*, 137–149. [CrossRef]

48. Sonnentag, O.; Hufkens, K.; Teshera-Sterne, C.; Young, A.M.; Friedl, M.; Braswell, B.H.; Milliman, T.; O'Keefe, J.; Richardson, A.D. Digital repeat photography for phenological research in forest ecosystems. *Agric. For. Meteorol.* **2012**, *152*, 159–177. [CrossRef]

49. White, M.A.; Thornton, P.E.; Running, S.W. A continental phenology model for monitoring vegetation responses to interannual climatic variability. *Glob. Biogeochem. Cycles* **1997**, *11*, 217–234. [CrossRef]

50. Zhang, X.Y.; Friedl, M.A.; Schaaf, C.B.; Strahler, A.H.; Hodges, J.C.F.; Gao, F.; Reed, B.C.; Huete, A. Monitoring vegetation phenology using MODIS. *Remote Sens. Environ.* **2003**, *84*, 471–475. [CrossRef]

51. Jönsson, P.; Eklundh, L. TIMESAT - A program for analyzing time-series of satellite sensor data. *Comput. Geoscis.* **2004**, *30*, 833–845. [CrossRef]

52. Fisher, J.I.; Mustard, J.F.; Vadeboncoeur, M.A. Green leaf phenology at Landsat resolution: Scaling from the field to the satellite. *Remote Sens. Environ.* **2006**, *100*, 265–279. [CrossRef]

53. Fisher, J.I.; Mustard, J.F. Cross-Scalar Satellite Phenology from Ground, Landsat, and MODIS Data. *Remote Sens. Environ.* **2007**, *109*, 261–273. [CrossRef]

54. Eklundh, L.; Jönson, P. TIMESAT: A Software Package for Time-Series Processing and Assessment of Vegetation Dynamics. In *Remote Sensing Time Series. Remote Sensing and Digital Image Processing*; Kuenzer, C., Dech, S., Wagner, W., Eds.; Springer: Cham, Switzerland, 2015; p. 22. [CrossRef]

55. Jönsson, P.; Eklundh, L. Seasonality extraction by function fitting to time-series of satellite sensor data. *IEEE Trans Geosci Remote Sens.* **2002**, *40*, 1824–1832. [CrossRef]

56. Rouse, J.W.; Haas, R.H.; Schell, J.A.; Deering, W.D. Monitoring vegetation systems in the Great Plains with ERTS. In Proceedings of the Third ERTS Symposium, NASA SP-351, Washington, DC, USA, 10–14 December 1973; pp. 309–317.

57. Gitelson, A.A.; Kaufman, Y.J.; Merzlyak, M.N. Use of a green channel in remote sensing of global vegetation from EOS-MODIS. *Remote Sens. Environ.* **1996**, *58*, 289–298. [CrossRef]

58. Huete, A.R. A soil-adjusted vegetation index. *Remote Sens. Environ.* **1988**, *25*, 295–309. [CrossRef]

59. Huete, A.R.; Didan, K.; Miura, T.; Rodriguez, E.P.; Gao, X.; Ferreria, L.G. Overview of the radiometric and biophysical performance of the MODIS vegetation indices. *Remote Sens. Environ.* **2002**, *83*, 195–213. [CrossRef]

60. Jiang, Z.Y.; Huete, A.R.; Didan, K.; Miura, T. Development of a two-band enhanced vegetation index without a blue band. *Remote Sens. Environ.* **2008**, *112*, 3833–3845. [CrossRef]

61. Dash, J.; Curran, P.J. 2004. The MERIS terrestrial chlorophyll index. *Int. J. Remote Sens.* **2004**, *25*, 5403–5413. [CrossRef]

62. Savitzky, A.; Golay, M.J.E. Smoothing and Differentiation of Data by Simplified Least-Squares Procedures. *Anal. Chem.* **1964**, *36*, 1627–1639. [CrossRef]

63. Chen, J.; Jönson, P.; Tamura, M.; Gu, Z.; Matsushita, B.; Eklundh, L. A simple method for reconstructing a high-quality NDVI time-series data set based on the Savitzky–Golay filter. *Remote Sens. Environ.* **2004**, *91*, 332–344. [CrossRef]

64. Jackson, J.E. *A User's Guide to Principal Components*; John Wiley and Sons: New York, NY, USA, 1991; p. 592.

65. Luo, Y.; El-Madany, T.S.; Filippa, G.; Ma, X.; Ahrens, B.; Carrara, A.; Migliavacca, M. Using near-infrared-enabled digital repeat photography to track structural and physiological phenology in Mediterranean tree-grass ecosystems. *Remote Sens.* **2018**, *10*, 1293. [CrossRef]

66. Richardson, A.D.; Hufkens, K.; Milliman, T.; Frolking, S. Intercomparison of phenological transition dates derived from the PhenoCam Dataset V1.0 and MODIS satellite remote sensing. *Sci. Rep.* **2018**, *8*, 1–12. [CrossRef]

67. Huete, A.R.; Liu, H.Q.; van Leeuwen, W.J.D. Use of vegetation indices in forested regions: Issues of linearity and saturation. *Int. Geosci. Remote Sens. Symp. (IGARSS)* **1997**, *4*, 1966–1968. [CrossRef]

68. Gu, Y.; Wylie, B.K.; Howard, D.M.; Phuyal, K.P.; Ji, L. NDVI saturation adjustment: A new approach for improving cropland performance estimates in the Greater Platte River Basin, USA. *Ecol. Indic.* **2013**, *30*, 1–6. [CrossRef]

69. Glenn, E.P.; Huete, A.R.; Nagler, P.L.; Nelson, S.G. Relationship between remotely-sensed vegetation indices and plant physiological processes: What vegetation indices can and cannot tell us about the landscape. *Sensors* **2008**, *8*, 2136. [CrossRef]

70. Rodriguez-Galiano, V.F.; Dash, J.; Atkinson, P.M. Intercomparison of satellite sensor land surface phenology and ground phenology in Europe. *Geophys. Res.* **2015**, *42*, 2253–2260. [CrossRef]

71. Sibanda, M.; Mutanga, O.; Rouget, M. Testing the capabilities of the new WorldView-3 space-borne sensor's red-edge spectral band in discriminating and mapping complex grassland management treatments. *Int. J. Remote Sens.* **2017**, *38*, 1–22. [CrossRef]

72. Browning, D.M.; Karl, J.W.; Morin, D.; Richardson, A.D.; Tweedie, C.E. Phenocams bridge the gap between field and satellite observations in an arid grassland ecosystem. *Remote Sens.* **2017**, *9*, 1071. [CrossRef]

73. Bolton, D.K.; Friedl, M.A. Forecasting crop yield using remotely sensed vegetation indices and crop phenology metrics. *Agric. For. Meteorol.* **2007**, *173*, 74–84. [CrossRef]

74. Hufkens, K.; Friedl, M.; Sonnentag, O.; Braswell, B.H.; Milliman, T.; Richardson, A.D. Linking near-surface and satellite remote sensing measurements of deciduous broadleaf forest phenology. *Remote Sens. Environ.* **2012**, *117*, 307–321. [CrossRef]

75. Klosterman, S.T.; Hufkens, K.; Gray, J.M.; Melaas, E.; Sonnentag, O.; Lavine, I.; Mitchell, L.; Norman, R.; Friedl, M.A.; Richardson, A.D. Evaluating remote sensing of deciduous forest phenology at multiple spatial scales using PhenoCam imagery. *Biogeosciences* **2014**, *11*, 4305–4320. [CrossRef]

76. Jolly, W.M.; Running, S.W. Effects of precipitation and soil water potential on drought deciduous phenology in the Kalahari. *Glob. Chang. Biol.* **2004**, *10*, 303–308. [CrossRef]

77. Allen, R.G. FAO Irrigation and Drainage Paper Crop. *Irrig. Drain.* **1998**, *300*, 300. [CrossRef]

78. Jones, M.O.; Kimball, J.S.; Nemani, R.R. Asynchronous Amazon. forest canopy phenology indicates adaptation to both water and light availability. *Environ. Res.* **2014**, *9*. [CrossRef]

MDPI

St. Alban-Anlage 66

4052 Basel

Switzerland

Tel. +41 61 683 77 34

Fax +41 61 302 89 18

www.mdpi.com

Remote Sensing Editorial Office

E-mail: remotesensing@mdpi.com

www.mdpi.com/journal/remotesensing